AF581643

Water Resource Management through the Lens of Planetary Health Approach

Water Resource Management through the Lens of Planetary Health Approach

Editors

Pankaj Kumar
Ram Avtar

MDPI • Basel • Beijing • Wuhan • Barcelona • Belgrade • Manchester • Tokyo • Cluj • Tianjin

Editors
Pankaj Kumar
Adaptation and Water
Institute for Global
Environmental Strategies
Hayama
Kanagawa
Japan

Ram Avtar
Faculty of Environmental
Earth Science
Hokkaido University
Sapporo
Japan

Editorial Office
MDPI
St. Alban-Anlage 66
4052 Basel, Switzerland

This is a reprint of articles from the Special Issue published online in the open access journal *Water* (ISSN 2073-4441) (available at: www.mdpi.com/journal/water/special_issues/water_planetary_health).

For citation purposes, cite each article independently as indicated on the article page online and as indicated below:

LastName, A.A.; LastName, B.B.; LastName, C.C. Article Title. *Journal Name* **Year**, *Volume Number*, Page Range.

ISBN 978-3-0365-6146-2 (Hbk)
ISBN 978-3-0365-6145-5 (PDF)

Contents

About the Editors vii

Pankaj Kumar and Ram Avtar
Water Resource Management through the Lens of Planetary Health Approach
Reprinted from: *Water* **2022**, *14*, 3490, doi:10.3390/w14213490 1

Rudra Mohan Pradhan, Ajit Kumar Behera, Sudhir Kumar, Pankaj Kumar and Tapas Kumar Biswal
Recharge and Geochemical Evolution of Groundwater in Fractured Basement Aquifers (NW India): Insights from Environmental Isotopes ($\delta^{18}O$, $\delta^{2}H$, and ^{3}H) and Hydrogeochemical Studies
Reprinted from: *Water* **2022**, *14*, 315, doi:10.3390/w14030315 7

Ajit Kumar Behera, Rudra Mohan Pradhan, Sudhir Kumar, Govind Joseph Chakrapani and Pankaj Kumar
Assessment of Groundwater Flow Dynamics Using MODFLOW in Shallow Aquifer System of Mahanadi Delta (East Coast), India
Reprinted from: *Water* **2022**, *14*, 611, doi:10.3390/w14040611 29

Abdul Kadir, Zia Ahmed, Md. Misbah Uddin, Zhixiao Xie and Pankaj Kumar
Integrated Approach to Quantify the Impact of Land Use and Land Cover Changes on Water Quality of Surma River, Sylhet, Bangladesh
Reprinted from: *Water* **2021**, *14*, 17, doi:10.3390/w14010017 47

Cheng-Wei Hung and Lin-Han Chiang Hsieh
Analysis of Factors Influencing the Trophic State of Drinking Water Reservoirs in Taiwan
Reprinted from: *Water* **2021**, *13*, 3228, doi:10.3390/w13223228 63

Arup Acharjee, Zia Ahmed, Pankaj Kumar, Rafiul Alam, M. Safiur Rahman and Jesus Simal-Gandara
Assessment of the Ecological Risk from Heavy Metals in the Surface Sediment of River Surma, Bangladesh: Coupled Approach of Monte Carlo Simulation and Multi-Component Statistical Analysis
Reprinted from: *Water* **2022**, *14*, 180, doi:10.3390/w14020180 77

Pankaj Kumar, Yukako Inamura, Pham Ngoc Bao, Amila Abeynayaka, Rajarshi Dasgupta and Helayaye D. L. Abeynayaka
Microplastics in Freshwater Environment in Asia: A Systematic Scientific Review
Reprinted from: *Water* **2022**, *14*, 1737, doi:10.3390/w14111737 93

Huynh Vuong Thu Minh, Van Pham Dang Tri, Vu Ngoc Ut, Ram Avtar, Pankaj Kumar and Trinh Trung Tri Dang et al.
A Model-Based Approach for Improving Surface Water Quality Management in Aquaculture Using MIKE 11: A Case of the Long Xuyen Quadangle, Mekong Delta, Vietnam
Reprinted from: *Water* **2022**, *14*, 412, doi:10.3390/w14030412 111

Megha Shyam, Gowhar Meraj, Shruti Kanga, Sudhanshu, Majid Farooq and Suraj Kumar Singh et al.
Assessing the Groundwater Reserves of the Udaipur District, Aravalli Range, India, Using Geospatial Techniques
Reprinted from: *Water* **2022**, *14*, 648, doi:10.3390/w14040648 133

Duc Cong Hiep Nguyen, Duc Canh Nguyen, Thi Tang Luu, Tan Cuong Le, Pankaj Kumar and Rajarshi Dasgupta et al.
Enhancing Water Supply Resilience in a Tropical Island via a Socio-Hydrological Approach: A Case Study in Con Dao Island, Vietnam
Reprinted from: *Water* **2021**, *13*, 2573, doi:10.3390/w13182573 . . . . **159**

Aparna Bera, Gowhar Meraj, Shruti Kanga, Majid Farooq, Suraj Kumar Singh and Netrananda Sahu et al.
Vulnerability and Risk Assessment to Climate Change in Sagar Island, India
Reprinted from: *Water* **2022**, *14*, 823, doi:10.3390/w14050823 . . . . **177**

About the Editors

Pankaj Kumar

Pankaj Kumar is working in the field of water resources and climate change adaptation at the Institute for Global Environmental Strategies (IGES), Hayama, Kanagawa Prefecture, Japan. His research work primarily focuses on socio-hydrology, water security, hydrological simulation, scenario analysis, and the water-health-food-energy nexus. To address his research objectives, he uses an integrated approach involving both qualitative and quantitative analysis. One of the important works that he is engaged in is 'Water quality assessments and scenario modeling for different water environment in different Asian countries', a transdisciplinary work that aims to enhance community resilience to address growing water demands under climate change and rapid urbanization, with the ultimate intention to give a solution for sustainable water resource management at the policy level both from adaptation and mitigation perspectives.

In parallel, he is also working for different global reports like Intergovernmental Panel on Climate Change (IPCC), Intergovernmental Science-Policy Platform on Biodiversity and Ecosystem Services (IPBES) and Global Environmental Outlook (GEO) in different capacities, and is currently working as lead author for nexus assessment of the interlinkages among biodiversity, water, food and health in the context of climate change under Intergovernmental Science-Policy Platform on Biodiversity and Ecosystem Services (IPBES) secretariat.

Ram Avtar

Dr Ram Avtar is working as Assistant Professor at the Faculty of Environmental Earth Science, Hokkaido University, Japan. He developed methods for forest mapping using multi-sensor remote sensing techniques and scenario analysis for sustainable forest management. His research interest is in applications of geospatial techniques to monitor terrestrial ecosystems including forest, agriculture, urban and disasters and disseminating the results to policy-makers. Currently, he is working on the synergistic use of remote sensing and unmanned aerial vehicles (UAVs) techniques to monitor the environment more precisely to solve environmental issues from global to local scale.

 water MDPI

Editorial

Water Resource Management through the Lens of Planetary Health Approach

Pankaj Kumar [1,*] and Ram Avtar [2]

[1] Institute for Global Environmental Strategies, Hayama 240-0115, Japan
[2] Faculty of Environmental Earth Science, Hokkaido University, Sapporo 060-0810, Japan
* Correspondence: kumar@iges.or.jp

Citation: Kumar, P.; Avtar, R. Water Resource Management through the Lens of Planetary Health Approach. *Water* **2022**, *14*, 3490. https://doi.org/10.3390/w14213490

Received: 14 October 2022
Accepted: 17 October 2022
Published: 1 November 2022

Publisher's Note: MDPI stays neutral with regard to jurisdictional claims in published maps and institutional affiliations.

For eternity, water resources have proven to be the key to inclusive social development and human well-being. However, the spatio-temporal variation of this finite resource over various landscapes makes it prone to misuse and mismanagement. Rapid global changes such as urbanization, population growth, socio-economic change, change in lifestyle, evolving energy needs, and climate change have put unprecedented pressure on this finite freshwater resource. Keeping this in mind, it is argued that achieving water security throughout the world is the key to achieving sustainable development in a comprehensive manner. However, scientific studies with a holistic point of view considering persistently changing dimensions are still in their embryonic stage. Broadly, water security evolves from ensuring reliable access to enough safe water for every person (at an affordable price where market mechanisms are involved) to lead a healthy and productive life, including future generations. Moreover, there is a need to transition from water scarcity towards water security for a water-secure present and future. This transition requires a look at this complex issue and interdependencies between water, environment, human health, and governance/institutions regulating it to be more inclusive. Despite recent progress in developing new strategies, practices, and technologies for water resource management, their dissemination and implementation have been limited. The nexus approach encompasses these interdependencies, and, to promote this idea, different global frameworks are there to address global health in holistic and comprehensive ways, like one earth, one health, eco-health, and planetary health. Planetary health is the most recent one advocated by the scientific communities as well as policymakers; however, very little has been conducted to present empirical scientific evidence from the ground.

Considering the above-mentioned information gap, this special issue aimed to capture the persistently changing dimensions and new paradigms of water security, providing a holistic view, including a wide range of sustainable solutions to address water security. It discussed gaps, opportunities, challenges, and lessons learned from past experiences for achieving water security in any particular landscape. It also highlighted how recent scientific innovations in the research methodologies had made progress in realizing a planetary health framework to address the water-food-energy-health-biodiversity nexus, urban-rural nexus, regional-circular-ecological-sphere approach, etc., and address the complex issue of water security. Finally, what are the ways forward for a better science-policy interface through the inclusion of every relevant stakeholder to codesign and codelivery of various adaptation and mitigation strategies needed to achieve global goals, e.g., SDGs at a local level in a timely manner? All the above-mentioned issues were reflected through ten articles included in this issue.

The first two articles (Pradhan et al. [1] and Behera et al. [2]) discussed the status of water resources and processing governing the evolution of its quality and quantity in the data-scarce region of India. Considering water as a limiting factor for socio-economic development, especially in arid/semi-arid regions, both scientific communities and policymakers are interested in groundwater recharge-related data. Therefore, Pradhan et al. [1] used an integrated approach of environmental isotopes and hydrogeochemical studies to understand the recharge processes and geochemical evolution of groundwater in the fractured basement terranes of Gujarat, NW India. Based on the major ionic distribution, results suggest that the chemical weathering of silicate minerals influences the groundwater chemistry in the aquifer system. Furthermore, the chemical composition of groundwater also reflects that the groundwater has interacted with distinct rock types (granites/granulites). The isotopic signature of the groundwater reveals that the local precipitation is the main source of recharge; however, it is affected by the evaporation process due to different geological conditions, irrespective of topographical differences in the study area.

Despite being a biodiversity hotspot, the Mahanadi delta in eastern India is facing groundwater salinization as one of the main environmental threats in the recent past. Hence, Behera et al. [2] attempt to understand the dynamics of groundwater and its sustainable management options through numerical simulation (MODFLOW) in the Jagatsinghpur deltaic region. The result shows that groundwater in the study area is extensively abstracted for agricultural activities, which also causes the depletion of groundwater levels. The hydraulic head value varies from 0.7 to 15 m above mean sea level (MSL), with an average head of 6 m in this low-lying coastal region. The horizontal hydraulic conductivity and the specific yield values in the area are found to vary from 40 to 45 m/day and 0.05 to 0.07, respectively. The interaction between the river and coastal unconfined aquifer system responds differently in different seasons. The net groundwater recharge to the coastal aquifer has been estimated and varies from 247.89 to 262.63 million cubic meters (MCM) in the years 2006–2007. The model further indicates a net outflow of 8.92–9.64 MCM of groundwater into the Bay of Bengal. Further, the outflow to the sea prevents the seawater from ingress into the shallow coastal aquifer system. The findings of both of these studies provided vital information for the decision-makers or policymakers to take appropriate measures to design water budgets as well as water management plans more sustainably.

The next paper by Kadir et al. [3] aims to assess the impacts of one of the direct drivers, i.e., land use and land cover (LULC) changes on the water quality of the Surma river in Bangladesh. For this, seasonal water quality (physico-chemical parameters) changes were assessed in comparison to the LULC changes recorded from 2010 to 2019. The obtained results from this study indicated that there is a significant seasonal pattern in the water quality changes, with relatively higher concentration found in the dry season. On the other hand, analysis of LULC revealed that agricultural and vegetation classes decreased, while built-up, waterbody, and barren lands increased. The correlation between LULC and water quality parameters showed a significant relationship between them. Built-up areas and waterbodies appeared to have the strongest effect on different water quality parameters. Scientific findings from this study will be vital for decision-makers in developing a more robust land use management plan at the local level.

The next three papers by Hung et al. [4], Acharjee et al. [5], and Kumar et al. [6] discussed the effect of water pollution and its impacts on different tropic levels, human health, and the ecosystem. Excessive nutrient enrichment or eutrophication is an environmental pollution problem that occurs in natural water bodies and causes a lot of socio-economic issues. Hung et al. [4] investigated the factors responsible for eutrophication in three reservoirs in Taiwan using regression analyses. The results indicate that the main factor influencing these reservoirs is total phosphorus, and the influence of total phosphorus when interacting with other factors on water quality trophic state is more serious than that of total phosphorus per se. This implies that the actual influence of total phosphorus on the eutrophic condition could be underestimated. Furthermore, there was no deterministic causality between climate and water quality variables. Moreover, it is found that the influencing patterns for the three reservoirs are different because the type, size, and background environment of each reservoir are different, which means it is difficult to predict eutrophication in reservoirs with a universal index or equation. However, the multiple linear regression model used in this study could be a suitable quick-to-use, case-by-case model option for this problem.

Acharjee et al. [5] examined river sediment as an environmental indicator to measure the pollution level in Surma River, Bangladesh. Further, it compares potential ecological risk index values using Hakanson Risk Index (RI) and Monte Carlo Simulation (MCS) approach to evaluate the environmental risks caused by these heavy metals. The obtained results using risk index values from RI and MCS provided valuable insights into the contamination profile of the river, indicating that the studied river is currently under low ecological risk for the studied heavy metals. This study can be utilized to assess the susceptibility of the river sediment to heavy metal pollution near an urban core and to have a better understanding of the contamination profile of a river.

Microplastics (MPs) are considered an emerging pollutant in the aquatic environment; however, there is a scientific knowledge gap in this regard in several Asian regions. Considering this aforementioned information, Kumar et al. [6] carried out a systematic review to provide an insightful understanding of the spatial distribution of scientific studies on MPs in freshwater conducted across the Asian region, utilized sampling methods, and a detailed assessment of the effects of MPs on different biotic components in freshwater ecosystems, with special focus on its potential risks on human health. The results of this review indicate that research on microplastics in Asia has gained attention since 2014, with a significant increase in the number in 2021 might be because of excessive plastic pollution during the COVID-19 situation. Moreover, these research works are concentrated in China, followed by India and South Korea. When talking about the type of research works, it was also found that most of the studies focused primarily on reporting the occurrence levels of MPs in freshwater systems, such as water and sediments, and aquatic organisms, with a lack of studies investigating the human intake of MPs and their potential risks to human health. Notably, comparing the results among different countries is a challenge because diverse sampling, separation, and identification methods were applied to estimate MPs. This review study suggests that further research on the dynamics and transport of microplastics in biota and humans is needed, as Asia is a major consumer of seafood products and contributes significantly to the generation of plastic litter in the marine environment. Moreover, there is a need for further research on policy and governance frameworks to address this emerging water pollutant more holistically.

The next four papers were based on different management methodologies for water resources and their interlinkages with other socio-ecological components. Considering the water-sensitive socio-economic growth in the Vietnamese Mekong Delta, Minh et al. [7] utilized MIKE 11 to quantify the spatio-temporal dynamics of water quality parameters. Results show that locations near cage culture areas exhibited higher BOD_5 values than sites close to pond/lagoon culture areas due to the effects of numerous point sources of pollution, including upstream wastewater and out-fluxes from residential and tourism activities in the surrounding areas, all of which had a direct impact on the quality of the surface water used for aquaculture. Moreover, as aquacultural effluents have intensified and dispersed over time, water quality in the surrounding water bodies has degraded. The findings suggest that the effective planning, assessment, and management of rapidly expanding aquaculture sites should be improved, including more rigorous water quality monitoring, to ensure the long-term sustainable expansion and development of the aquacultural sector in the Long Xuyen Quadrangle in particular and the Vietnamese Mekong Delta as a whole.

Considering the water resources as a limiting factor for socio-economic development, especially in arid/semi-arid regions, Shyam et al. [8] carried out the assessment of groundwater reserves, which was carried out in the Udaipur district, Aravalli range, India using an integrated approach of remote sensing, GIS, and field-based spatial modeling. Results show that the principal aquifer for the availability of groundwater in the studied area is quartzite, phyllite, gneisses, schist, and dolomitic marble, which occur in unconfined to semi-confined zones. Furthermore, all primary chemical ingredients were found within the permissible limit, including granum. We also found that the dynamic GW reserves of the area are 637.42 mcm/annum, and the total groundwater draft is 639.67 mcm/annum. The deficit GW reserves are 2.25 mcm/annum from an average rainfall of 627 mm; hence, the stage of groundwater development is 100.67% and categorized as over-exploited. However, as per the relationship between reserves and rainfall events, surplus reserves are available when rainfall exceeds 700 mm. It was concluded that enough static GW reserves are available in the studied area to sustain the requirements of the drought period. For the long-term sustainability of groundwater use, controlling groundwater abstraction by optimizing its use, managing it properly through techniques such as sprinkler and drip irrigation, and achieving more crop-per-drop schemes, will go a long way to conserving this essential reserve and creating maximum groundwater recharge structures.

Considering the emergence of different nexus approaches to understanding the dynamics and co-evolution of water and human systems, Nguyen et al. [9] developed a social-hydrological approach to enhance the water supply resilience in Con Dao Island, Vietnam. It used a water-balance model involving the Water Evaluation and Planning (WEAP) tool to conduct a scenario-based evaluation of water demands. In doing so, we assessed the impacts of socio-economic development, such as population growth and climate change, on increasing water demand. The modeling results showed that the existing reservoirs—the main sources to recharge the groundwater—play a critical role in enhancing water supply resilience on the island, particularly during the dry season. In addition, future water shortages can be solved by investment in water supply infrastructures in combination with the use of alternative water sources, such as rainwater and desalinated seawater. The findings further indicate that while the local actors have a high awareness of the role of natural resources, they seem to neglect the impacts of climate change. To meet the future water demands, this paper also gave some potential suggestions like upgrading and constructing new reservoirs, mobilizing resources for freshwater alternatives, and investing in water supply facilities as among the most suitable roadmaps for the island. In addition, strengthening adaptive capacity, raising awareness, and building professional capacity for both local people and officials are strongly recommended. The research concludes with a roadmap that envisages the integration of social capacity to address the complex interaction and co-evolution of the human–water system to foster water-supply resilience in the study area.

Inhabitants of low-lying islands face increased threats due to climate change as a result of their higher exposure and lesser adaptive capacity. Sagar Island, the largest inhabited estuarine island of Sundarbans, is experiencing severe coastal erosion, frequent cyclones, flooding, storm surges, and breaching of embankments, resulting in land, livelihood, and property loss, and the displacement of people at a huge scale. Hence, Bera et al. [10] assessed climate change-induced vulnerability and risk for Sagar Island, India, using an integrated geostatistical and geoinformatics-based approach. Based on the IPCC AR5 framework, the proportion of variance of 26 exposure, hazard, sensitivity, and adaptive capacity parameters was measured and analyzed. The results showed that 19.5% of mouzas (administrative units of the island), with 15.33% of the population in the southern part of the island, i.e., Sibpur–Dhablat, Bankimnagar–Sumatinagar, and Beguakhali–Mahismari, are at high risk (0.70–0.80). It has been concluded that the island has undergone tremendous land system transformations and changes in climatic patterns. Therefore, there is a need to formulate comprehensive adaptation strategies at the policy- and decision-making levels to help the communities of this island deal with the adverse impacts of climate change. The findings of this study will help adaptation strategies based on site-specific information and sustainable management for the marginalized populations living on similar islands worldwide.

Author Contributions: Conceptualization, P.K.; writing—original draft preparation, P.K. and R.A. All authors have read and agreed to the published version of the manuscript.

Funding: This research received no external funding.

Data Availability Statement: Not applicable.

Conflicts of Interest: The authors declare no conflict of interest.

References

1. Pradhan, R.M.; Behera, A.K.; Kumar, S.; Kumar, P.; Biswal, T.K. Recharge and Geochemical Evolution of Groundwater in Fractured Basement Aquifers (NW India): Insights from Environmental Isotopes ($\delta^{18}O$, δ^2H, and ^3H) and Hydrogeochemical Studies. *Water* **2022**, *14*, 315. [CrossRef]
2. Behera, A.K.; Pradhan, R.M.; Kumar, S.; Chakrapani, G.J.; Kumar, P. Assessment of Groundwater Flow Dynamics Using MODFLOW in Shallow Aquifer System of Mahanadi Delta (East Coast), India. *Water* **2022**, *14*, 611. [CrossRef]
3. Kadir, A.; Ahmed, Z.; Uddin, M.M.; Xie, Z.; Kumar, P. Integrated Approach to Quantify the Impact of Land Use and Land Cover Changes onWater Quality of Surma River, Sylhet, Bangladesh. *Water* **2022**, *14*, 17. [CrossRef]
4. Hung, C.W.; Chiang Hsieh, L.H. Analysis of Factors Influencing the Trophic State of Drinking Water Reservoirs in Taiwan. *Water* **2021**, *13*, 3228. [CrossRef]
5. Acharjee, A.; Ahmed, Z.; Kumar, P.; Alam, R.; Rahman, M.S.; Simal-Gandara, J. Assessment of the Ecological Risk from Heavy Metals in the Surface Sediment of River Surma, Bangladesh: Coupled Approach of Monte Carlo Simulation and Multi-Component Statistical Analysis. *Water* **2022**, *14*, 180. [CrossRef]
6. Kumar, P.; Inamura, Y.; Bao, P.N.; Abeynayaka, A.; Dasgupta, R.; Abeynayaka, H.D.L. Microplastics in Freshwater Environment in Asia: A Systematic Scientific Review. *Water* **2022**, *14*, 1737. [CrossRef]
7. Thu Minh, H.V.; Tri, V.P.D.; Ut, V.N.; Avtar, R.; Kumar, P.; Dang, T.T.T.; Hoa, A.V.; Ty, T.V.; Downes, N.K. A Model-Based Approach for Improving SurfaceWater Quality Management in Aquaculture Using MIKE 11: A Case of the Long Xuyen Quadangle, Mekong Delta, Vietnam. *Water* **2022**, *14*, 412. [CrossRef]
8. Shyam, M.; Meraj, G.; Kanga, S.; Sudhanshu; Farooq, M.; Singh, S.K.; Sahu, N.; Kumar, P. Assessing the Groundwater Reserves of the Udaipur District, Aravalli Range, India, Using Geospatial Techniques. *Water* **2022**, *14*, 648. [CrossRef]
9. Nguyen, D.C.H.; Nguyen, D.C.; Luu, T.T.; Le, T.C.; Kumar, P.; Dasgupta, R.; Nguyen, H.Q. Enhancing Water Supply Resilience in a Tropical Island via a Socio-Hydrological Approach: A Case Study in Con Dao Island, Vietnam. *Water* **2021**, *13*, 2573. [CrossRef]
10. Bera, A.; Meraj, G.; Kanga, S.; Farooq, M.; Singh, S.K.; Sahu, N.; Kumar, P. Vulnerability and Risk Assessment to Climate Change in Sagar Island, India. *Water* **2022**, *14*, 823. [CrossRef]

Article

Recharge and Geochemical Evolution of Groundwater in Fractured Basement Aquifers (NW India): Insights from Environmental Isotopes ($\delta^{18}O$, $\delta^{2}H$, and ^{3}H) and Hydrogeochemical Studies

Rudra Mohan Pradhan [1,*], Ajit Kumar Behera [2], Sudhir Kumar [3], Pankaj Kumar [4,*] and Tapas Kumar Biswal [1]

1 Department of Earth Sciences, Indian Institute of Technology Bombay, Powai 400076, India; tkbiswal@iitb.ac.in
2 Marine Geoscience Group, National Centre for Earth Science Studies, Thiruvananthapuram 695011, India; ajitgeol.89@gmail.com
3 Hydrological Investigations Division, National Institute of Hydrology Roorkee, Roorkee 247667, India; sudhir.nih@gmail.com
4 Institute for Global Environmental Strategies, Kamiyamaguchi, Hayama 2108-11, Kanagawa, Japan
* Correspondence: rmp.geol@gmail.com (R.M.P.); kumar@iges.or.jp (P.K.); Tel.: +81-046-855-3858 (P.K.)

Abstract: Considering water as a limiting factor for socio-economic development, especially in arid/semi-arid regions, both scientific communities and policymakers are interested in groundwater recharge-related data. India is fast moving toward a crisis of groundwater due to intense abstraction and contamination. There is a lack of understanding regarding the occurrence, movement, and behaviors of groundwater in a fractured basement terrane. Therefore, integrated environmental isotopes ($\delta^{18}O$, $\delta^{2}H$, and ^{3}H) and hydrogeochemical studies have been used to understand the recharge processes and geochemical evolution of groundwater in the fractured basement terranes of Gujarat, NW India. Our results show that the relative abundance of major cations and anions in the study basin are $Ca^{2+} > Na^{+} > Mg^{2+} > K^{+}$ and $HCO_3^{-} > Cl^{-} > SO_4^{2-} > NO_3^{-}$, respectively. This suggests that the chemical weathering of silicate minerals influences the groundwater chemistry in the aquifer system. A change in hydrochemical facies from Ca-HCO_3 to Na-Mg-Ca-Cl. HCO_3 has been identified from the recharge to discharge areas. Along the groundwater flow direction, the presence of chemical constituents with different concentrations demonstrates that the various geochemical mechanisms are responsible for this geochemical evolution. Furthermore, the chemical composition of groundwater also reflects that the groundwater has interacted with distinct rock types (granites/granulites). The stable isotopes ($\delta^{18}O$ and $\delta^{2}H$) of groundwater reveal that the local precipitation is the main source of recharge. However, the groundwater recharge is affected by the evaporation process due to different geological conditions irrespective of topographical differences in the study area. The tritium (^{3}H) content of groundwater suggests that the aquifer is mainly recharged by modern rainfall events. Thus, in semi-arid regions, the geology, weathering, and geologic structures have a significant role in bringing chemical changes in groundwater and smoothening the recharge process. The findings of this study will prove vital for the decision-makers or policymakers to take appropriate measures to design water budgets as well as water management plans more sustainably.

Keywords: groundwater; fractured rock; hydrogeochemistry; geochemical evolution; environmental isotopes ($\delta^{18}O$; $\delta^{2}H$; and ^{3}H); Ambaji Basin; NW India; socio-economic development; water resource management

Citation: Pradhan, R.M.; Behera, A.K.; Kumar, S.; Kumar, P.; Biswal, T.K. Recharge and Geochemical Evolution of Groundwater in Fractured Basement Aquifers (NW India): Insights from Environmental Isotopes ($\delta^{18}O$, $\delta^{2}H$, and ^{3}H) and Hydrogeochemical Studies. *Water* **2022**, *14*, 315. https://doi.org/10.3390/w14030315

Academic Editors: Pankaj Kumar, Ram Avtar and Domenico Cicchella

Received: 7 December 2021
Accepted: 18 January 2022
Published: 21 January 2022

1. Introduction

Globally, groundwater resources serve one-third of freshwater supplies, accounting for nearly 36% for domestic purposes, 42% for agricultural use, and 27% for industrial activities [1]. About 70% of the land area in India is underlain by crystalline basement rocks and approximately 30% of them are merely covered with Precambrian basement or hard rocks [2]. In the last few decades, the demand for groundwater and overexploitation of groundwater resources has significantly amplified, particularly in semi-arid and arid areas due to an increase in population, urbanization, and growth of the worldwide economy.

In semi-arid environments, understanding the groundwater evolution, flow path, and recharge processes mechanisms are very much needed for the optimal usage of groundwater resources [3,4]. Furthermore, groundwater resources in basement hard rock terranes are often restricted to the top weathered fractured zones that usually spread up to 50 m depth, and below that, groundwater movement is mainly controlled by deeply fractured zones [2,5–8]. In these terranes, the geochemical properties of groundwater are essentially a function of the mineral composition of the rock through which it flows [9,10]. In addition, the concentration of chemical species varies along its flow path [9,11–14], and several hydrogeochemical processes impact groundwater geochemistry viz. topography, precipitation intensity, water–rock interaction, mixing, dissolution, ion exchange, and oxidation–reduction process [15–17]. These processes are mainly depending on the physicochemical and biological properties of bedrocks with climatic conditions. Furthermore, hydrogeochemical properties are also used as a proxy for determining the recharge areas and sources [18]. Additionally, physical methods are also used to assess the recharge process. However, the interpretation remains equivocal due to variation in weathered fractured zones as well as seasonal variability of rainfall and groundwater level [19]. Hence, integrating the hydrogeochemical datasets with stable isotopes ($\delta^{18}O$ and δ^2H) and radiogenic isotopes (hydrogen-3, i.e., ^3H) can aid significantly in tracing the hydrogeochemical processes and in identifying the recharge environments of groundwater systems [20].

Generally, the stable isotopic compositions are preserved in the groundwater body until it mixed with other water with altered stable isotopic signatures [21,22]. Several studies have used stable isotopes in tracing the sources of groundwater around the world viz. groundwater–surface water interaction studies [23,24], groundwater recharge assessment along different flow paths [22,25–27], isotope tracing of paleo groundwater [28], and evaporation effects on groundwater resources [27,29]. Among the other environmental isotopes, tritium (^3H) is a unique isotope used in hydrogeological studies to understand the flow direction and groundwater age, and it has been applied in several regions (Northern China [30], South Florida [31], California Basin [32], Banana Plain [14], Punjab state [33]).

Identifying the recharge processes in semi-arid regions has gained substantial attention in recent years. However, most of the work focused mainly on sedimentary or alluvial terranes and less on crystalline fractured rock terranes with varied climatic zones [4,34–37]. Especially in NW India, no such research has been carried out, although the communities are heavily dependent on groundwater resources for domestic and agricultural activities.

Therefore, the present study is aimed at understanding the complex fractured basement aquifers using hydrogeochemical, stable isotopes ($\delta^{18}O$ and δ^2H), as well as radiogenic isotopes (^3H). The specific goal is to identify the geochemical evolution and groundwater recharge processes that occurred within the fractured crystalline basement aquifers of Ambaji basin (Gujarat), NW India. The outcome of the work will lead to better groundwater management and practices for sustainable water use especially in fractured basement terrane.

2. Description of the Study Area

2.1. General Characteristics

The study area (Ambaji basin) lies in the northern end part of the Banaskantha district (North Gujarat), NW India. The area is bounded by latitude 24°10′–24°22′ N and longitude 72°30′–72°50′ E (Figure 1). The areal extent is ≈450 km^2 and is divided into two Talukas

(Amirgarh and Danta taluka). The area is mostly dominated by hilly terrane and a smaller portion of the low-lying flat terrane. The altitude varies from 250 to 650 m a.m.s.l. The area comes under a semi-arid climatic type and is characterized by extreme temperature in the summer months (May–July), erratic rainfall, and high evapotranspiration rates [10,38]. The average annual rainfall is ≈771 mm and is typically received through the southwest monsoon, and the temperature varies between 15 and 42 °C in this area.

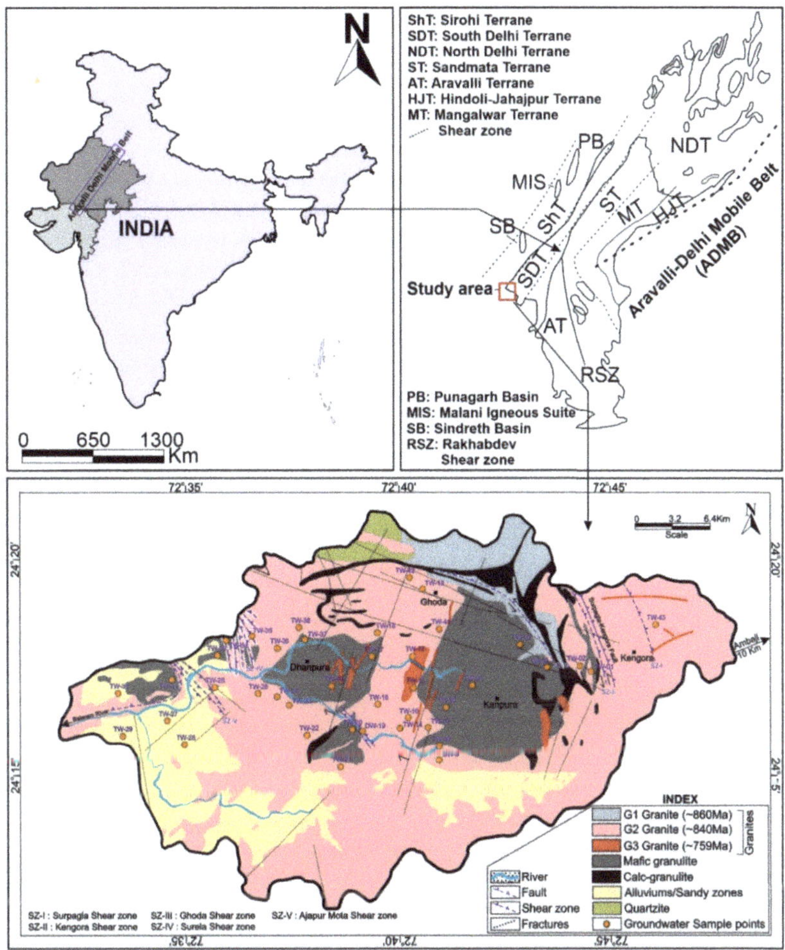

Figure 1. Geological map of the study area (Ambaji Basin, NW India) and groundwater sampling locations (Pre-monsoon, December 2017). Figure 1 inset showing the major geological terranes of Aravalli Delhi Mobile Belt (ADMB), NW India after GSI, [38].

2.2. Geological and Geomorphological Settings

Geologically, the Ambaji basin belongs to the South Delhi Terrane (SDT) of Aravalli-Delhi Mobile Belt (ADMB), NW India (Figure 1). The ADMB comprises several geological terranes including SDT (Figure 1 inset) [39]. Furthermore, several granitic intrusions have occurred in the ADMB i.e., the Berach, the Jasrapur, the Sendra, the Ambaji (study area), the mount Abu Granite, and the Erinpura granites. The different geological units in the study area were demarcated through extensive fieldwork and the existing geological map of the Geological Survey of India. As the study basin comes under the Meso-proterozoic age of SDT, the rocks present in this terrane were mainly of pelitic-, calcareous-, and basic granulites, where three phases of granite intruded namely G1 (gneissic), G2 (medium to coarse-grained, highly fractured), and G3 granites (fine-grained or micro-granite) [38,39]. Figure 1 shows that G2 granite is mostly dominated in this area, which is followed by basic granulite, alluviums, G1 granite, and G3 granites. Furthermore, the G2 granite is highly weathered in different regions. The area is predominantly comprised of different erosional and depositional hydrogeomorphic units viz. structural hills, denudational hills, residual hills, shallow to deep buried pediments, and valley fills. The eastern part mainly consists of hilly terranes with highly undulating surfaces, whereas the southwestern end areas are of gently undulating surfaces. The upper part of the basin mainly consists of structural hills, whereas the lower part of the basin is covered with alluviums [38].

2.3. Structural and Hydrogeological Scenarios

The basin has witnessed several faults, fractures, and shear zones. The fractures mostly show three sets of orientations i.e., NE-SW, NNW-SSE, and NW-SE (Figure 1). The study area comprises several criss-crossed fractures in the northern part and northwestern part due to transtensional settings and is characterized by both extensional and compressional structures [38]. These transtensional settings lead to multiple phases of deformation, which caused several criss-crossed fractures or lineaments. The major percentage of the study basin is covered by crystalline basement rocks. Due to the absence of primary porosity, secondary porosity such as fractures, faults, and shear zones serve as the main source of groundwater resources. Furthermore, groundwater resources in the shallow part were mostly covered with weathered zones/topsoil, while the deeper part mostly consists of fault and fracture zones and shear zones followed by massive rocks. The groundwater water levels in this area range from 2.5 to 49 m below ground level (b.g.l.) in the pre-monsoon (May 2017) and 0.65–47.0 m b.g.l in the post-monsoon (December 2017). The regional groundwater flows from the northeast to southwest direction and is primarily influenced by the secondary porosities (faults/fractures/joints) and surface topography. The area has largely three types for hydrogeological formations: (1) the top weathered zones range from 1 to 30 m, (2) the fault and fractured zones comprised of granite range from 20 to 150 m b.g.l., and (3) massive granites (Figure 2). The area belongs to a semi-arid climate, and the rivers flowing through it are mostly ephemeral and the major drainage network is constituted by the Banas River flowing in the NE–SW direction and its tributaries i.e., Balaram River and Teliya Nadi.

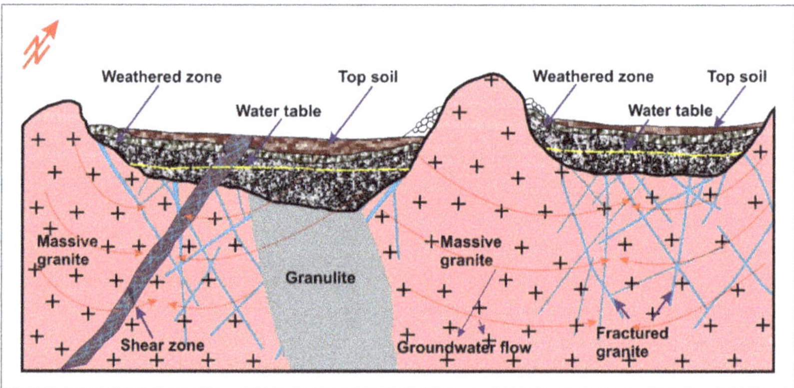

Figure 2. A schematic diagram shows different hydrogeological zones (top-soil, weathered zones, fractured zones, and massive zones) and groundwater circulations in the subsurface.

3. Materials and Methods

3.1. Groundwater Sampling and Geochemical Analysis

A total of forty (n = 40) groundwater samples were collected for this study for stable isotopes (oxygen and hydrogen), radiogenic isotope (tritium), and major cations–anions analyses during December 2017 (post-monsoon) from different tube wells, dug wells, and bore wells with variable depths (<30 m) (Figure 1). Pre-cleaned high-density polyethylene (HDPE) bottles (500 mL) were used to collect the groundwater samples. Water was pumped out for 10–15 min before the sampling. For stable isotopes ($\delta^{18}O$ and $\delta^{2}H$) analyses, the water samples were collected in 10 mL HDPE bottles, and for radiogenic isotopes (^{3}H), water samples are collected in 1 L HDPE bottles and sealed tightly. The groundwater samples collected for major cations and anions were filtered using 0.45 μm Millipore filter paper in the field. In situ parameters such as pH, electrical conductivity (EC), temperature (°C), and total dissolved solids (TDS) were measured in the field through HANNA USA-made (Model: HI98130) portable meter. Bicarbonate (HCO_3) of groundwater samples was measured in the field by the titration method. The collected samples were preserved in a cool place after sampling and then transferred to a refrigerator for preservation until the geochemical analysis. The major cations such as Ca^{2+}, Mg^{2+}, Na^{+}, and K^{+} and anions Cl^{-}, SO_4^{2-}, and NO_3^{-} of groundwater samples were measured through a UV visible spectrophotometer at IIT Bombay and ion chromatography (Metrohm 883 Basic IC Plus) at the SEOCS, IIT Bhubaneswar with appropriate standards.

3.2. Stable Isotope Data ($\delta^{18}O$ and $\delta^{2}H$) Analysis

The refrigerated water samples collected for stable isotopes ($\delta^{18}O$ and $\delta^{2}H$) were analyzed at the Nuclear Hydrology Laboratory, NIH Roorkee. For measuring $\delta^{2}H$, a dual inlet stable isotope-ratio mass spectrometer (DISIRMS) was used, and a continuous-flow stable isotope-ratio mass spectrometer (CFSIRMS) was used to quantify the $\delta^{18}O$. The water samples were equilibrated with CO_2 and H_2 to quantity $\delta^{18}O$ and $\delta^{2}H$ values respectively using the standard method [40]. Then, the instrument was calibrated to determine the $\delta^{18}O$ and $\delta^{2}H$ composition by analyzing IAEA standards i.e., Vienna standard mean ocean water (VSMOW) [41] with precision range ±1.0‰ for $\delta^{2}H$ and ±0.1‰ for $\delta^{18}O$. The results of the isotopes are expressed in terms of per mil (‰) relative to VSMOW using the 'δ' notation and Equation (1).

$$\delta(‰) = \left(\frac{R_{sample} - R_{standard}}{R_{standard}}\right) \times 1000 \quad (1)$$

Here, R_{sample} is the ratio of $^{18}O/^{16}O$ and $^{2}H/H$ isotopes for the collected groundwater sample, and $R_{reference}$ is the ratio of $^{18}O/^{16}O$ and $^{2}H/H$ isotopes for the standard water sample. The reference standard is usually considered IAEA VSMOW, and the measurement precision is ±0.1‰ and ±1‰ for $\delta^{18}O$ and $\delta^{2}H$, respectively. The isotope data reported in this paper correspond to VSMOW.

3.3. Radiogenic Tritium (^{3}H) Analysis

A total of twenty-two identical groundwater samples were collected for tritium (^{3}H) analysis. Usually, one liter of sample is enough for radiogenic ^{3}H analysis. The water samples were collected in unfiltered condition, and no preservatives were added. The samples were stored in HDPE bottles with air-tight caps. The samples were analyzed at the Nuclear Hydrology Laboratory, NIH Roorkee. For tritium analysis, three steps were followed: (a) sample distillation, (b) fractionation by electrolytic enrichment (for removing ^{1}H and ^{2}H), and (c) measurement of tritium on Ultra Low-Level Liquid Scintillation spectrometry. The WinQ and QuickStart software on the QUANTULUS system was used to process the data. The measured tritium concentrations are stated in tritium units (TU), and the error varies between ±0.12 and ±0.24 TU.

4. Results and Discussion

4.1. Hydrogeochemical Studies

4.1.1. General Hydrogeochemistry

The physicochemical parameters and statistical data of the groundwater samples used in this study have been provided in Table 1. Detailed information about the water chemistry data is provided in Supplementary Table S1. The pH range of groundwater varies from 6.60 to 7.38 with an average value of 6.97, indicating that the groundwater is slightly acidic, which may be due to the mixture of carbonic acid in the water of the aquifer system. The electrical conductivity (EC) ranges between 360 and 2980 μS/cm, whereas the TDS varies from 216 to 1788 mg/L. The respective average value of EC and TDS is 1231.50 μS/cm and 738 mg/L, which suggests that most of the groundwater samples can be used for drinking water purposes, as it follows the WHO guidelines. However, few groundwater samples (Table S1) with high NO_3^- (>45 mg/L, WHO) are not suitable for drinking purposes. The order of relative abundance of the major cations are $Ca^{2+} > Na^+ > Mg^{2+} > K^+$ and anions $HCO_3^- > Cl^- > SO_4^{2-} > NO_3^-$ (Figure 3).

Table 1. Physicochemical and isotopic results of groundwater samples collected during the post-monsoon season from Ambaji Basin (NW India) (December 2017).

Parameter	Units	Minimum	Maximum	Average
pH		6.6	7.38	6.97
EC	(μS)	360	2980	1231.5
TDS	(mg/L)	216	1788	738.9
Ca^{2+}	(mg/L)	21.92	183.6	87.04
Mg^{2+}	(mg/L)	11.94	111.28	48.97
Na^+	(mg/L)	28.7	260.15	74.19
K^+	(mg/L)	0.5	7.27	2.69
SO_4^{2-}	(mg/L)	12.29	195.94	40.61
Cl^-	(mg/L)	16.08	425.7	105.01
NO_3^-	(mg/L)	0.61	182.48	34.34
HCO_3^-	(mg/L)	125	535	359.25
δ^2H	(‰)	−41.9	−24.51	−34.03
$\delta^{18}O$	(‰)	−6.17	−3.23	−4.93
d-excess	(‰)	1.3	8.11	5.44
^3H	(TU)	1.97	28.05	5.1

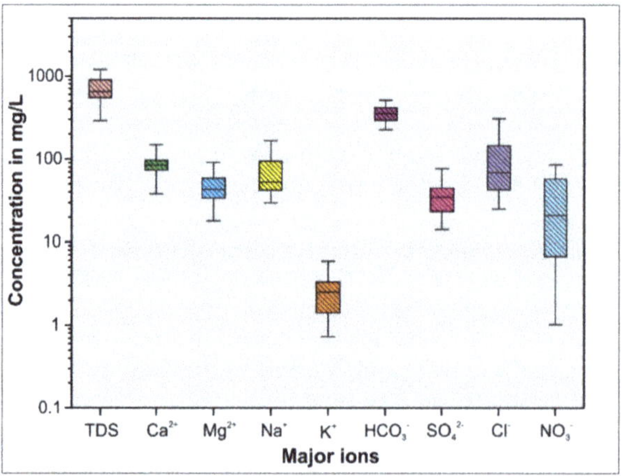

Figure 3. Major ion concentration variation in the study area.

Bicarbonates (HCO_3^-) represent the alkalinity that ranges from 125 to 535 mg/L with an average of 359.25 mg/L. Such a high concentration of bicarbonates in groundwater is observed due to the chemical weathering of carbonate minerals and calcite dissolution [42,43]. Similarly, chloride (Cl^-) with an average concentration of 105.01 mg/L indicates that there is an interaction between freshwater from the recharge area and highly dissolved water from the discharge area [44–46]. Apart from the chloride (Cl^-) content, a high concentration of nitrate (NO_3^-) (>45 mg/L) [47] is also observed in 30% of the groundwater samples due to agricultural activities and semi-arid climatic conditions. However, the sulfate (SO_4^{2-}) concentration is between 12.29 and 195.94 mg/L with an average of 40.61 mg/L, indicating that the various sources such as pesticides used for agricultural productivity, evaporate dissolution, and domestic sewages are responsible for the high concentration of sulfate [48,49]. Calcium (Ca^{2+}) is the most abundant ion; it ranges from 21.92 to 183.60 mg/L (average 87.04 mg/L), and it is derived from calc-silicate minerals due to chemical weathering and the dissolution of calcium carbonate minerals [50]. Sodium (Na^+) is the second most abundant cation, which varies from 28.70 to 260.15 mg/L. This illustrates that the source of sodium (Na^+) in groundwater is mostly plagioclase feldspar, which is the common silicate mineral of granite [9,11] as covered in most of the study regions. The magnesium concentration ranges from 11.94 to 111.28 mg/L with an average of 48.97 mg/L, which indicates that the groundwater has significantly interacted with granulites as it consists of Mg-calcite, biotite, and amphibole minerals [51]. Potassium (K^+) is the lowest; it is the most abundant in the groundwater samples ranging from 0.50 to 7.27 mg/L (average 2.69 mg/L), which leached out from the clay-bearing minerals being produced by the chemical weathering of silicate minerals [52].

4.1.2. Hydrochemical Facies Variation

A piper trilinear plot is very useful in analyzing and understanding the geochemical evolution and chemical relations of groundwater. This plot is further used to assess the recharge flow paths by plotting the concentrations of major cations and anions [53]. The plot (Figure 4) shows that the groundwater samples are categorized by major water types such as (I) Ca-HCO_3, (II) Na-Cl, (III) Mixed Ca-Na-HCO_3, (IV) Mixed Ca-Mg-Cl, (V) Ca-Cl, and (VI) Na-HCO_3. Most of the samples are collected from granite (G2) and Mafic granulite of the shallow aquifer (depth < 50 m) except for four (04) samples (TW-27, TW-28, TW-29, and TW-30), which are collected from alluvial/sandy zones, as shown in Figure 1. Around 77% of the total groundwater samples show Ca-HCO_3 type, out of which 36% samples are derived from the recharge area, which is supposed to be recharged only by rainfall [2,11]. Furthermore, the low groundwater salinity with the Ca-HCO_3 dominated type of water indicates the rapid recharge of groundwater [18]. This infers that the samples (TW-13, TW-18, TW-38, TW-41) with a TDS value <500 mg/L collected from the recharge area of granite aquifer (Figure 1) are in fresh condition, which is due to the early stage of the rainfall recharge process. However, TW-1 and TW-2 (TDS > 500 mg/L) and TW-43 (TDS > 1000 mg/L) are located in the same recharge region, suggesting that the infiltration through fractures leads to the dilution of groundwater with existing groundwater and subsequently increases the TDS value along its flow direction [54]. Furthermore, the hydrochemical facies shift from Ca-HCO_3 to Ca-Na-HCO_3 due to the release of sodium ions from clay minerals, which are formed due to the chemical weathering of silicate minerals. Thus, ion exchange plays a significant role in this geochemical evolution during the initial stage of groundwater flow and is expressed in Equations (2) and (3).

$$2Na(AlSi_3)O_8 + 2H^+ + 9H_2O \rightarrow Al_2Si_2O_5(OH)_4 + 2Na^+ + 4H_4SiO_4 \quad (2)$$

$$Ca(Al_2Si_2)O_8 + 2H^+ + H_2O \rightarrow Al_2Si_2O_5(OH)_4 + Ca^{2+} \quad (3)$$

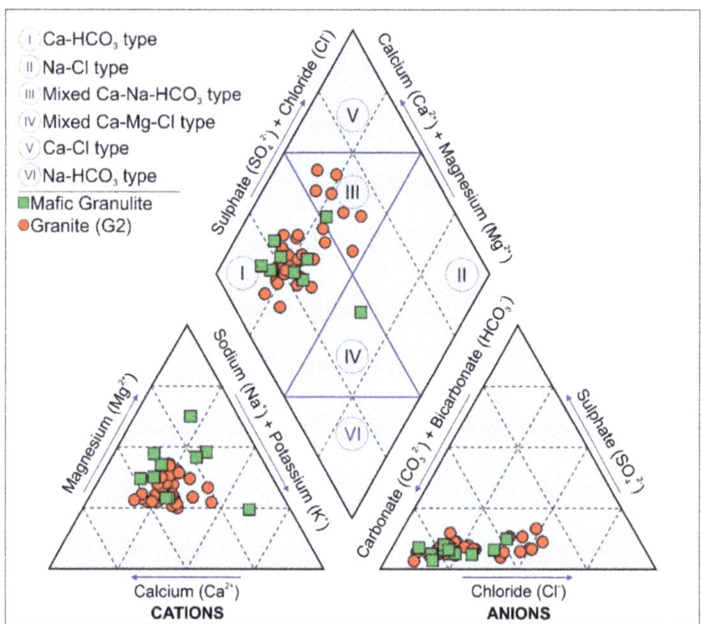

Figure 4. Piper trilinear diagram showing major groundwater types and geochemical evolution.

The samples (TW-3 to TW-6) collected from the granulite shallow aquifer are showing TDS > 500 mg/L, which illustrates that there is a prolonged groundwater interaction with the aquifer matrix [2,11]. The incongruent dissolution of ferromagnesian minerals, hornblende, and pyroxene minerals from mafic granulite rock is attributed to the Mg(Ca)-Na-HCO_3 hydrochemical facies in these samples. The presence of Mg^{2+} in groundwater is driven by the leaching of pyroxene from the granulite by Equation (4).

$$(Mg_{0.7}CaAl_{0.3})(Al_{0.3}Si_{1.7})O_6 + 3.4H^+ + 1.1H_2O \rightarrow 0.3Al_2Si_2O_5(OH)_4 + Ca^{2+} + 0.7Mg^{2+} + 1.1H_4SiO_4 \quad (4)$$

The rest of the groundwater samples have been categorized under type III water i.e., Mixed Ca-Na-HCO_3, which are mostly from the granitic body. The hydrochemical facies shown by these samples is Ca-Mg-Na-HCO_3. This is due to the groundwater chemistry driven by the weathering of ferromagnesian silicate minerals, as there must be an interaction with granulite rock during its flow. In addition, the samples collected from a low topography area or discharge area with a TDS value >1000 mg/L show the mixed-type of water facies viz. Ca-Na (Mg)-HCO_3(Cl), Ca-Mg (Na)-HCO_3, Na-Mg-Ca-Cl HCO_3, etc. The presence of a high concentration of chloride (Cl^-) in groundwater implies the sluggish movement of groundwater with minimal flushing capacity of the aquifer [11].

4.1.3. Mechanisms Controlling the Groundwater Chemistry

Gibb's plot of groundwater samples in the Ambaji Basin is plotted in Figure 5. The plot shows that most of the groundwater is confined in the rock dominance area. Around 95% of the groundwater samples show ratios of $Na^+/(Na^+ + Ca^{2+})$ and $Cl^-/(Cl^- + HCO_3^-)$ that are less than 0.5 with a TDS value level <1000 mg/L. This suggests that the groundwater significantly interacted with fractured granite or mafic granulite during its movement. Chemical weathering is the controlling factor for releasing Ca into the groundwater, as calcium silicate and ferromagnesian minerals are the major constituents in these rock types [52,55]. Very few numbers (around 5%) of samples show the ionic ratio of $Na^+/(Na^+ + Ca^{2+})$ and $Cl^-/(Cl^- + HCO_3^-)$ more than 0.5 having a TDS value >1000 mg/L, with an indication of shifting of samples from rock dominance to evaporation dominance (Figure 5). This may be due to the chemical weathering and anthropogenic activities that are attributed to evaporation [56], which leads to changes in the groundwater chemistry in Ambaji Basin. In this case, the release of Na^+ ions from clay minerals or fertilizers and increase in residence time may be responsible for a higher ratio of $Na^+/(Na^+ + Ca^{2+})$ in groundwater. No samples belong to the precipitation dominance category, illustrating a limited supply of ions from the atmosphere [57].

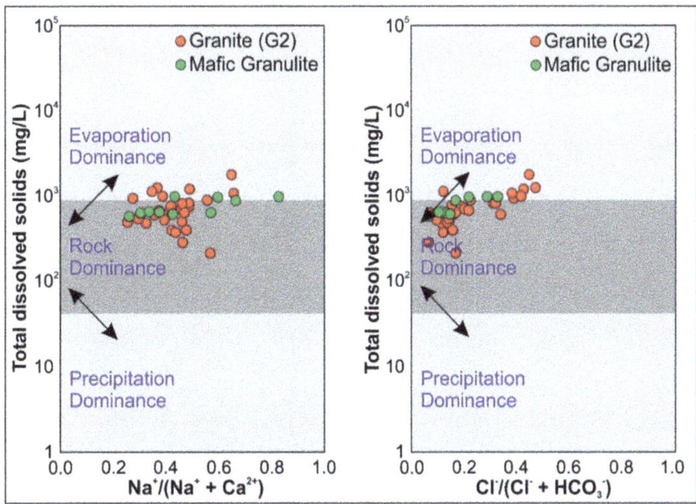

Figure 5. Gibbs plot showing geochemical evolution and rock–water interaction mechanisms.

4.1.4. Ionic Ratios and Hydrogeochemical Evolution

Groundwater chemistry mainly depends on the geochemical reactions and processes that occur within the groundwater system [58]. Various binary plots (Figure 6) can explain the various controlling mechanisms for the geochemical evolution of groundwater. In Figure 6a, the samples lie in between reference lines 1:1 and 2:1 on the bivariate plot of total cations versus alkali earth metals (Ca + Mg), indicating that the contribution of Ca and Mg ions has a significant role in the derivation of groundwater chemistry. This may be due to carbonate dissolution [59] or the weathering of calc-silicate and ferromagnesian minerals/carbonate minerals (calcite and dolomite) that are likely to be present in fractured rock bodies (both in granite and granulite) [11]. However, the scatter plot (Figure 6b) between total cations versus alkali metals (Na + K) shows that nearly all samples are plotted below the reference line (2:1). This infers that the contribution of alkalis metal is relatively less than alkali earth metals, which can be explained by the enrichment of Ca^{2+} in groundwater due to the Ca-Na exchange process.

The ion exchange and silicate weathering are the major controlling factors for the geochemical evolution of groundwater in this aquifer system. Furthermore, the presence of fractures or faults may also be responsible for the ion exchange process [60], which can bring additional Ca^{2+} ions to the aquifer system. The scatter plot between Ca^{2+} and HCO_3^- as shown in Figure 6d illustrates that most of the samples are lying above the 1:1 ratio reference line, whereas very few numbers of samples fall below the reference line. This indicates that the relative enrichment of HCO_3^- with respect to Ca^{2+} is due to silicate weathering. Furthermore, an excess of Ca^{2+} (some samples falling below the line) infers that the cation exchange process is one of the driving mechanisms for the chemistry of groundwater. The weathering of ferromagnesian silicate minerals in the groundwater can also be elucidated through the correlation between $Ca^{2+} + Mg^{2+}$ and HCO_3^-, as shown in

Figure 6c. In this case, the majority of the samples are plotted very close to the equilibrium line (1:1 ratio reference line) as opposite to Figure 6d. Thus, adding Mg^{2+} brings the major chemical variation in the groundwater system, which originates from clay minerals. During chemical weathering, the feldspar gets altered to clay minerals that mainly constitute Ca^{2+}, Mg^{2+}, Na^{+}, K^{+}, etc. as major ions. As the groundwater interacts with the aquifer matrix, an exchange of ions between Ca^{2+} or Mg^{2+} and Na^{+} takes place, which is also known as the reverse ion exchange process [59,61] and seemed to be one of the controlling factors.

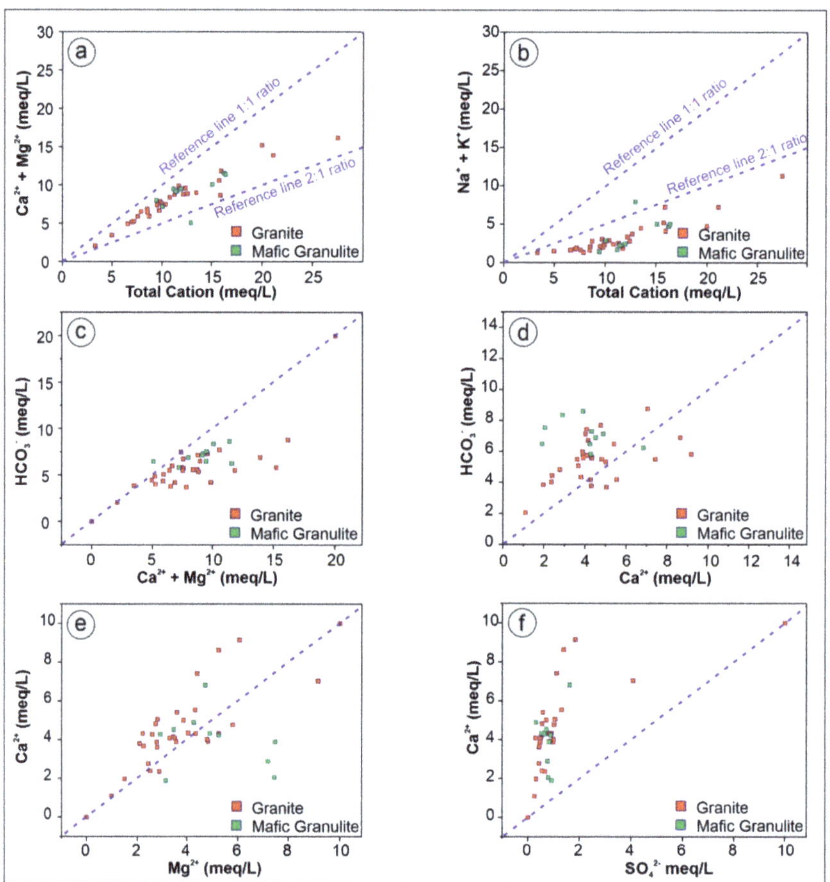

Figure 6. Bivariate plots of (**a**) Ca^{2+} + Mg^{2+} vs. Total cation, (**b**) Na^{+} + K^{+} vs. Total cation, (**c**) HCO_3^{-} vs. Ca^{2+} + Mg^{2+}, (**d**) HCO_3^{-} vs. Ca^{2+}, (**e**) Ca^{2+} vs. Mg^{2+}, (**f**) Ca^{2+} vs. SO_4^{2-}.

The relationship between Ca^{2+} and Mg^{2+} can be derived from the scatter plot, as shown in Figure 6e. This graph shows that the samples collected from granite terrane fall both above the reference line, whereas the samples collected from granulite terrane lies below the reference line. This suggests that the samples above and below the reference line are Ca and Mg-rich, respectively. The presence of calc-silicate and ferromagnesian minerals in different lithology influences the groundwater system and is an attribute of the chemical variation in the groundwater. However, to identify any influence of the carbonate weathering process in the groundwater system, a scatter plot between Ca^{2+} and SO_4^{2-} has been plotted (Figure 6f). All the samples lie above the reference line with an excess of Ca^{2+} relative to SO_4^{2-} ions inferring the existence of carbonate weathering [42].

4.1.5. Major Geochemical Processes

Silicate weathering is the main controlling mechanism for the geochemical evolution of groundwater in the study basin. The study area is predominantly composed of granites and granulites, which consist of silicate minerals such as feldspars, pyroxenes, biotite, quartz, etc. The bivariate plots Ca^{2+}/Na^{+} versus Mg^{2+}/Na^{+} and Ca^{2+}/Na^{+} versus Na^{+}/HCO_3^{-} are used to classify three important processes that govern the geochemistry of water i.e., carbonate dissolution, evaporite dissolution, and silicate weathering [17,62]. As shown in Figure 7a, Ca/Na > 1 is an indication of Ca-rich groundwater, and HCO_3/Na > 1 suggests the dissolution of silicate minerals. In this study area, granites consist of Ca feldspar, which has undergone a weathering process, and the dissolution of feldspar releases Ca^{2+} to the groundwater system. While the groundwater moves from higher to lower topography (from NE to SW direction) in the investigated area (Figure 1), it shows increasing TDS value with an increase in Ca^{2+}. However, there is a sharp change in groundwater chemistry along its flow path when it interacts with mafic granulite rock, consisting of ferromagnesian minerals. In this particular region, Mg^{2+} slightly exceeds Ca^{2+} and gives rise to the Mg-Ca-Na-HCO_3 type of water, as discussed in the earlier section of the paper. Thus, the weathering of Mg-rich silicate minerals releases Mg^{2+} into the groundwater and replaces Ca^{2+} on participating in the cation exchange process. The Mg/Na > 1 reveals that Mg^{2+} concentration is more than that of Na^{+} in groundwater (Figure 7b), whereas the Mg/Na < 1 shows the dominancy of Na^{+} ion in some groundwater samples, suspecting the normal ion exchange process.

To understand this, a plot, Na^{+} + K^{+}-(Cl^{-}) versus Ca^{2+} + Mg^{2+}-(SO_4^{2-} + HCO_3^{-}), has been plotted, in which almost all samples fall on the straight line with a slope of –1.31 (Figure 8a). This suggests that the reverse ion exchange process is one of the significant factors for the enrichment of Ca or Mg in deeply fractured aquifer relative to Na and also influences the groundwater chemistry in the semi-arid region [60,61,63], as in the case of Ambaji Basin. During this process, Na+ gets adsorbed in favorable exchange sites and replaced with Ca^{2+} or Mg^{2+}. This can be explained by Equation (5).

$$2Na^{+} + Ca^{2+}\left(Mg^{2+}\right) - Clay \leftrightarrow Na^{+} - Clay + Ca^{2+}\left(Mg^{2+}\right). \quad (5)$$

The scattered plot between Na/Cl and Cl illustrates that the Na concentration in groundwater is due to the ion exchange and dissolution process (Figure 8b). The ratio of Na/Cl > 1 indicates the excess of Na over Cl in groundwater, as driven by the ion exchange process in a continuous groundwater flow system. However, some samples show Na/Cl < 1 with rising in salinity (Cl^{-}) concentration, suggesting the effect of dissolution processes, as observed in groundwater discharge points. The subsequent decrease in groundwater flow is attributed to a high Cl^{-} concentration, which also indicates the influence of dissolution on groundwater chemistry. Additionally, the prevailing semi-arid climatic conditions over this region enhance the evaporation effect and may cause an increase in ionic concentration in groundwater [11].

4.1.6. Saturation Index (SI) and Geochemical Modeling

For this study, the PHREEQC code [64] has been performed to compute the saturation indices (SI) of dominant carbonate mineral species such as calcite and dolomite, which are the common fracture-filling minerals. As the concentration of Ca and Mg is observed in groundwater samples, the saturation indices with respect to calcite and dolomite can be used as a proxy for identifying the geochemical processes. In Figure 9, the samples have been classified as saturated (SI = 0), undersaturated (SI < 0 or negative), and oversaturated (SI > 0, positive) [50,65]. Most of the samples collected from granitic rocks are plotted below the equilibrium line (SI = 0) (Figure 9a). This infers that groundwater is undersaturated to calcite and dolomite with an excess of Ca and Mg, which are mainly derived from the dissolution of silicate minerals. The higher dissolution rate may be due to the continuous flow of groundwater through fractured granite. However, few samples (BW-9, DW-19,

TW-10, TW-11, TW-36, TW-8, TW-42) are distributed close to the saturated line (Figure 9b), acquiring the saturation or over-saturation condition with the increasing of Ca and Mg. These samples are located near the groundwater discharge locations with an average TDS value >1000 mg/L. The relatively slower movement of the groundwater and the evaporation process may be the driving factors for excess ionic concentration.

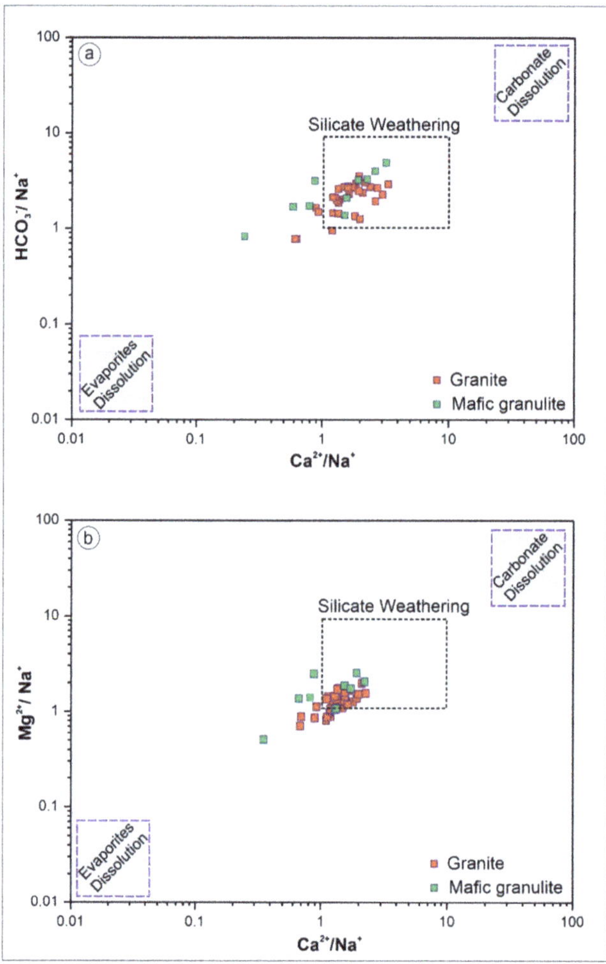

Figure 7. Ionic ratio plot (**a**) HCO_3^-/Na^+ vs. Ca^{2+}/Na^+, (**b**) Mg^{2+}/Na^+ vs. Ca^{2+}/Na^+ showing the silicate weathering as controlling mechanisms.

In the case of mafic granulite, the majority of the samples are near the equilibrium line (Figure 9b), indicating that groundwater is in a nearly saturated state with regard to calcite and dolomite. Groundwater is oversaturated to dolomite in mafic granulite compared to granite. Thus, the sluggish movement of groundwater may lead to an over-saturation condition of calcite and dolomite in granulite rocks, unlike granitic rocks. The concentration of alkali earth metals (Ca + Mg) increases as the groundwater is subjected to saturated or oversaturated conditions to both calcite and dolomite (Figure 9b) due to an incongruent dissolution of silicates.

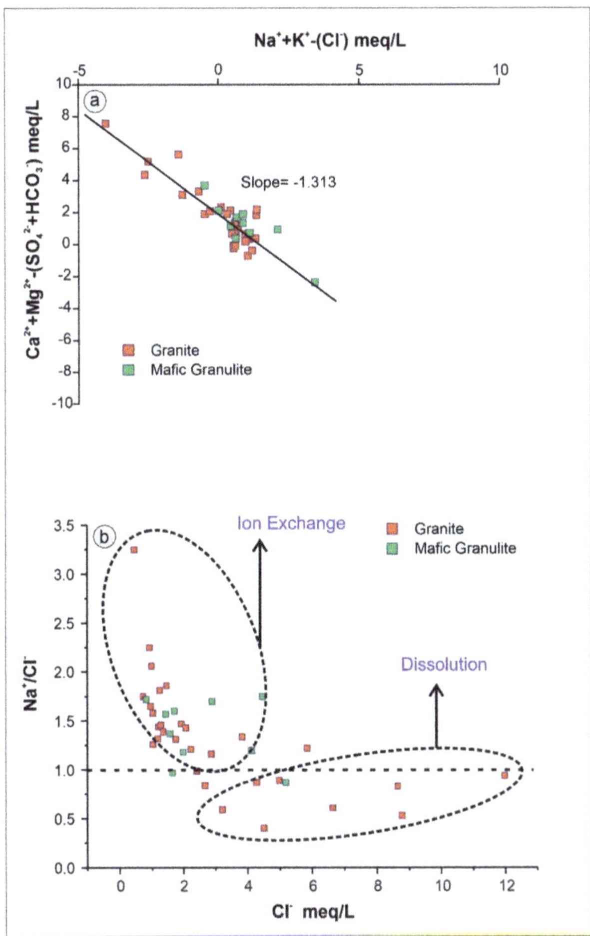

Figure 8. Plot (a) $Ca^{2+}Mg^{2+}$-(SO_4^{2-} + HCO_3^- vs. Na^++ K^+-(Cl^-), (b) Na^+/Cl^- vs. Cl^- showing the reversible ion exchange and evaporation process as the major contributors of Na ions in groundwater at different stages of groundwater evolution.

4.2. Stable Isotopic ($\delta^{18}O$ and δ^2H) Signatures and Recharge Process

The stable isotopic ($\delta^{18}O$ and δ^2H) composition of the groundwater in the study area varies from −6.17‰ to −3.23‰ in $\delta^{18}O$ and −41.90‰ to −24.51‰ in δ^2H (Table 1) with the mean value of −4.93‰ for $\delta^{18}O$ and −34.03‰ for δ^2H. The Local Meteoric Water Line (LMWL) was calculated based on the concept of Global Meteoric Water Line (GWML) using Equation (6) as shown below [66,67] to understand the relation between ^2H and ^{18}O.

$$\delta^2H = 8\delta^{18}O + 10 \quad (6)$$

Around 17% and 83% of the total groundwater samples are plotted slightly below and above the LMWL [68] (δ^2H = (7.6 ± 0.6) × $\delta^{18}O$-(2.9 ± 2.2)) respectively, suggesting that groundwater is of meteoric origin (Figure 10).

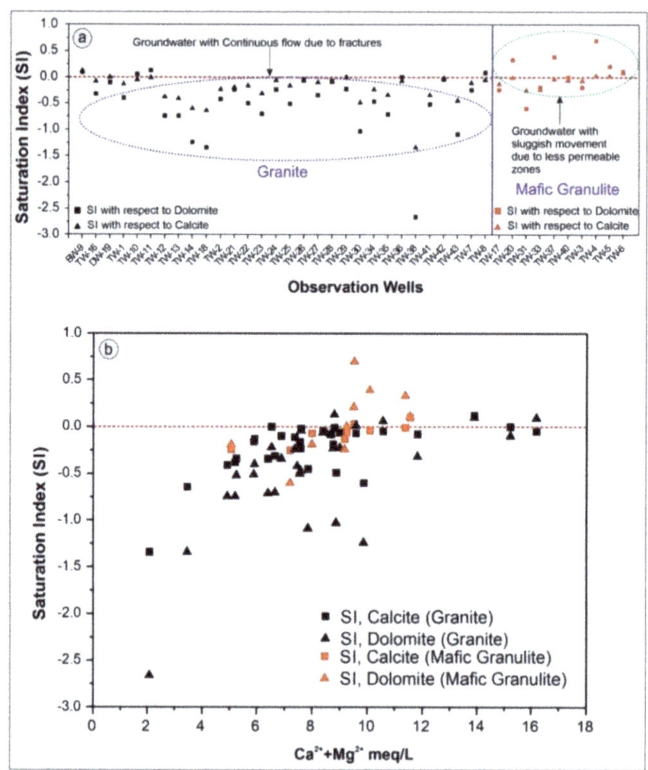

Figure 9. Plot (**a**) spatial variation of saturation index (SI) with respect to calcite and dolomite, (**b**) saturation index (SI) vs. Ca^{2+} + Mg^{2+} showing the contribution of silicate minerals in fractured granite and mafic granulite through geochemical modeling.

Furthermore, there is a variation in isotopic composition in groundwater, which is collected from granite and mafic granulite, suspecting the influence of geology on the recharge process. Figure 10 illustrates that the regression line, $\delta^2H = 5.89116 \times \delta^{18}O- 4.73697$, which was plotted for the shallow fractured granite aquifer, has a lower slope (approximately 5.9) and intercept (−4) than LMWL. This is due to the fractionation, which leads to a depletion of isotopic composition in groundwater with a positive and strong correlation ($R^2 = 0.9$) between $\delta^{18}O$ and δ^2H. Moreover, the d-excess value of groundwater samples collected from fractured aquifer systems varies from 1.81 to 8.11‰ with an average value of 5.93‰, which also indicates the evaporation process. Similarly, the regression line, $\delta^2H = 5.57 \times \delta^{18}O\text{-}7.0672$ fits positively ($R^2 = 0.83$) between $\delta^{18}O$ and δ^2H of groundwater samples extracted from mafic granulite. The slope (5.57) and intercept (−7.06) of the regression line is less than that of LMWL, indicating the evaporation effect prior to recharging the groundwater in the mafic granulite aquifer. This is also supported by the d-excess value ranging from 1.30 to 7.10‰ with the mean value of 3.98‰. The slope (<7.6) and d-excess (<10‰) values of the groundwater from both granitic and mafic granulites suggest that the groundwater is influenced by the evaporation process before recharge.

In order to understand the recharge source, a flow direction map of groundwater has been prepared (Figure 11).

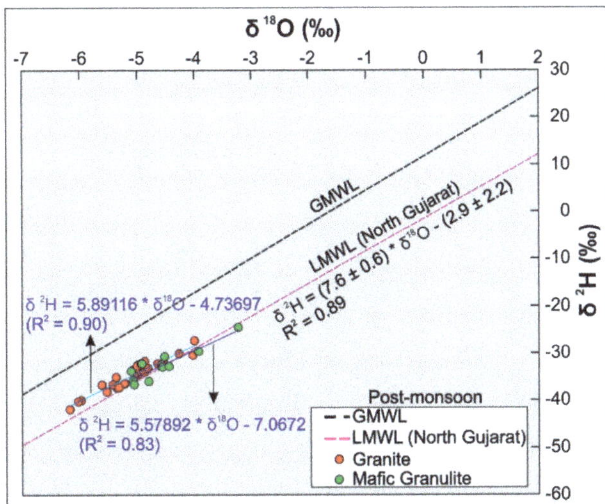

Figure 10. A plot of δ^2H (‰) versus δ^{18}O (‰) for groundwater resources in Ambaji Basin, North Gujarat (NW India). The Local Meteoric Water Line (LMWL) of North Gujarat and the Global Meteoric Water Line (GMWL) are also shown.

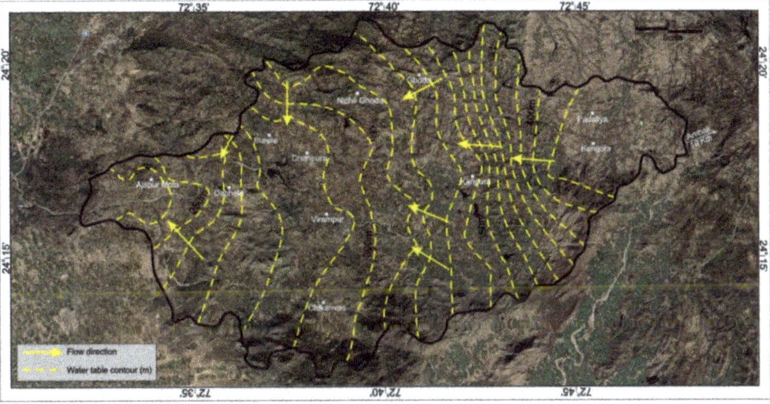

Figure 11. Groundwater flow direction map of the study area.

Sampling locations such as TW-1, TW-43, T-13, TW-42, and TW-38 show depleted isotopic compositions (average δ^{18}O < −5.00‰) with the higher hydraulic heads and are located in the Kengora and Ghoda area. Similarly, groundwater from TW-3, TW-5, TW-4, and TW-6 in the Kanpura area shows slightly enriched isotopic values (average δ^{18}O > −5.00‰) along the flow direction with a decrease in the hydraulic head (Figure 11), which indicates that a higher topography region is depleted with isotopic composition as compared to a low topography region, although they originate from the same source of recharge i.e., local precipitation. However, the depleted isotopic composition is also observed in TW-7, TW-10, TW-21, TW-22, TW-23, TW-25, TW-27, and TW-29 sampling locations with lower hydraulic heads, which are also falling on granitic rocks of the lower topography region. Remarkably, it reveals that groundwater in fractured granite shows the depleted isotopic composition and higher d-excess value as compared to mafic granulite (Figure 12), although they have the same source of recharge. This may be due to the kinetic evaporation of soil moisture that affects the groundwater more effectively in mafic granulite than fractured granite before it recharges. Furthermore, the sluggish movement of groundwater due to less permeability in mafic granulite is attributed to the enrichment of isotopic composition as compared to the fractured permeable aquifer.

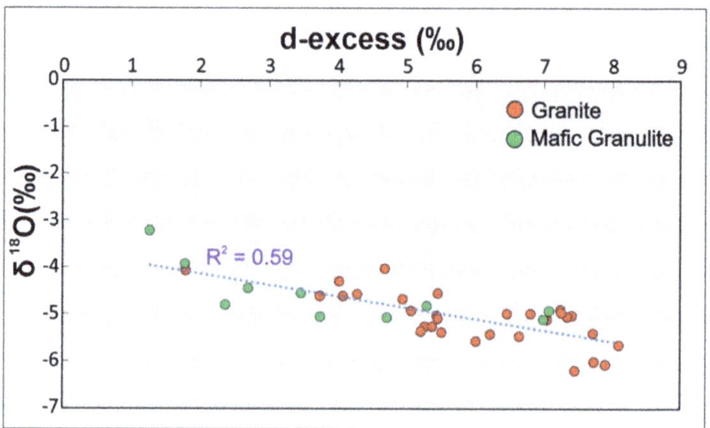

Figure 12. Plot of d-excess vs. $\delta^{18}O$ (‰) of groundwater samples.

4.3. Radiogenic (^3H) Isotopic Signatures

In this study, tritium (^3H) dating is used to measure the age of groundwater. The half-life of ^3H is 12.32 ± 0.02 years. The ^3H concentration in the hydrosphere can be cosmogenic and anthropogenic, and the concentration of ^3H in rain due to cosmic ray production is about 6–8 tritium units (TU) in Indian rainwaters [69]. Based on the tritium unit (TU) values, the semi-quantitative age of groundwater can be inferred [70,71]. Twenty-two (22) representative groundwater samples were collected from different aquifers during December 2017 and investigated for measuring the ^3H in the study area in (Table 1). The measured range of TU varies from 1.97 ± 0.13 TU to 28.05 ± 0.70 TU with an average of 5.10 TU. Usually, the ^3H content in groundwater starts decreasing from the recharge zone to discharge zones with time. The area also shows similar results, i.e., relatively high TU in the recharge areas and low TU at the discharge areas (southwest part) (Figure 13). The results show that the measured tritium values (TU) in groundwater can be categorized broadly into (1) a mixture of sub-modern and modern water (0.8–4.0 TU, 30–40 years), (2) modern water (5–15 TU, <5–10 years), and (3) recent recharge (>15 TU). The ^3H values show that most of the samples show a mixture between sub-modern and modern recharge followed by modern and recent recharge. Figure 13 shows the semi-quantitative groundwater ages and exemplifies that groundwater is dynamically recharged through weathered zones, secondary fractures, faults, and shear zones. The existence of recent and modern groundwater in the deeper depth specifies a link between shallow and deeper aquifers. Furthermore, it directs that the groundwater level will be shallow even if pumping is more in these areas. Among the entire area, one sample shows 28.05 TU (TW-04, Kanpura area), representing the recent age groundwater, which might be due to a river being close to the well location. In this study basin, the identified recharge locations are Kengora (TW-1), Padaliya (TW-43), Ghoda (TW-13), Nichli Ghoda (TW-18), Surela (TW-34, TW-35), Virampur (TW-21, TW-22), and Yogdadi areas (TW-27) (>5 TU). These areas are also a good pact with the presence of structural components i.e., major faults, fractures, and shear zones that control the groundwater circulations in the study basin.

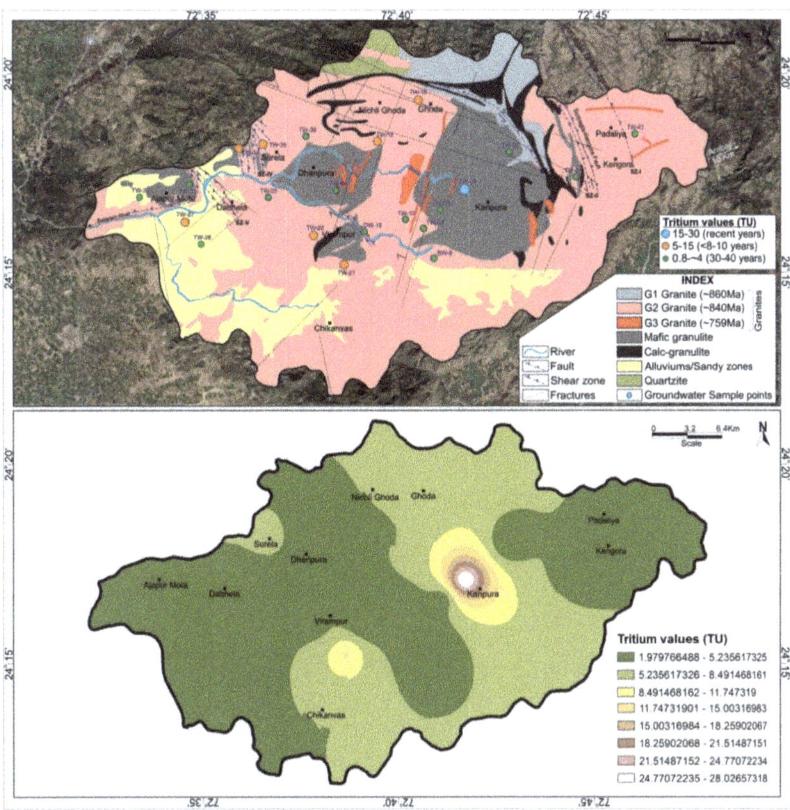

Figure 13. Distribution and variation of environmental tritium (^3H) content in Ambaji Basin (NW India).

More than 50% of the samples show the TU values between 0.8 and 4.0, which indicates that the groundwater is mixed type i.e., modern and sub-modern age (30–40 years). This implies that the study area is moderately rechargeable in these areas even if few fracture and lineament traces are present. The area consists of several mafic granulite patches where fractures and weathered zones are relatively less as compared to granitic rock. The mafic granulite is massive in nature, which results in the sluggish behavior of groundwater movement in the subsurface. Therefore, these moderately rechargeable areas are expected to show a depletion of groundwater levels if pumping is more.

5. Scientific Outcome and Its Policy Relevance for Sustainable Water Resources Management

This study has not only given mere value for scientific investigation but also an open door for policymakers' or decision-makers' interventions by achieving different objectives, as shown in Figure 14. One of the dimensions is to hasten the process of SDGs targets in a timely manner. Although there are 16 sustainable development goals (SDGs), all the goals are interlinked, and accessing water management is vital for achieving all the goals. For the second dimension, robust science is necessary for designing for better policy in arid/semi-arid regions; that is why this kind of scientific evidence will open pathways for better science–policy interlinkages. For the third dimension, water always proved to be a limiting factor for socio-economic growth in a holistic manner, especially in water-scarce or water-shortage regions such as arid/semi-arid areas. This study will be proved crucial for laying a good foundation for management policies. Furthermore, this study also helps to achieve a better bio-diversity management plan or global Aichi target.

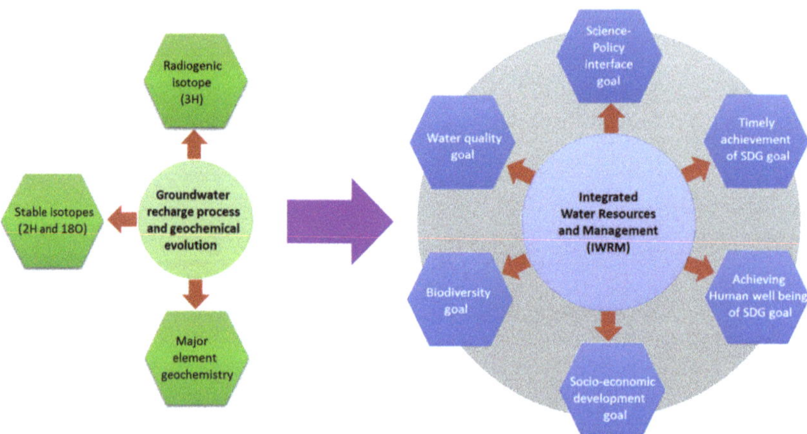

Figure 14. Interactive link between groundwater recharge process and geochemical evolution with IWRM and SDG.

6. Conclusions

Based on hydrogeochemical parameters and isotopic composition, the study was carried out to understand the various geochemical processes that were subjected to hydrogeochemical evolution in a structurally controlled area. The environmental isotopes ($\delta^{18}O$, δ^2H, and ^3H) of groundwater samples were analyzed to identify the source of groundwater and the hydrological process for groundwater recharge. In groundwater, Ca^{2+} is the dominated cation followed by Na^+, Mg^{2+}, and K^+, whereas from the anionic category, HCO_3^- is dominant with a gradual decrease in the concentration of Cl^-, SO_4^{2-}, and NO_3^-. The interaction of groundwater with the aquifer matrix is a common phenomenon for the different ionic distribution in the different aquifer systems. To understand the hydrogeochemical pattern from recharge to discharge areas, hydrochemical facies was plotted. All samples are majorly represented by two types of hydrochemical facies, i.e., Ca-HCO_3 and Mixed Ca-Na-HCO_3, which are collected from fractured granite and mafic granulite. The chemical behavior of groundwater is different, as it moves from one point to another. Initially, the groundwater shows Ca-HCO_3 facies in both granites and granulites as it gets recharged by rainfall. As soon as it moves further, mixed Ca-Na-HCO_3 type hydrochemical facies are observed. In fractured granite and mafic granulite, groundwater is represented by Ca-Na-HCO_3 and Mg (Ca)-Na-HCO_3, respectively, which is due to the chemical weathering of silicate minerals. It is also supported by Gibb's rock–water interaction plot. Along with chemical weathering, ion exchange is also responsible for controlling the groundwater chemistry, in which the Na releases with the replacement of Ca during the early stage of groundwater movement. Gradually, the Ca and Mg replace Na by a reversible ion exchange process.

The stable isotopic ($\delta^{18}O$ and δ^2H) composition of groundwater reveals that rainfall is the primary source of the groundwater recharge and shows more depletion value in the recharge area as compared to discharge. Furthermore, the enriched isotopic composition $\delta^{18}O > 5‰$ and low d-excess value ($<10‰$) indicate that the groundwater is subjected to evaporation before it recharges. The radiogenic isotope (^3H) concentration shows that the study basin is predominantly modern groundwater (<8–10 years) and mixed-type sub-modern to modern groundwater age (30–40 years). Furthermore, the modern recharge is linked to the presence of different structural components i.e., secondary faults, fractures, and shear zones that circulate the precipitation into the subsurface groundwater system. As a way forward, the study recommends hydrological simulation-related studies or detailed policy-relevant studies to achieve water security in the present study area or areas with similar climatic and geographical characteristics.

Supplementary Materials: The following supporting information can be downloaded at: https://www.mdpi.com/article/10.3390/w14030315/s1, Table S1. Physicochemical and isotopic results of groundwater samples collected during post-monsoon season from Ambaji Basin (NW India) (December 2017)

Author Contributions: Writing—original draft, R.M.P.; reviewing and editing, R.M.P., A.K.B., S.K., P.K. and T.K.B.; methodology, R.M.P. and A.K.B.; supervision, T.K.B., S.K. and P.K.; data collection and conceptualization, R.M.P. All authors have read and agreed to the published version of the manuscript.

Funding: This work was supported by the Ministry of Earth Sciences (MoES), Govt. of India [MoES/P.O (Geosci)/50/2015].

Institutional Review Board Statement: Not applicable.

Informed Consent Statement: Not applicable.

Data Availability Statement: The data presented in this study are available on request from the corresponding author.

Acknowledgments: Rudra Mohan Pradhan thanks the IRCC IIT Bombay and Ministry of Earth Sciences (MoES), Govt. of India for the financial support to carry out this work as part of the Ph.D. Thesis. TK Biswal is thankful to the MoES, Govt. of India for the sponsored Aquifer mapping project (Project no. MoES/P.O (Geosci)/50/2015). We are thankful to the Department of Earth Sciences, IIT Bombay and Hydrological Investigation Division, NIH Roorkee for providing the necessary laboratory facilities to carry out the analyses.

Conflicts of Interest: The authors declare no conflict of interest.

References

1. Taylor, R.G.; Todd, M.C.; Kongola, L.; Maurice, L.; Nahozya, E.; Sanga, H.; MacDonald, A.M. Evidence of the dependence of groundwater resources on extreme rainfall in East Africa. *Nat. Clim. Chang.* **2013**, *3*, 374–378. [CrossRef]
2. Singhal, B.; Gupta, R. *Applied Hydrogeology of Fractured Rocks*, 2nd ed.; Springer: Berlin/Heidelberg, Germany, 2010.
3. Sun, Z.; Ma, R.; Wang, Y.; Ma, T.; Liu, Y. Using isotopic, hydrogeochemical-tracer and temperature data to characterize recharge and flow paths in a complex karst groundwater flow system in northern China. *Hydrogeol. J.* **2016**, *24*, 1393–1412. [CrossRef]
4. Sreedevi, P.D.; Sreekanth, P.D.; Reddy, D.V. Recharge environment and hydrogeochemical processes of groundwater in a crystalline aquifer in South India. *Int. J. Environ. Sci. Technol.* **2021**, 1–18. [CrossRef]
5. Wright, E.P. The hydrogeology of crystalline basement aquifers in Africa. *Geol. Soc. Lond. Spec. Publ.* **1992**, *66*, 1–27. [CrossRef]
6. Banks, D.; Odling, N.E.; Skarphagen, H.; Rohr-Torp, E. Permeability and stress in crystalline rocks. *Terra Nova* **1996**, *8*, 223–235. [CrossRef]
7. Dewandel, B.; Lachassagne, P.; Wyns, R.; Maréchal, J.C.; Krishnamurthy, N.S. A generalized 3-D geological and hydrogeological conceptual model of granite aquifers controlled by single or multiphase weathering. *J. Hydrol.* **2006**, *330*, 260–284. [CrossRef]
8. Guihéneuf, N.; Boisson, A.; Bour, O.; Dewandel, B.; Perrin, J.; Dausse, A.; Viossanges, M.; Chandra, S.; Ahmed, S.; Maréchal, J.C. Groundwater flows in weathered crystalline rocks: Impact of piezometric variations and depth dependent fracture connectivity. *J. Hydrol.* **2014**, *511*, 320–334. [CrossRef]
9. Elango, L.; Kannan, R.; Kumar, M.S. Major ion chemistry and identification of hydrogeochemical processes of ground water in a part of Kancheepuram district, Tamil Nadu, India. *Environ. Geosci.* **2003**, *10*, 157–166.
10. Pradhan, R.M.; Biswal, T.K. Fluoride in groundwater: A case study in Precambrian terranes of Ambaji region, North Gujarat, India. *Proc. Int. Assoc. Hydrol. Sci.* **2018**, *379*, 351–356. [CrossRef]
11. Freeze, R.A.; Cherry, J.A. *Groundwater (No. 629.1 F7)*; Prentice-Hall: Englewood Cliffs, NJ, USA, 1979.
12. Domenico, P.A.; Schwartz, F.W. *Physical and Chemical Hydrogeology*; Wiley: New York, NY, USA, 1998.
13. Kortatsi, B.K. Hydrochemical framework of groundwater in the Ankobra Basin, Ghana. *Aquat. Geochem.* **2007**, *13*, 41–74. [CrossRef]
14. Ako, A.A.; Shimada, J.; Hosono, T.; Ichiyanagi, K.; Nkeng, G.E.; Eyong, G.E.T.; Roger, N.N. Hydrogeochemical and isotopic characteristics of groundwater in Mbanga, Njombe and Penja (Banana Plain)–Cameroon. *J. Afr. Earth Sci.* **2012**, *75*, 25–36. [CrossRef]
15. Rosen, M.R.; Jones, S. Controls on the groundwater composition of the Wanaka and Wakatipu basins, Central Otago, New Zealand. *Hydrogeol. J.* **1998**, *6*, 264–281.

16. Tirumalesh, K.; Shivanna, K.; Sriraman, A.K.; Tyagi, A.K. Assessment of quality and geochemical processes occurring in groundwaters near central air conditioning plant site in Trombay, Maharashtra, India. *Environ. Monit. Assess.* **2010**, *163*, 171–184. [CrossRef] [PubMed]
17. Roy, A.; Keesari, T.; Mohokar, H.; Pant, D.; Sinha, U.K.; Mendhekar, G.N. Geochemical evolution of groundwater in hard-rock aquifers of South India using statistical and modelling techniques. *Hydrol. Sci. J.* **2020**, *65*, 951–968. [CrossRef]
18. He, J.; Ma, J.; Zhang, P.; Tian, L.; Zhu, G.; Edmunds, W.M. Groundwater recharge environments and hydrogeochemical evolution in the Jiuquan Basin, Northwest China. *J. Appl. Geochem.* **2012**, *27*, 866–878. [CrossRef]
19. Maréchal, J.C.; Selles, A.; Dewandel, B.; Boisson, A.; Perrin, J.; Ahmed, S. An observatory of groundwater in crystalline rock aquifers exposed to a changing environment: Hyderabad, India. *Vadose Zone J.* **2018**, *17*, 1–14. [CrossRef]
20. Sami, K. Recharge mechanisms and geochemical processes in a semi-arid sedimentary basin, Eastern Cape, South Africa. *J. Hydrol.* **1992**, *139*, 27–48. [CrossRef]
21. Gat, J. Oxygen and hydrogen isotopes in the hydrologic cycle. *Annu. Rev. Earth Planet. Sci.* **1996**, *24*, 225–262. [CrossRef]
22. Jeelani, G.; Lone, S.A.; Nisa, A.U.; Deshpande, R.D.; Padhya, V. Use of stable water isotopes to identify and estimate the sources of groundwater recharge in an alluvial aquifer of Upper Jhelum Basin (UJB), western Himalayas. *Hydrol. Sci. J.* **2021**, *66*, 2330–2339. [CrossRef]
23. Kalbus, E.; Reinstorf, F.; Schirmer, M. Measuring methods for groundwater–surface water interactions: A review. *Hydrol. Earth Syst. Sci.* **2006**, *10*, 873–887. [CrossRef]
24. Mahlangu, S.; Lorentz, S.; Diamond, R.; Dippenaar, M. Surface water-groundwater interaction using tritium and stable water isotopes: A case study of Middelburg, South Africa. *J. African Earth Sci.* **2020**, *171*, 103886. [CrossRef]
25. De Vries, J.J.; Simmers, I. Groundwater recharge: An overview of processes and challenges. *Hydrogeol. J.* **2002**, *10*, 5–17. [CrossRef]
26. Sukhija, B.S.; Reddy, D.V.; Nagabhushanam, P.; Bhattacharya, S.K.; Jani, R.A.; Kumar, D. Characterisation of recharge processes and groundwater flow mechanisms in weathered-fractured granites of Hyderabad (India) using isotopes. *Hydrogeol. J.* **2006**, *14*, 663–674. [CrossRef]
27. Oiro, S.; Comte, J.C.; Soulsby, C.; Walraevens, K. Using stable water isotopes to identify spatio-temporal controls on groundwater recharge in two contrasting East African aquifer systems. *Hydrol. Sci. J.* **2018**, *63*, 862–877. [CrossRef]
28. Sukhija, B.S.; Reddy, D.V.; Nagabhushanam, P. Isotopic fingerprints of paleoclimates during the last 30,000 years in deep confined groundwaters of Southern India. *Quat. Res.* **1998**, *50*, 252–260. [CrossRef]
29. Negrel, P.; Pauwels, H.; Dewandel, B.; Gandolfi, J.M.; Mascré, C.; Ahmed, S. Understanding groundwater systems and their functioning through the study of stable water isotopes in a hard-rock aquifer (Maheshwaram watershed, India). *J. Hydrol.* **2011**, *397*, 55–70. [CrossRef]
30. Chen, J.Y.; Tang, C.Y.; Shen, Y.J.; Sakura, Y.; Kondoh, A.; Shimada, J. Use of water balance calculation and tritium to examine the dropdown of groundwater table in the piedmont of the North China Plain (NCP). *J. Environ. Geol.* **2003**, *44*, 564–571. [CrossRef]
31. Price, R.M.; Top, Z.; Happell, J.D.; Swart, P.K. Use of tritium and helium to define groundwater flow conditions in Everglades National Park. *Water Resour. Res.* **2003**, *39*, 1267. [CrossRef]
32. Moran, J.E.; Hudson, G.B. *Using Groundwater Age and Other Isotopic Signatures to Delineate Groundwater Flow and Stratification (No. UCRL-PROC-215146)*; Lawrence Livermore National Lab. (LLNL): Livermore, CA, USA, 2005.
33. Krishan, G.; Kumar, B.; Sudarsan, N.; Rao, M.S.; Ghosh, N.C.; Taloor, A.K.; Bhattacharya, P.; Singh, S.; Kumar, C.P.; Sharma, A.; et al. Isotopes ($\delta^{18}O$, δD and ^3H) variations in groundwater with emphasis on salinization in the State of Punjab, India. *Sci. Total Environ.* **2021**, *789*, 148051. [CrossRef]
34. Kohfahl, C.; Sprenger, C.; Herrera, J.B.; Meyer, H.; Chacon, F.F.; Pekdeger, A. Recharge sources and hydrogeochemical evolution of groundwater in semiarid and karstic environments: A field study in the Granada Basin (Southern Spain). *J. Appl. Geochem.* **2008**, *23*, 846–862. [CrossRef]
35. Joshi, S.K.; Rai, S.P.; Sinha, R.; Gupta, S.; Densmore, A.L.; Rawat, Y.S.; Shekhar, S. Tracing groundwater recharge sources in the northwestern Indian alluvial aquifer using water isotopes ($\delta^{18}O$, δ^2H and ^3H). *J. Hydrol.* **2018**, *559*, 835–847. [CrossRef]
36. Khayat, S.; Marei, A.; Hippler, D.; Barghouthi, Z.; Dietzel, M. Using environmental isotopes to investigate the groundwater recharge mechanisms and dynamics in the North-eastern Basin, Palestine. *Hydrol. Sci. J.* **2020**, *65*, 583–596. [CrossRef]
37. Kumar, S.; Joshi, S.K.; Pant, N.; Singh, S.; Chakravorty, B.; Saini, R.K.; Kumar, V.; Singh, A.; Ghosh, N.C.; Mukherjee, A.; et al. Hydrogeochemical evolution and groundwater recharge processes in arsenic enriched area in central Gangetic plain, India. *J. Appl. Geochem.* **2021**, *131*, 105044. [CrossRef]
38. Pradhan, R.M.; Guru, B.; Pradhan, B.; Biswal, T.K. Integrated multi-criteria analysis for groundwater potential mapping in Precambrian hard rock terranes (North Gujarat), India. *Hydrol. Sci. J.* **2021**, *66*, 961–978. [CrossRef]
39. Singh, Y.K.; De Waele, B.; Karmakar, S.; Sarkar, S.; Biswal, T.K. Tectonic setting of the Balaram-Kui-Surpagla-Kengora granulites of the South Delhi Terrane of the Aravalli Mobile Belt, NW India and its implication on correlation with the East African Orogen in the Gondwana assembly. *Precambrian Res.* **2010**, *183*, 669–688. [CrossRef]
40. Epstein, S.; Mayeda, T. Variation of O18 content of waters from natural sources. *Geochim. Cosmochim. Acta* **1953**, *4*, 213–224. [CrossRef]
41. Behera, A.K.; Chakrapani, G.J.; Kumar, S.; Rai, N. Identification of seawater intrusion signatures through geochemical evolution of groundwater: A case study based on coastal region of the Mahanadi delta, Bay of Bengal, India. *Nat. Hazards* **2009**, *97*, 1209–1230. [CrossRef]

42. Ahmed, A.; Clark, I. Groundwater flow and geochemical evolution in the Central Flinders Ranges, South Australia. *Sci. Total Environ.* **2016**, *572*, 837–851. [CrossRef]
43. Pant, N.; Rai, S.P.; Singh, R.; Kumar, S.; Saini, R.K.; Purushothaman, P.; Nijesh, P.; Rawat, Y.S.; Sharma, M.; Pratap, K. Impact of geology and anthropogenic activities over the water quality with emphasis on fluoride in water scarce Lalitpur district of Bundelkhand region, India. *Chemosphere* **2021**, *279*, 130496. [CrossRef]
44. Marei, A.; Khayat, S.; Weise, S.; Ghannam, S.; Sbaih, M.; Geyer, S. Estimating groundwater recharge using the chloride mass-balance method in the West Bank, Palestine. *Hydrol. Sci. J.* **2010**, *55*, 780–791. [CrossRef]
45. Naranjo, G.; Cruz-Fuentes, T.; Cabrera, M.D.C.; Custodio, E. Estimating natural recharge by means of chloride mass balance in a volcanic aquifer: Northeastern Gran Canaria (Canary Islands, Spain). *Water* **2015**, *7*, 2555–2574. [CrossRef]
46. Karlović, I.; Marković, T.; Vujnović, T. Groundwater Recharge Assessment Using Multi Component Analysis: Case Study at the NW Edge of the Varaždin Alluvial Aquifer, Croatia. *Water* **2022**, *14*, 42. [CrossRef]
47. BIS. *Bureau of Indian Standards Specification for Drinking Water*; IS: 10500:91. Revised 2003; Bureau of Indian Standards: New Delhi, India, 2003.
48. Jakóbczyk-Karpierz, S.; Sitek, S.; Jakobsen, R.; Kowalczyk, A. Geochemical and isotopic study to determine sources and processes affecting nitrate and sulphate in groundwater influenced by intensive human activity-carbonate aquifer Gliwice (southern Poland). *J. Appl. Geochem.* **2017**, *76*, 168–181. [CrossRef]
49. Gómez-Alday, J.J.; Hussein, S.; Arman, H.; Alshamsi, D.; Murad, A.; Elhaj, K.; Aldahan, A. A multi-isotopic evaluation of groundwater in a rapidly developing area and implications for water management in hyper-arid regions. *Sci. Total Environ.* **2022**, *805*, 150245. [CrossRef]
50. Appelo, C.A.J.; Postma, D. *Geochemistry, Groundwater and Pollution*, 2nd ed.; Appelo, C.A.J., Postma, D., Eds.; CRC Press: Boca Raton, FL, USA, 2005.
51. Lasaga, A.C. Chemical kinetics of water-rock interactions. *J. Geophys. Res. Solid Earth* **1984**, *89*, 4009–4025. [CrossRef]
52. Lee, B.D.; Oh, Y.H.; Cho, B.W.; Yun, U.; Choo, C.O. Hydrochemical properties of groundwater used for Korea bottled waters in relation to geology. *Water* **2019**, *11*, 1043. [CrossRef]
53. Piper, A.M. A graphic procedure in the geochemical interpretation of water-analyses. *Eos Trans. Am. Geophys. Union* **1944**, *25*, 914–928. [CrossRef]
54. Gascoyne, M. Hydrogeochemistry, groundwater ages and sources of salts in a granitic batholith on the Canadian Shield, southeastern Manitoba. *J. Appl. Geochem.* **2004**, *19*, 519–560. [CrossRef]
55. Srinivasamoorthy, K.; Gopinath, M.; Chidambaram, S.; Vasanthavigar, M.; Sarma, V.S. Hydrochemical characterization and quality appraisal of groundwater from Pungar sub basin, Tamilnadu, India. *J. King Saud Univ. Sci.* **2014**, *26*, 37–52. [CrossRef]
56. Rao, N.S.; Rao, V.G.; Gupta, C.P. Groundwater pollution due to discharge of industrial effluents in Venkatapuram area, Visakhapatnam, Andhra Pradesh, India. *Environ. Geol.* **1998**, *33*, 289–294.
57. Luo, W.; Gao, X.; Zhang, X. Geochemical processes controlling the groundwater chemistry and fluoride contamination in the Yuncheng Basin, China—An area with complex hydrogeochemical conditions. *PLoS ONE* **2018**, *13*, e0199082. [CrossRef] [PubMed]
58. Cederstrom, D.J. Genesis of ground waters in the Coastal Plain of Virginia. *Econ. Geol.* **1946**, *41*, 218–245. [CrossRef]
59. Rajmohan, N.; Elango, L. Identification and evolution of hydrogeochemical processes in the groundwater environment in an area of the Palar and Cheyyar River Basins, Southern India. *Environ. Geol.* **2004**, *46*, 47–61. [CrossRef]
60. Abdalla, F.A.; Scheytt, T. Hydrochemistry of surface water and groundwater from a fractured carbonate aquifer in the Helwan area, Egypt. *J. Earth Syst. Sci.* **2012**, *121*, 109–124. [CrossRef]
61. Sunkari, E.D.; Abu, M.; Zango, M.S. Geochemical evolution and tracing of groundwater salinization using different ionic ratios, multivariate statistical and geochemical modeling approaches in a typical semi-arid basin. *J. Contam. Hydrol.* **2021**, *236*, 103742. [CrossRef] [PubMed]
62. Mukherjee, A.; Fryar, A.E.; Rowe, H.D. Regional-scale stable isotopic signatures of recharge and deep groundwater in the arsenic affected areas of West Bengal, India. *J. Hydrol.* **2007**, *334*, 151–161. [CrossRef]
63. Nandakumaran, P.; Balakrishnan, K. Groundwater quality variations in Precambrian hard rock aquifers: A case study from Kerala, India. *Appl. Water Sci.* **2020**, *10*, 1–13. [CrossRef]
64. Parkhurst, D.L.; Appelo, C.A.J. *User's Guide to PHREEQC (Version 2)—A Computer Program for Speciation, Batch-Reaction, One-Dimensional Transport, and Inverse Geochemical Calculations*; Water Resources Investigations Report 99-4259; United States Geological Survey: Washington, DC, USA, 1999.
65. Drever, J.I. *The Geochemistry of Natural Waters: Surface and Groundwater Environments*, 3rd ed.; Prentice Hall: New York, NY, USA, 1997.
66. Craig, H. Isotopic variation in meteoric waters. *Science* **1961**, *133*, 1702–1703. [CrossRef]
67. Kumar, P.; Kumar, A.; Singh, C.K.; Saraswat, C.; Avtar, R.; Ramanathan, A.L.; Herath, S. Hydrogeochemical evolution and appraisal of groundwater quality in Panna District, Central India. *Expo. Health* **2016**, *8*, 19–30. [CrossRef]
68. Gupta, S.K.; Deshpande, R.D. Groundwater isotopic investigations in India: What has been learned? *Curr. Sci.* **2005**, *89*, 825–835.
69. Rao, S.M. *Practical Isotope Hydrology*; New India Publishing: New Delhi, India, 2006.
70. Schlosser, P.; Stute, M.; Sonntag, C.; Münnich, K.O. Tritiogenic 3He in shallow groundwater. *Earth Planet. Sci. Lett.* **1989**, *94*, 245–256. [CrossRef]
71. Clarke, I.D.; Fritz, P. *Environmental Isotopes in Hydrogeology*; Lewis Publishers: New York, NY, USA, 1997.

MDPI

Article

Assessment of Groundwater Flow Dynamics Using MODFLOW in Shallow Aquifer System of Mahanadi Delta (East Coast), India

Ajit Kumar Behera [1,2,*], Rudra Mohan Pradhan [3], Sudhir Kumar [4], Govind Joseph Chakrapani [1] and Pankaj Kumar [5,*]

1 Department of Earth Sciences, Indian Institute of Technology Roorkee, Roorkee 247 667, India; govind.chakrapani@es.iitr.ac.in
2 Marine Geoscience Group, National Centre for Earth Science Studies, Thiruvananthapuram 695 011, India
3 Department of Earth Sciences, Indian Institute of Technology Bombay, Powai 400 076, India; rmp.geol@gmail.com
4 Hydrological Investigations Division, National Institute of Hydrology Roorkee, Roorkee 247 667, India; sudhir.nih@gmail.com
5 Institute for Global Environmental Strategies, Kamiyamaguchi, Hayama 2108-11, Kanagawa, Japan
* Correspondence: ajitgeol.89@gmail.com (A.K.B.); kumar@iges.or.jp (P.K.); Tel.: +91-471-2511717 (A.K.B.)

Abstract: Despite being a biodiversity hotspot, the Mahanadi delta is facing groundwater salinization as one of the main environmental threats in the recent past. Hence, this study attempts to understand the dynamics of groundwater and its sustainable management options through numerical simulation in the Jagatsinghpur deltaic region. The result shows that groundwater in the study area is extensively abstracted for agricultural activities, which also causes the depletion of groundwater levels. The hydraulic head value varies from 0.7 to 15 m above mean sea level (MSL) with an average head of 6 m in this low-lying coastal region. The horizontal hydraulic conductivity and the specific yield values in the area are found to vary from 40 to 45 m/day and 0.05 to 0.07, respectively. The study area has been calibrated for two years (2004–2005) by using these parameters, followed by the validation of four years (2006–2009). The calibrated numerical model is used to evaluate the net recharge and groundwater balance in this study area. The interaction between the river and coastal unconfined aquifer system responds differently in different seasons. The net groundwater recharge to the coastal aquifer has been estimated and varies from 247.89 to 262.63 million cubic meters (MCM) in the year 2006–2007. The model further indicates a net outflow of 8.92–9.64 MCM of groundwater into the Bay of Bengal. Further, the outflow to the sea is preventing the seawater ingress into the shallow coastal aquifer system.

Keywords: groundwater; MODFLOW; groundwater modeling; hydraulic conductivity; coastal aquifer; Mahanadi delta

Citation: Behera, A.K.; Pradhan, R.M.; Kumar, S.; Chakrapani, G.J.; Kumar, P. Assessment of Groundwater Flow Dynamics Using MODFLOW in Shallow Aquifer System of Mahanadi Delta (East Coast), India. *Water* **2022**, *14*, 611. https://doi.org/10.3390/w14040611

Academic Editors: Dimitrios E. Alexakis and Zbigniew Kabala

Received: 20 December 2021
Accepted: 15 February 2022
Published: 17 February 2022

1. Introduction

The coastal aquifer is one of the most important water resources in coastal regions that supplies water to more than a billion people worldwide [1–3] and connects the world's oceanic and hydrologic systems [4]. The general hydrogeological characteristics of the coastal aquifers are influenced by geologic environments, mixing zones, long and short-term sea fluctuations, and the density gradients due to differences in salinity [1]. In general, groundwater is an attractive source of water ($15{,}300 \times 10^3$ km^3) as it is fresh and readily available [5,6]. However, the groundwater of coastal regions is more susceptible to deterioration due to several factors, such as rapid urbanization, intensified agricultural development, economic development, climate change, sea-level rise, and lack of sufficient surface water resources [7–9]. The effect of groundwater withdrawal on coastal aquifer systems with the smallest topographic gradient is more pronounced than the impact of sea-level rise and variation in groundwater recharge due to the dynamic nature of surface water–groundwater interaction [3]. Consequently, the coastal aquifer will become more saline due to saltwater intrusion and make this precious water resource unfit for consumption without any sophisticated treatment [10]. Therefore, it is essential to set up a diligent monitoring system, e.g., using numerical models for evaluating the maximum viable pumping rates to protect from seawater intrusion in the coastal aquifers [11].

Various tools and techniques viz. statistical analysis, water quality index development, hydrochemical analysis, hydrological modeling, etc. are being used to assess and monitor water resources in the coastal aquifers. The groundwater models are the conceptual description that describes the physical systems through mathematical equations [12,13]. Groundwater modeling is based on three techniques i.e., finite element method [14–16], analytical element method [17], and the finite difference method [18–20]. Essink 2001 [21] established a density-dependent groundwater flow model to examine the effect of seawater intrusion in the coastal aquifer system of the Netherlands. Similarly, the numerical model (MOCDENS3D) study helped to calculate the fluctuations in coastal groundwater flow [10,22,23]. Numerical models have been used by several researchers to estimate the regional groundwater budget in different aquifer systems [24–27], to predict the consequences of proposed development actions, to link any connection between locations and aquifer boundaries, and to assess the groundwater quality within the aquifer system and the amount of natural recharge to a particular aquifer [12]. A visual MODFLOW software package has been used worldwide for groundwater flow simulation as it is user-friendly and robust [28,29]. A SEAWAT model is a transport and density-dependent groundwater flow model generally used to understand the saltwater intrusion processes [30]. However, a 3D finite element model is more advanced technology and can be used to simulate the saltwater intrusion for single as well as multiple complex coastal aquifer systems [31]. The development of groundwater models along with management models is useful to make proper decisions in optimal usage and management of groundwater resources [32]. The advantages of applying these models or codes lie in simplification of the aquifer system with certain limitations. Accordingly, we can make future plans or decisions on the usage of groundwater resources and predict the groundwater condition. In the case of a complex aquifer system, execution of the model takes a lot of time, which is one of the major disadvantages of using the modeling technique.

Odisha state on the East coast of India is home to a rich ecosystem and biodiversity but is very vulnerable to rapid global changes due to poor adaptive capacity [33]. Jagatsinghpur is considered to be a disaster-prone region as it experiences floods and cyclones almost every year. Despite this, the local habitants of this region still depend on agriculture for their survival and use groundwater for drinking water and irrigation purposes. Further, the heavy abstraction of groundwater from the deep aquifer system leads to groundwater salinity in some parts of this region. Again, this situation forces the villagers to depend on shallow fresh groundwater for daily usage. Freshwater resources act as a limiting factor for human well-being and sound environmental development. Data scarcity is one of the biggest challenges here to design any robust management plans. Among a few, one of the studies focuses on participatory coastal land-use management (PCLM) that was introduced in the coastal aquifers of the Brahmani River basin of Odisha for the sustainable management of water resources based on simulated water quality [33]. Further, one study reported numerous ecosystem services this area provides and highlights the need for preservation of the mangroves of Bhitarkanika and Mahanadi delta, which reduces the coastal degradation and protects the coastal aquifers from salinity [34].

So far, no such numerical simulation has been conducted on groundwater flow dynamics in this area. Therefore, this study has been attempted for the first time using a modeling technique to understand the groundwater flow pattern and behavior of the shallow aquifer of this coastal region influenced by hydrological components based on calculated water flux. We have also used MODFLOW software packages for this study because of their flexible modular units to represent hydrogeological conditions. Besides, MODFLOW is an easily available software package in the public domain, which can be used to calculate water balance for small-scale aquifer systems. This will help to estimate the optimal use of shallow groundwater to prevent seawater inflow. Considering this first of its kind of study, the result will prove to be a milestone for the decision-makers in designing better water management plans.

2. Study Area

The coastal aspects of Odisha, consisting of alluvial formation for agricultural activities and fresh water in the coastal aquifer system, are the key factors that attract the people to live in these regions. There are six coastal districts along the coastal tract (480 km) of Odisha [35]. Jagatsinghpur district is one of the districts in the coastal belts with a geographical area of 1668 km^2 and nearly 1.1 million population residing in these regions [35]. Geographically, the Jagatsinghpur area is located between longitude 86°03′ to 86°45′ E and latitude 19°53′ to 20°23′ N (Figure 1). This coastal region is a part of the Mahanadi delta, surrounded by two rivers i.e., the Mahanadi River (flowing from west to east) and the Devi River (flowing from north-northwest to south-southeast) forming the northern boundary, and the southern and western boundary of the district, respectively, and Bay of Bengal in the eastern part [36]. The study area comprises the central and middle part of the Mahanadi delta with a thick deposition of quaternary sediments. As it belongs to the coastal region, possible vulnerabilities, e.g., sea-level rise, saltwater intrusion, and frequent climate variations, may affect the study region. The average annual rainfall in this region is 1436 mm and is received mainly from the southwest monsoon. As the study area is a part of a deltaic region, it mainly consists of thick sediments supplied by the rivers, such as Mahanadi, Birupa, Kathjodi, Devi, and Kuakhai. Moreover, it has a gentle slope towards the Bay of Bengal [37,38].

Agricultural sectors are the major activities in these regions, and the key crops of the district are paddy, turmeric, sugarcane, cotton, and jute. The population of the district is largely dependent on the monsoon for irrigation, which is very erratic. Due to its geographical situation, the regions face acute natural calamities, e.g., floods, cyclones, and droughts. Almost all blocks of the Jagatsinghpur coastal district were severely affected by the super cyclone, with a wind speed of above 200 km/h on 29 October 1999 [39]. The places like Ersama, Kujang, and Balikuda were submerged due to the tidal wave (Height > 7 m) of the Bay of Bengal and this super cyclonic storm brought destruction to homes, human life, livestock, and other property [39,40]. The incidence of drought has been predicted due to the increase of surface air temperature at the rate of 1.1 °C per century over the Mahanadi Basin and reduced effective rainfall [41]. Hence, agricultural production is affected by soil salinity, waterlogging, and natural disasters. Further, different industries, such as fisheries, manufacturing, and processing, also contribute to economic development and the Jagatsinghpur district is one of the leading districts in the state in terms of industrialization, housing many industries related to fertilizers and petroleum products.

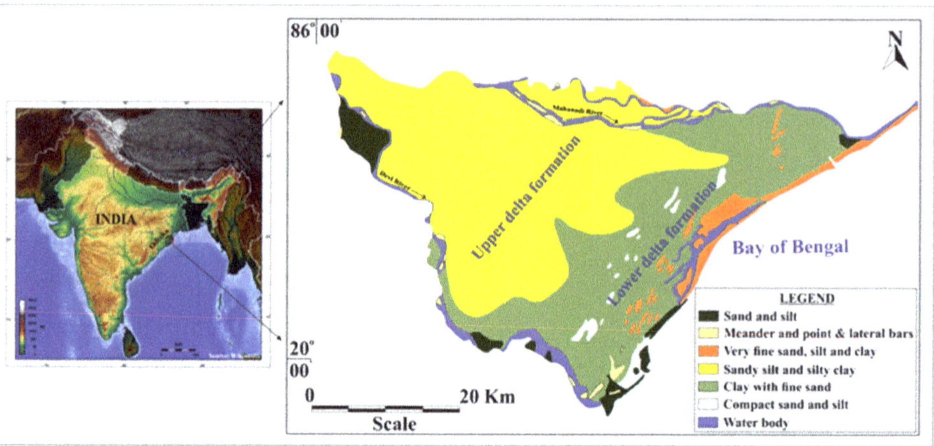

Figure 1. Location map of Jagatsinghpur coastal area, Odisha, India [36,42].

3. Hydrogeology

The deltaic region of Jagatsinghpur belongs to the quaternary formation that covers recent sediments of flood plain deposits of the Mahanadi River and the Devi River. These are mainly comprised of gravel, sand (fine to coarse grain), silt, and clay (black, red, and yellow) materials [42]. These unconsolidated to semi-consolidated materials act as a good repository of groundwater resources. The lower part of this study area lying close to the coast is characterized by low lying wet plains, fine-grained sediments, tidal infected rivers, tidal creeks, swamps, ill drainage of land, and non-development of levees [37]. This coastal tract acts as a favorable zone of groundwater availability due to the large thickness of sediments of varying sizes deposited in this part of the study area. The aquifer system of the Jagatsinghpur area is divided mainly into two zones, i.e., a shallow aquifer zone (<50 m thickness) and deeper aquifer zone (50–300 m thickness) below ground level [35].

4. Methodology

The groundwater flow simulation can be performed through Visual MODFLOW, which integrates the modular 3D finite-difference groundwater flow code [20]. Several numerical codes are used to simulate groundwater flow both in local and regional groundwater systems [43]. The groundwater flow equation (Equation (1)) is used in MODFLOW to know the groundwater flow in three different directions for this study [12].

$$\frac{\partial}{\partial x}\left(K_x\frac{\partial h}{\partial x}\right)+\frac{\partial}{\partial y}\left(K_y\frac{\partial h}{\partial y}\right)+\frac{\partial}{\partial z}\left(K_z\frac{\partial h}{\partial z}\right)=S_s\frac{\partial h}{\partial t}-R \quad (1)$$

where K_x, K_y, and K_z refer to hydraulic conductivity in three different directions; S_s, h, and R represent specific storage, hydraulic head, and sink or source, respectively. Visual MODFLOW is based on the finite-difference mathematical equation (Equation (2)) with assumptions of constant density and viscosity of groundwater flow under transient state conditions [44].

$$K_x\frac{\partial^2 h}{\partial x^2}+K_y\frac{\partial^2 h}{\partial y^2}+K_z\frac{\partial^2 h}{\partial z^2}\pm W=S_s\frac{\partial h}{\partial t} \quad (2)$$

Hydraulic head value does not change with time under steady-state conditions. This condition expresses as (Equation (3)).

$$K_x\frac{\partial^2 h}{\partial x^2}+K_y\frac{\partial^2 h}{\partial y^2}+K_z\frac{\partial^2 h}{\partial z^2}=0 \quad (3)$$

Visual MODFLOW has several solvers, such as Preconditioned Conjugate Gradient (PCG), Strongly Implicit Procedure (SIP) package, WHS solver for Visual MODFLOW package (WHS), Slice Successive Over Relaxation (SOR) package, and Geometric Multigrid solver (GMG) package, are used to solve the numerical equation for groundwater flow simulation purposes. For this study, the WHS solver package with Bi-Conjugate Gradient Stabilized (Bi-CGSTAB) accelerator us used to resolve the partial differential equations through iterative procedures.

4.1. Development of the Model

The groundwater model starts with the development of a groundwater flow model for a particular study area, which represents its physical condition. Similarly, for this study, a model consisting of a single layer has been conceptualized based on geology, lithologs, river boundary conditions, and groundwater level data sets [45]. The shallow unconfined aquifer composed of unconsolidated formation up to 50 m depth has been considered as the modeled single layer. After preparing a conceptual model, it is translated into a numerical model with the help of the Visual MODFLOW software package. The numerical model is developed through several steps. The input parameters in the conceptual model are presented in Table 1.

Table 1. Input model parameters.

Sl No.	Parameters	Inputs
1.	Cell	
1.1	Active	White Cells (600 m × 600 m)
1.2	Inactive	Green Cells (600 m × 600 m)
2.	Model Boundaries	
2.1	Constant Head	Head = 0 m (Bay of Bengal-SW to NE)
2.2	Recharge	Variable
2.3	Evapotranspiration	Rate = 1400 mm/year
		Extinction Depth = 3.0 m
3.	Layer	
3.1	Layer No.	1
3.2	Layer Type	Unconfined
4.	Aquifer Parameters	
4.1	Hydraulic Conductivity (K)	$K_x = K_y$ = 40 to 45 m/d
		K_z = 4 to 4.5 m/d
	Specific Yield (S_y)	0.05 to 0.07
5.	Wells	
5.1	Observation Wells	11 nos.
6.	Aquifer Stresses	Data for individual pumping wells is not available, the same has been included in net recharge
7.	Simulation Period	
7.1	Steady State	1 January 2004 (1 day)
7.2	Transient State	2004 to 2009

4.1.1. Discretization of the Study Area

The model study area covers 1668 km^2 and is gridded into 9394 cells with 77 rows (I = 77) and 122 columns (J = 122) and each cell consists of 600 m × 600 m blocks (Figure 2). The modeled layer thickness varies approximately from 30 to 50 m in the study area. The layer elevation and ground elevation data are imported in Visual MODFLOW through an ASCII file.

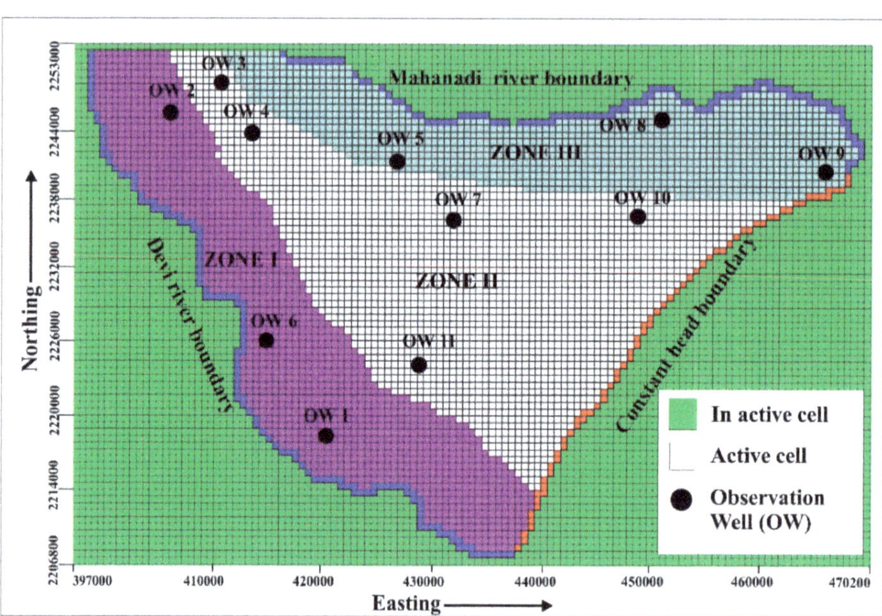

Figure 2. Discretization of the study area into grids.

4.1.2. Hydraulic Head Data

Hydraulic head data of eleven (11) different observation wells were collected from the Central Ground Water Board (CGWB) in the Jagatsinghpur coastal aquifer system for this study. The hydraulic head varies from 0.7 m near the sea coast to 15 m away from the shoreline. For groundwater flow simulation, 1 January 2004 has been taken as the initial time. Annually, four different periods of head data (from the year 2004 to 2009) have been used for both calibration and validation of the model. These head data have been categorized as post-monsoon (Rabi), pre-monsoon, monsoon, and post-monsoon (Kharif).

4.1.3. Boundary Conditions

The Visual MODFLOW simulates the groundwater flow followed by different types of boundary conditions. In the Jagatsinghpur coastal aquifer system, two types of boundaries have been used (Figure 2). The Bay of Bengal is considered as a constant head boundary or Dirichlet boundary [46]. Two large perennial rivers, i.e., the Mahanadi River and its distributary the Devi River, flowing along the two flanks of the study area are considered as the Cauchy boundary or head-dependent flux boundary. The river conductance value can be determined from Equation (4) [29].

$$C_{RIVER} = \frac{K_r \times L \times W_r}{B} \quad (4)$$

where K_r = hydraulic conductivity of the river bed (m/day), L = length of the reach/grid size (m) W_r = width of the river (m), and B = thickness of the river bed (m). The river bed conductance of two rivers is approximately the same, i.e., 30,000–35,000 m^2/day [35].

4.2. Hydrological Parameters

The model domain is classified into three hydraulic conductivities and specific yield zones for this single unconfined aquifer system (Figure 2), which also belongs to the alluvial formation of the Mahanadi delta. The horizontal hydraulic conductivities (K_h) are 40 m/day, 42 m/day, and 45 m/day for ZONE I, ZONE II, and ZONE III, respectively, whereas the corresponding vertical conductivity (K_v) of the three zones is 4, 4.2, and 4.5 m/day. The different specific yield values of 0.05, 0.06, and 0.07 for three respective zones I–III were taken during the calibration of the groundwater model (Table 2). As the water table is very close to the ground surface, some groundwater is extracted through the evapotranspiration process. Hence the evapotranspiration data have been taken into consideration for the groundwater simulation model. Further, the study area has been divided into eleven different recharge zones, in which monthly rainfall recharge values have been assigned.

Table 2. Aquifer parameters in the study area.

Zones	Horizontal Hydraulic Conductivity (K_h) in m/Day	Vertical Hydraulic Conductivity (K_v) in m/Day	Specific Yield
I	40	4	0.05
II	42	4.2	0.06
III	45	4.5	0.07

4.3. Calibration and Validation of Model

Calibration is the process through which the calculated head value is subjected to match with the observed head value. The head values from the year 2004 to the year 2005 are taken for calibrating the model for this study. In the present study, the calibration of the model is done through trial and error in which the unknown hydrogeologic parameters are set to be fixed to minimize the head difference between the calculated head and observed head at steady-state conditions. Then, the model is run for 2 years from January 2004 to December 2005 under transient state conditions. The groundwater simulation is said to be a good fit when the computed head value is very close to the observed head value and this good match can be analyzed through calibration criteria, e.g., the mean error (ME) (Equation (5)), the mean absolute error (MAE) (Equation (6)), and the root mean squared error (RMSE) (Equation (7)) [12,47,48]. After calibration, the validation of the model is performed by taking the hydraulic head values from January 2006 to December 2009.

$$\text{Mean Error (ME)} = 1/n \sum_{i=1}^{n} (h_o - h_c)_i \quad (5)$$

$$\text{Mean Absolute Error (MAE)} = 1/n \sum_{i=1}^{n} [(h_o - h_c)_i] \quad (6)$$

$$\text{Root Mean Squared Error (RMSE)} = \sqrt{[1/n \sum_{i=1}^{n} (h_o - h_c)_i^{\,2}]} \quad (7)$$

where h_o refers to the observed head value, h_c the calculated head value, and n the total number of observed data. A statistical analysis of calibrated model under steady-state conditions has been given (Figure 3). Under transient state conditions, the calibrated and validated model indicates a good correlation between the observed head and calculated head in the study area (Figure 4a,b). The correlation coefficient values for calibrated and validated models are 0.994 and 0.988 respectively.

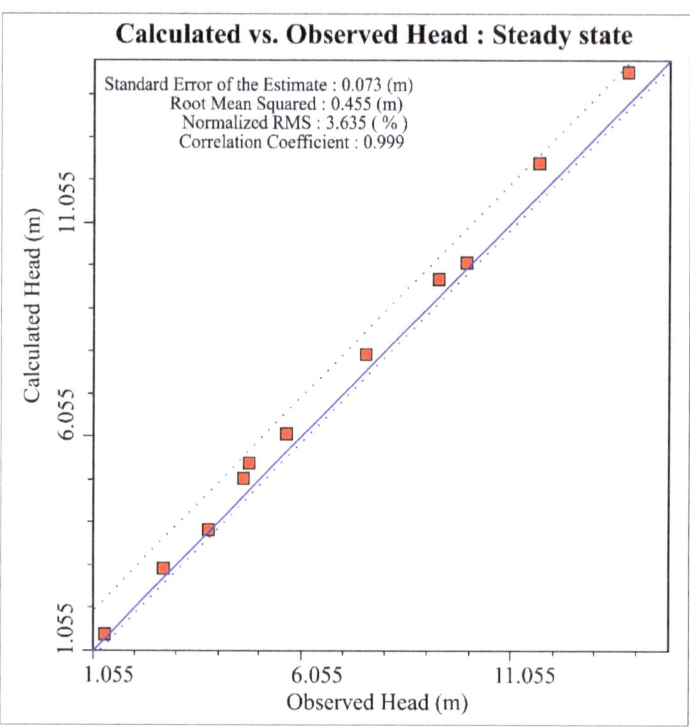

Figure 3. Steady-state condition: calculated head versus observed head during calibration (2004–2005) of the model.

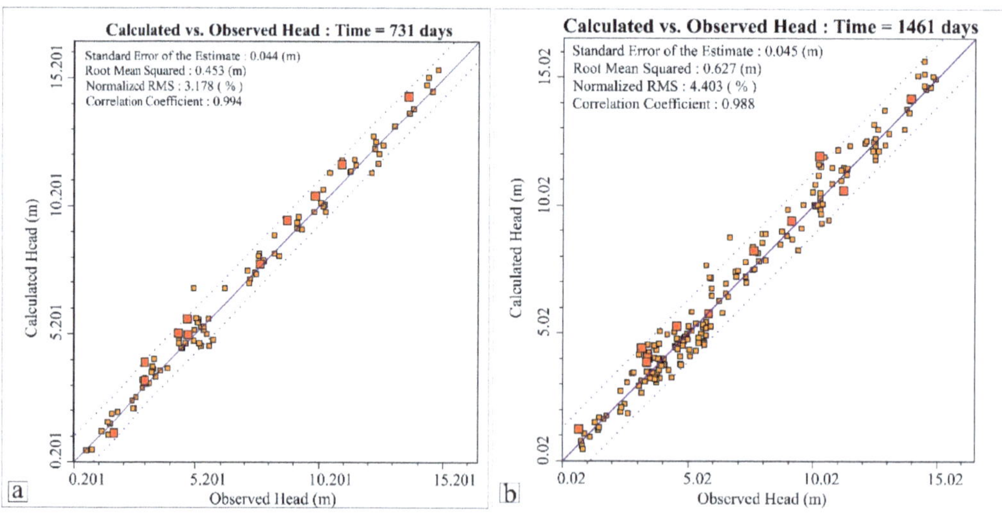

Figure 4. Transient state condition (**a**) calculated head versus observed head during calibration of the model (**b**) calculated head versus observed head during validation of the model.

4.4. Parameter Estimation (PEST) Model

The groundwater model parameters are also estimated by using PEST [49]. Knowling and Adrian (2016) used PEST to minimalize the weighted least squares objective function based on Tikhonov regularization [50]. In this study, the groundwater model has been auto-calibrated with the help of the PEST module of MODFLOW to optimize the aquifer parameters (hydraulic conductivity and specific yield). The calibrated model shows a good correlation between the observed head value and calculated head value with a correlation coefficient value of 0.993 (Figure 5). The automated calibrated (PEST) aquifer parameters (conductivity and specific yield) have been compared to the manually calibrated aquifer parameters, as shown in Table 3. The uncertainty is the process through which the uncertainty on the estimated parameters is quantified to understand the risk associated with different groundwater management models [51].

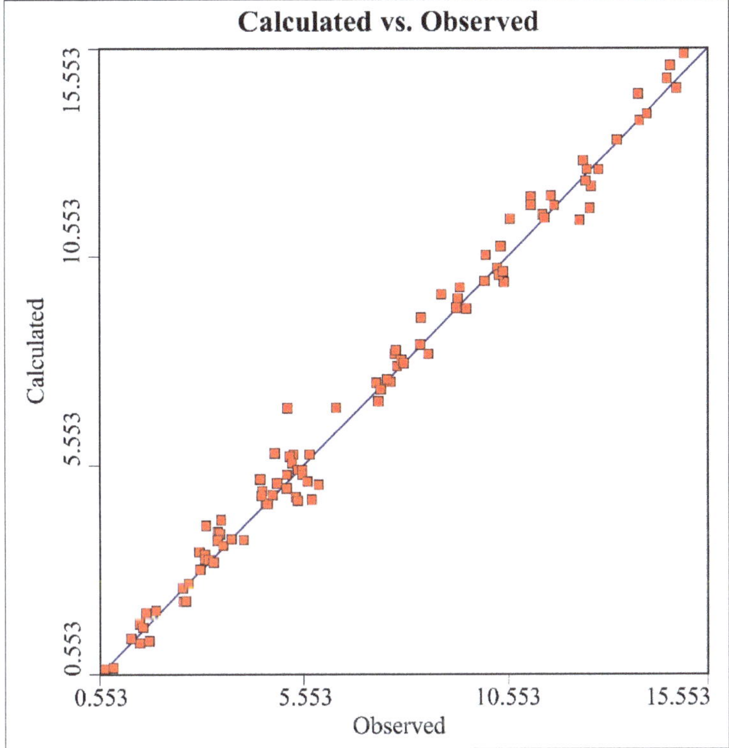

Figure 5. Correlation graph between observed head and calculated head by PEST.

Table 3. Initial and PEST hydraulic parameters.

Zones	Initial Hydraulic Parameters		PEST Estimated Parameters	
	Hydraulic Conductivity in m/Day	Specific Yield	Hydraulic Conductivity in m/Day	Specific Yield
I	40	0.05	36.85	0.058
II	42	0.06	44.39	0.075
III	45	0.07	44.01	0.053

The estimated parameters viz. specific yield and hydraulic conductivity by the PEST tool have been subjected to uncertainty analysis, which shows a 95% confidence interval between 40.19 and 44.96 m/day (Table 4).

Table 4. Uncertainty analysis (95% confidence interval) by Parameter estimation (PEST) techniques.

Zones	Hydraulic Conductivity (K) in m/Day	Specific Yield (S_y)
I	30.746 < K < 44.18	$0.046 < S_y < 0.074$
II	40.81 < K < 48.30	$0.048 < S_y < 0.116$
III	39.70 < K < 48.78	$0.043 < S_y < 0.067$

5. Results and Discussion

5.1. Interaction between Aquifer and River

Both inflow from the rivers into the aquifer system and outflow from the aquifer system to the rivers have been observed in a different time period. This unconfined coastal aquifer receives water from rivers in the pre-monsoon and post-monsoon period, whereas the excess amount of groundwater in the form of base flow discharges to the rivers during monsoon season, as shown in Figure 6a,b. The inflow from the river boundary has been estimated as 34 MCM in the post-monsoon and pre-monsoon period in the year 2006–2007. In the monsoon period, the unconfined coastal aquifer system supplies around 23 to 27 MCM of groundwater to the river system after irrigation (Figure 6a,b). This deltaic aquifer system is mostly recharged by rainfall during wet days. Figure 6c shows that the extraction of groundwater is different in different time periods. In the post-monsoon time period, withdrawal of groundwater is more than that of the pre-monsoon period to provide water for post-monsoon crop, though there is available of adequate amount of water in coastal areas to meet the monsoon period Kharif crop [52]. The extracted groundwater for the sustainability of agricultural productivity and livelihoods in the post-monsoon season has been estimated at around 180 MCM in the year 2006–2007 (Figure 6c). This implies that the heavy abstraction of groundwater for agriculture activity and other domestic uses declines the groundwater level of the aquifer system, which is also a respondent of river inflow into the aquifer system. The resultant river inflow of 33.92 MCM of water entering into this coastal aquifer is due to the pumping of groundwater. Similarly, during the pre-monsoon time, the groundwater is extracted to fulfill the water demand for Rabi crops, but the amount of water required is less than that of the post-monsoon season. It is calculated that about 100 MCM of groundwater is pumped out for agricultural activity and other needs, which also causes the inflow of water through the river boundary. When there is groundwater stress, whether less or more, it also affects the river system.

The river–aquifer interaction indicates a good relationship between the river stage and the outflow/inflow from the river boundary (Figure 7a,b).

As the outflow from the river boundary increases, the river stage also increases. The estimated base flow (7.81 MCM) to the river could be one of the factors contributing to the highest river stage value of 8 m in August of the monsoon period. In contrast, the inflow from river boundary to aquifer system during pre- and post-monsoon time causes declination of river stage from 8 to 3 m. According to the estimation, about 10 MCM of water from the river system enter into the aquifer systems during this season. The interpreted result shows the influence of groundwater flux on the river stage and also suggests a good interaction between the river and the coastal aquifer system [52].

5.2. Fluctuation in Groundwater Level

The spatiotemporal variation in groundwater level has been identified in this coastal aquifer system (Figure 8).

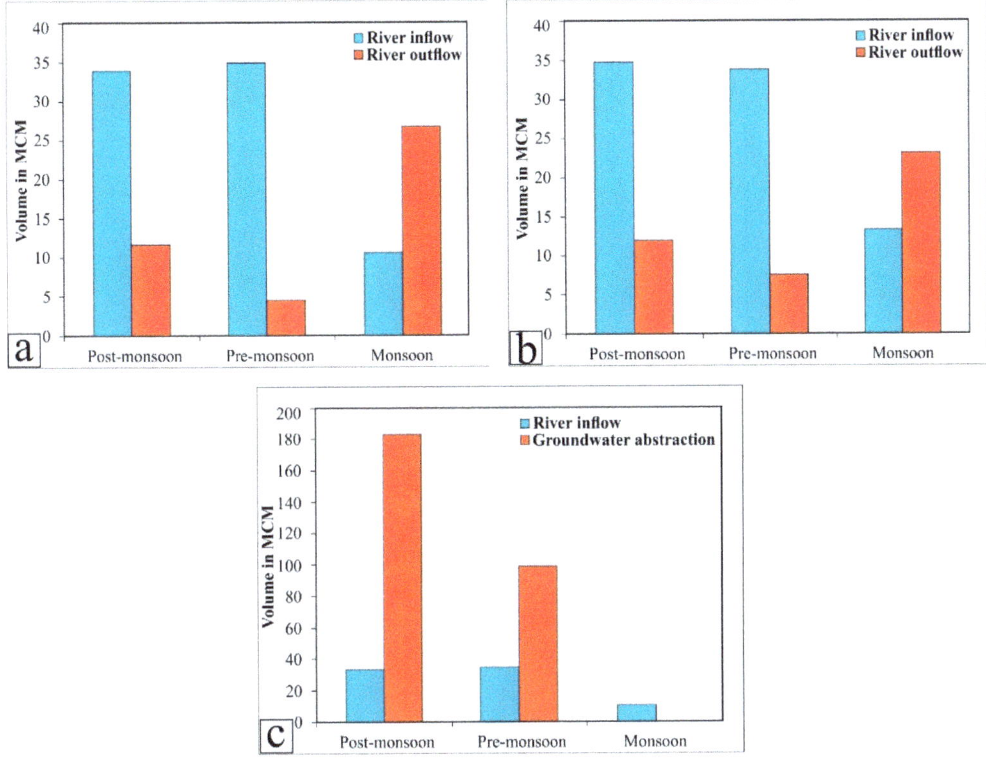

Figure 6. Interaction between river and aquifer (**a**) River inflow/outflow into/from the aquifer system in 2006 (**b**) River inflow/outflow into/from the aquifer system in 2007 (**c**) Graph showing the influence of groundwater abstraction on the river systems in seasons.

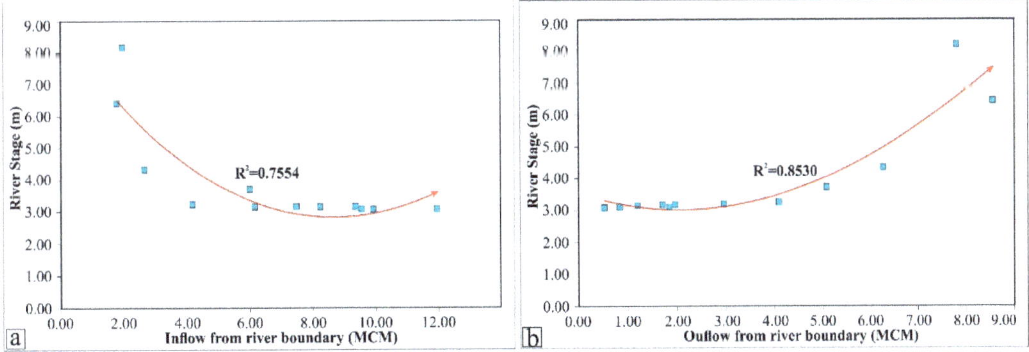

Figure 7. A correlation Graph between the river stage and groundwater flux (**a**) the river stage vs. inflow from the river boundary (**b**) the river stage vs. outflow the river boundary.

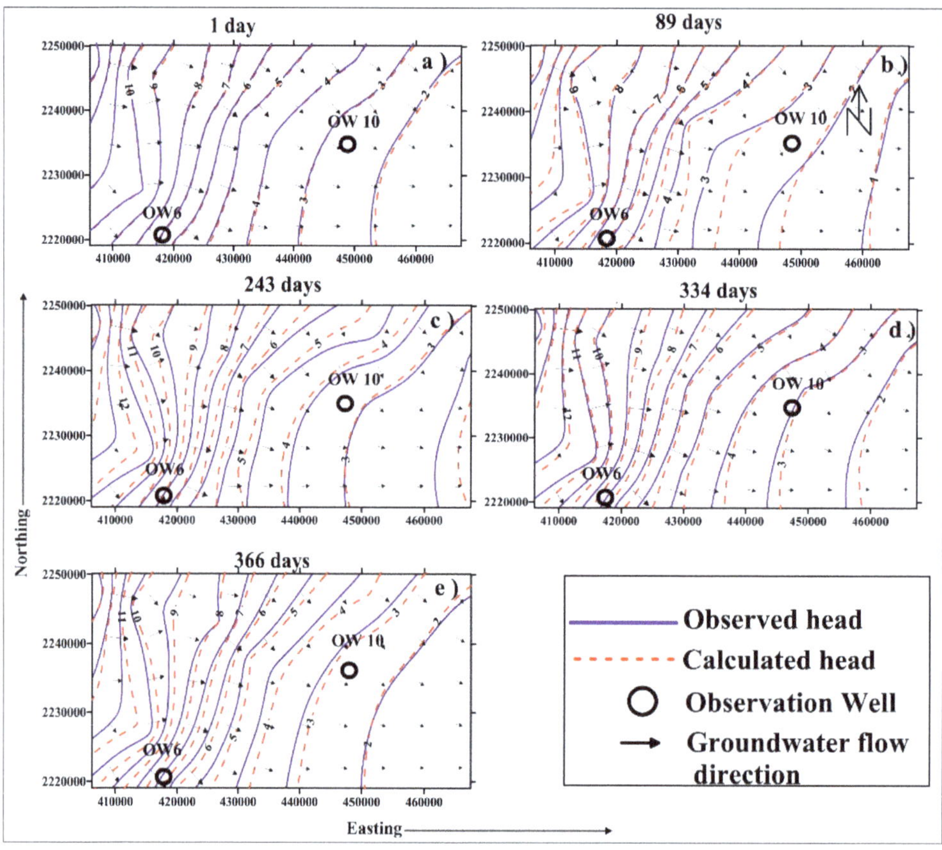

Figure 8. Spatio-temporal variation of the hydraulic head (m) in the study area.

The observation well (OW 6) is showing the highest hydraulic head rise of 3.07 m, situated away from the coastal tract of the study area. The lowest hydraulic head rise of 0.75 m has been observed in the observation well (OW 10), which is close to the Bay of Bengal. This spatial variation in hydraulic head rise depends on the topography, rainfall intensity, type of soil, and land-use patterns. Groundwater stress has been observed during the pre-monsoon period. This also results in a decline in groundwater levels due to the absence of rainfall events in between days 1 and 89, as shown in Figure 8. In the case of monsoon and post-monsoon periods (243 and 334 days), the hydraulic head increases from place to place, suggesting that the groundwater is being mostly recharged by rainfall and regained the groundwater potentiality in the study area. The contour lines in the upper part of the study area are very close, which reveals the presence of a high hydraulic gradient and encounters high groundwater movement [53]. Gradually, large spacing between two equipotential lines has been observed close to the sea shoreline, indicating the sluggish movement of groundwater as the presence of a low hydraulic gradient (Figure 8). However, there is an average rise of 1.84 m of hydraulic head in the monsoon period as compared to the pre-monsoon period, which implies that the shallow aquifer of this coastal region is recharged quickly due to rainfall events. This suggests that sustainable management of coastal aquifers is required during dry periods as there is an absence of a primary recharge source (i.e., rainfall).

5.3. Groundwater Recharge Estimation

In this area, rainfall, seepage from the riverbed, and irrigation return flow are the major sources of groundwater recharge. Groundwater recharge estimation plays a vital role in the optimal development and efficient management of fresh groundwater resources in coastal areas. The study area of the Jagatsinghpur district is divided into eleven recharge zones (Figure 9) in the form of Thiesen polygons to estimate the groundwater recharge [43].

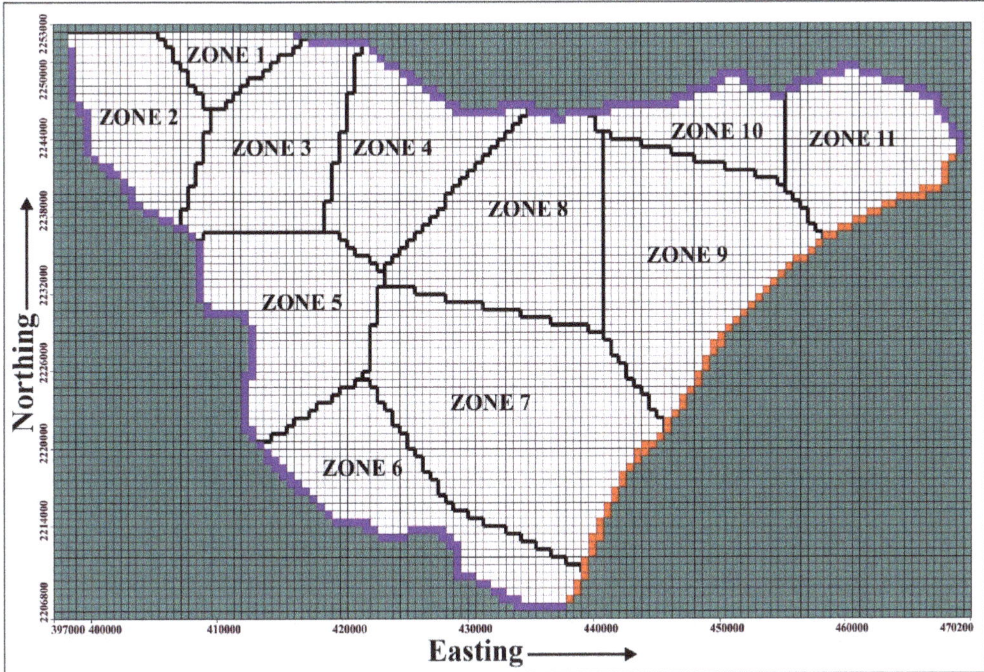

Figure 9. Recharge zones in the study area.

In the groundwater simulation model, the recharge was manually optimized to minimize the head difference between the observed head and calculated head [46]. Modeled recharge rates vary across the recharge zones and also from year to year due to large variations of rainfall over the simulation period. The temporal and spatial distribution of annual rainfall from the year 2004 to 2009 varies from 875 to 1229 mm. The average net recharge (rainfall recharge–groundwater draft) in the area varies from 247.89 to 262.63 million cubic meters (MCM) in the year 2006–2007. The net recharge in post-monsoon is estimated to be less than that of pre-monsoon and monsoon periods (Figure 10). Around 6–15 MCM of water recharge the aquifer system during the post-monsoon period, whereas 33–54 MCM of water percolates into the aquifer system in the pre-monsoon period. Huge extraction of groundwater leads to less net recharge in the post-monsoon season, though natural rainfall happens to have occurred in the study area. As compared to the post and pre-monsoon season, the coastal aquifer system gets highly recharged by rainfall in the monsoon period, estimated in between 180.26 and 223.08 MCM.

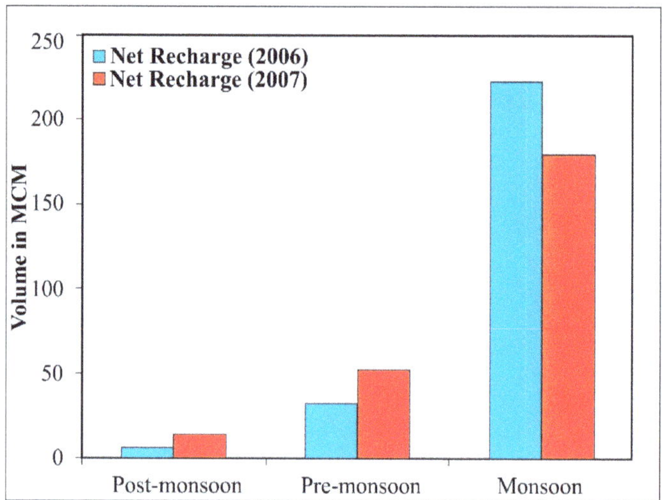

Figure 10. Estimated net recharge in the study area in different time periods for the year 2006–2007.

5.4. Groundwater Outflow to the Bay of Bengal

The groundwater flow direction in the study area is towards the Bay of Bengal as shown in Figure 8. The model simulation results show that the outflow to the Bay of Bengal varies from 8.92 to 9.64 MCM on an annual basis (2006–2007) (Figure 11). The unconfined coastal aquifer of the studied area discharges nearly 50% of its groundwater during the monsoon period, while the rest discharges during dry periods [29]. This is due to the groundwater recharge by rainfall and reduction of groundwater abstraction during the monsoon season. The resultant outflow from the Bay of Bengal also prevents seawater ingress into the aquifer system.

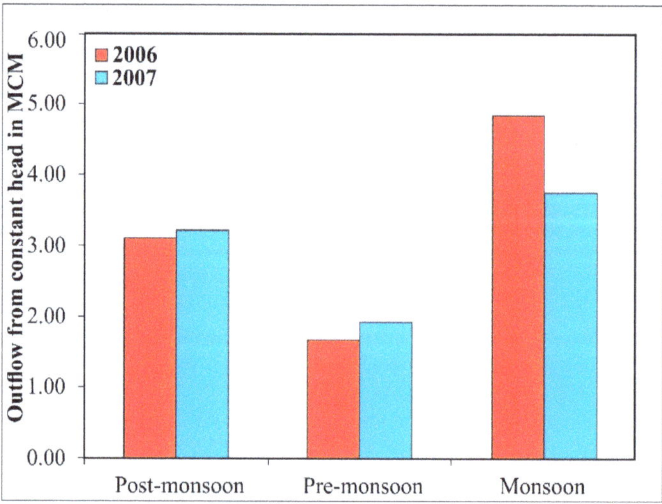

Figure 11. Groundwater outflow from the constant head in different time periods for the year 2006–2007.

6. Conclusions

The groundwater dynamics in the coastal area of the Mahanadi delta was studied using a modeling technique with the help of Visual MODFLOW. The simulated heads matched significantly with the observed heads characterized by different statistical parameters, showing the development of a robust model for this study area. The aquifer parameters (hydraulic conductivity and specific yield) estimated by the PEST tool and trial-and-error method through modeling are approximately the same. Further, the parameters have been subjected to uncertainty analysis, yielding a 95% confidence interval between 39.70 and 48.43 m/day for a particular zone.

The shallow coastal aquifer was influenced by the river system. The estimated groundwater abstraction (180 MCM) led to inflow from the river during the non-monsoon time. Further, the excess amount of groundwater as base flow during the monsoon period recharges the river. This is one of the controlling factors that also affects the river stage. The good connectivity between river and shallow aquifer indicates the regular water circulation, exchange from groundwater to surface water or vice versa. This can improve the aquatic environmental condition. On the other hand, the decline of groundwater was due to the heavy extraction of groundwater to grow non-monsoon agricultural products. Furthermore, the temporal variation of the hydraulic head reveals that the shallow coastal aquifer is very sensitive to rainfall as it quickly responds to rainfall events.

Based on the estimated aquifer parameters, i.e., hydraulic conductivity and specific yield, the net groundwater recharge to this coastal aquifer was estimated to vary between 247.89 and 262.63 MCM. The results derived from the groundwater modeling indicate that there is a net outflow of groundwater into the Bay of Bengal. The outflow varies between 8.92 and 9.64 MCM, which may prevent the seawater ingress into the coastal aquifer of Jagatsinghpur, Odisha. Further, this scientific evidence may prove to be significant for the development of resilient water development plans for dynamic coastal aquifers, so that future generations can utilize the precious resource against climate-related hazards.

Author Contributions: Writing—original draft, A.K.B.; reviewing and editing, A.K.B., R.M.P., S.K., G.J.C. and P.K.; methodology, A.K.B. and S.K.; supervision, S.K. and G.J.C.; data collection and conceptualization, A.K.B. All authors have read and agreed to the published version of the manuscript.

Funding: This research received no external funding.

Institutional Review Board Statement: Not available.

Informed Consent Statement: Not available.

Data Availability Statement. The data presented in this study are available on request from the corresponding author.

Acknowledgments: We would like to acknowledge CGWB SE Region, Bhubaneswar for providing the necessary data to research groundwater modeling in the Jagatsinghpur area, Odisha. University Grant Commission (Govt. of India), New Delhi is acknowledged for the award of Junior Research Fellowship to Ajit Kumar Behera. We are thankful to the editor and three anonymous reviewers for their constructive comments and suggestions which has greatly improved the present version of the manuscript.

Conflicts of Interest: The authors declare no conflict of interest.

References

1. Post, V.E.A. Fresh and saline groundwater interaction in coastal aquifers: Is our technology ready for the problems ahead? *Hydrogeol. J.* **2005**, *13*, 120–123. [CrossRef]
2. Antonellini, M.; Mollema, P.; Giambastiani, B.; Bishop, K.; Caruso, L.; Minchio, A.; Pellegrini, L.; Sabia, M.; Ulazzi, E.; Gabbianelli, G. Salt water intrusion in the coastal aquifer of the southern Po Plain, Italy. *Hydrogeol. J.* **2008**, *16*, 1541–1556. [CrossRef]
3. Ferguson, G.; Gleeson, T. Vulnerability of coastal aquifers to groundwater use and climate change. *Nat. Clim. Chang.* **2012**, *2*, 342–345. [CrossRef]
4. Moore, W. Large groundwater inputs to coastal waters revealed by 226Ra enrichments. *Nature* **1996**, *380*, 612–614. [CrossRef]

Article

Integrated Approach to Quantify the Impact of Land Use and Land Cover Changes on Water Quality of Surma River, Sylhet, Bangladesh

Abdul Kadir [1], Zia Ahmed [1], Md. Misbah Uddin [2], Zhixiao Xie [3] and Pankaj Kumar [4,*]

1 Department of Geography and Environment, Shahjalal University of Science and Technology, Sylhet 3114, Bangladesh; abdulkadir.sust@gmail.com (A.K.); ziaahmed-gee@sust.edu (Z.A.)
2 Department of Civil and Environmental Engineering, Shahjalal University of Science and Technology, Sylhet 3114, Bangladesh; mun-cee@sust.edu
3 Department of Geoscience, Florida Atlantic University, Boca Raton, FL 33431, USA; xie@fau.edu
4 Institute for Global Environmental Strategies, Hayama, Kanagawa 240-0115, Japan
* Correspondence: kumar@iges.or.jp

Abstract: This study aims to assess the impacts of land use and land cover (LULC) changes on the water quality of the Surma river in Bangladesh. For this, seasonal water quality changes were assessed in comparison to the LULC changes recorded from 2010 to 2019. Obtained results from this study indicated that pH, electrical conductivity (EC), and total dissolved solids (TDS) concentrations were higher during the dry season, while dissolved oxygen (DO), 5-day biological oxygen demand (BOD_5), temperature, total suspended solids (TSS), and total solids (TS) concentrations also changed with the season. The analysis of LULC changes within 1000-m buffer zones around the sampling stations revealed that agricultural and vegetation classes decreased; while built-up, waterbody and barren lands increased. Correlation analyses showed that BOD_5, temperature, EC, TDS, and TSS had a significant relationship (5% level) with LULC types. The regression result indicated that BOD_5 was sensitive to changing waterbody (predictors, R^2 = 0.645), temperature was sensitive to changing waterbodies and agricultural land (R^2 = 0.889); and EC was sensitive to built-up, vegetation, and barren land (R^2 = 0.833). Waterbody, built-up, and agricultural LULC were predictors for TDS (R^2 = 0.993); and waterbody, built-up, and barren LULC were predictors for TSS (R^2 = 0.922). Built-up areas and waterbodies appeared to have the strongest effect on different water quality parameters. Scientific finding from this study will be vital for decision makers in developing more robust land use management plan at the local level.

Keywords: water quality; buffer zone; land use/land cover; Bangladesh

Citation: Kadir, A.; Ahmed, Z.; Uddin, M.M.; Xie, Z.; Kumar, P. Integrated Approach to Quantify the Impact of Land Use and Land Cover Changes on Water Quality of Surma River, Sylhet, Bangladesh. *Water* **2022**, *14*, 17. https://doi.org/10.3390/w14010017

Academic Editor: František Petrovič

Received: 16 November 2021
Accepted: 19 December 2021
Published: 22 December 2021

Publisher's Note: MDPI stays neutral with regard to jurisdictional claims in published maps and institutional affiliations.

1. Introduction

Water is a vital resource for the maintenance of life, ecological functioning, biological diversity, and social well-being. Despite its importance for life, in recent decades, excessive human land use has severely harmed the quality and quantities of available water resources [1]. In particular, it is well known that rivers function as integrators of land-water connections, receiving pollutants from the surrounding landscapes [2], and river water quality could be negatively impacted. Surma River, an important river in Bangladesh, has been collecting pollutants from a wide variety of point and non-point sources along its course from agricultural wastes, industrial effluents, menage wastes, and municipal sewage [3]. Due to population growth, urbanization, and industrialization, the surrounding landscape of the Surma river has been changing, and the riverside has experienced tremendous development in terms of commercial, human settlement, and industrial development. The population and urban sprawl have adverse effects on the quality of the Surma river water; the increased urban area is responsible for generating large amounts of nonpoint source pollution through runoff and degraded the river water quality.

Land use and land cover (LULC) within the immediate environments of a waterbody have a direct impact on the physicochemical and microbiological properties of water, and such impact varies with the type, extent, location of human land uses, and the inputs from the watershed. Water quality problems arise when the type and extent of human land use exceed the natural ability of the watershed to mitigate accumulated land-use-related stress [4]. Numerous studies have shown that human activities lead to landscape pattern changes, which in turn had significant impacts on the conditions of river water [2,5,6]. LULC change, especially urbanization, has a major impact on hydrology, affecting water quality and quantity on a range of spatial and temporal scales [7,8]. Zhu [9] found that water quality degrading was particularly affected by alteration from farmland to commercial and residential land, and the expansion in an urban area causes streamflow increase, carrying more sediment, bank erosion, and nutrients in streams. Hossain [10] reported that water quality variables are correlated with LULC change. As a consequence of the spatiotemporal LULC change, the concentration of diffuse pollutants in streams varies as well as vegetation types, and watershed climate are responsible for stream water quality change.

Landscape pattern has a complex, space- and scale-dependent effect on water quality [11,12], and different landscape characteristics play different roles in receiving water at different spatial scales varying from the local to eco-regional scales [13]. Bhaduri et al. [7] noted that the most significant human impacts on the hydrologic system and water resources are caused by land-use changes on local, regional, and global scales, driven by a rise in urban areas. Pollution from nonpoint sources (NPS) is a challenging problem to solve as it comes from a variety of origins difficult to pinpoint, and it occurs in a variety of environments; however, Geographical Information Systems (GIS) software provides a more comprehensive description of land cover patterns and the spatial distribution of NPS pollution [14] and has been commonly used. To explore the landscape pattern's impacts on the lakes and rivers water quality several studies have been conducted [15,16], and it seems that the riparian buffer zone landscape patterns are more powerful in explaining water quality variations [17]. Land use types have an influence on surface water quality which can be analyzed by using statistical methods, remote sensing (RS), and geographic information system (GIS) [18–21]. Li et al. [12] stated that in riparian zones, the landscape category has a significant impact on water quality, and the alterations in the landscape through urban spread have put a lot of pressure on undeveloped land. Ribeiro et al. [22] found that water quality was worse in the sub-basin, also characterized by the presence of more agriculture, permanent conservation area, lesser natural forest, and a greater drainage area; however, the existence of agriculture negatively affect water quality in the riparian area. As proven by numerous studies, water quality is closely correlated to landscape pattern, which includes the landscape structure and spatial configuration [23–26].

So, it is essential to find out the spatiotemporal and future potential effect on water quality by LULC change. Therefore, this study examines the LULC change in the riparian buffer zone of the Surma river, over the period 2010 to 2019. The seasonal surface water quality variation is analyzed next. The study aims to assess the impacts of land use and land cover (LULC) changes on the Surma river surface water quality.

The outcomes of this study will make it helpful to acquire sustainability of land and water resources. Additionally, it will facilitate other researchers' pursuit of studies about LULC change impact on water quality in the study area and similar regions.

2. Materials and Methods

2.1. Design of the Study

We have conducted this research work systematically and scientifically; the major methods and techniques, which were followed very carefully, are illustrated in Figure 1.

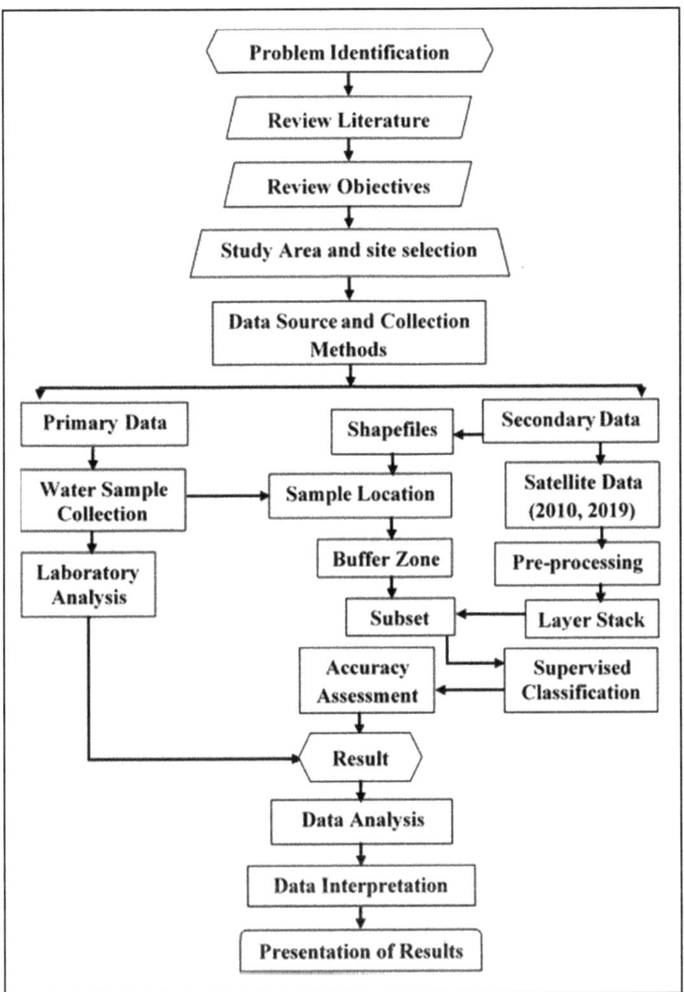

Figure 1. Research Design.

2.2. Study Area

In Bangladesh, Surma is an important river, as a part of the Surma–Meghna river system, which originates when the Barak River in northeastern India and then splits into two branches at the Bangladesh border as the Surma (a northern branch that flows west and then runs southwest to the town of Sylhet) and the Kushiyara rivers [27]. Sylhet Sadar Upazila and Dakshin Surma Upazila are two Upazilas of the Sylhet district, which are located in the country's north-eastern region. Upazila is an administrative region in Bangladesh, equivalent to a county of Western countries. The study was conducted in the Surma river portion, situated between Sylhet Sadar Upazila and Dakshin Surma Upazila land, extending 1000 m toward both sides of the riverbank. It is situated between 24°51′39.1″ N to 24°54′37.07″ N latitude and between 91°49′40.9″ E to 91°55′51.9″ E longitude (Figure 2).

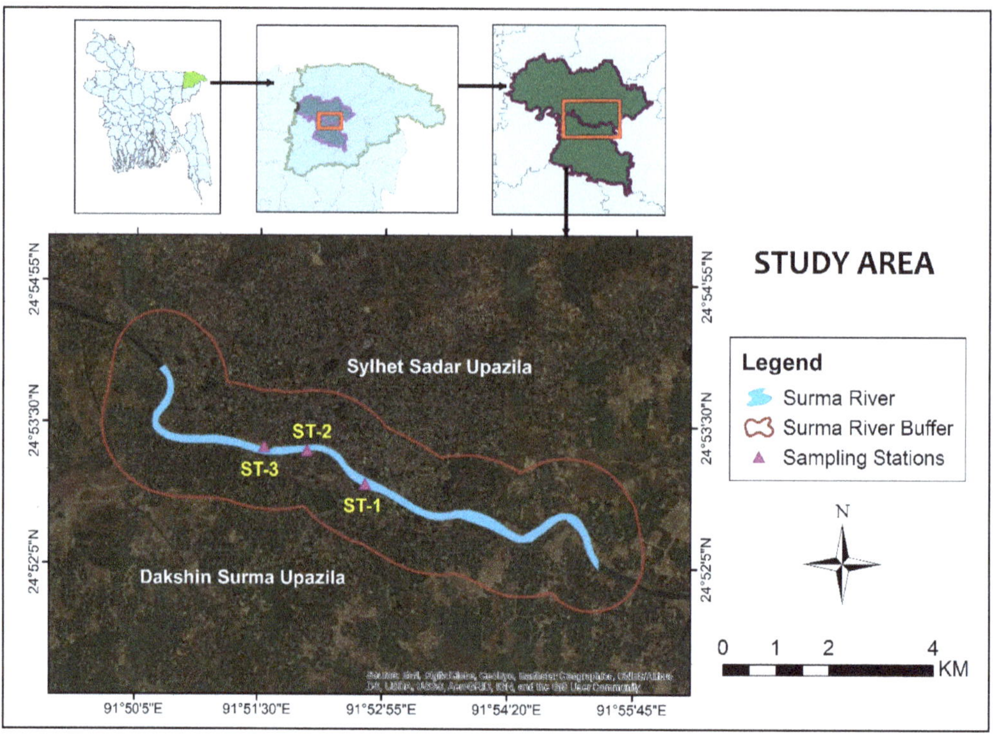

Figure 2. The study area and sampling stations.

2.3. Water Sampling and Analytical Methods

This study considered both primary and secondary data on water quality. Water quality data for the dry and wet season of 2010 was collected from the Department of Environment (DoE), Sylhet Divisional Office, Sylhet, Bangladesh.

Water samples were collected from three (ST-1, ST-2, and ST-3) separate Surma river sampling stations, each with a different geographic location, as shown in Table 1. Following the collection, water samples were kept in an ice box and bought to the laboratory. Sampling was performed in the dry season (2019) and the wet season (2019).

Table 1. Sampling stations with location.

Station No.	Station Zone	Location		Flow Direction
		Latitude	Longitude	
ST-1	Shahjalal Bridge	24°52′54.408″ N	91°52′42.78″ E	⇩
ST-2	Keane Bridge	24°53′14.244″ N	91°52′3.252″ E	
ST-3	Kazir Bazar	24°53′16.26″ N	91°51′33.696″ E	

Eight water quality parameters were selected, including dissolved oxygen (DO), 5-day biological oxygen demand (BOD_5), pH, electrical conductivity (EC), temperature, total suspended solids (TSS), total dissolved solids (TDS), and total solids (TS). All the water samples of 2019 were analyzed in the Water Supply and Sewerage Engineering Laboratory of Civil and Environmental Engineering Department, and Environmental Laboratory of Geography and Environment Department, Shahjalal University of Science and Technology, Sylhet, Bangladesh. DO and BOD_5 were measured by the Winkler titration method, pH and temperature were measured by electrometric methods (pH meter, HANNA-HI 9125), electrical conductivity (EC) was measured using an EC meter (HANNA-HI 98192); and TDS, TSS, and TS were measured in the laboratory following standard methods [28].

2.3.1. LULC Change Analysis

For land use and land cover (LULC) change analysis, satellite images of 2010 (Landsat 5 Thematic Mapper) and 2019 (Sentinel 2 MSI), acquired from the United States Geological Survey (USGS), were used to produce LULC classification maps for both years using remote sensing (ERDAS IMAGINE 2014) and geographic information system (ArcGIS 10.8) software. LULC classes are categorized into five major categories including, waterbody (river, canals, pond, lakes, reservoirs), built-up (urban areas, human settlements, road networks. Commercial and industrial areas), agricultural land (cropland, pasture, herb, shrub, fallow land, permeable surface), vegetation (canopy, mixed forest, evergreen forest), and barren land (bare soil, sand, rocks without vegetation). Satellite image preprocessing, as well as geometrical rectification, registration of image, corrections viz. atmospheric and radiometric, were conducted by ERDAS IMAGINE 2014. Supervised classification was conducted to create LULC maps [29]. Accuracy assessment was conducted, indicating that the overall classification accuracy of the 2010 image was 81.65% and a kappa statistics of 0.7524, and overall classification accuracy of the 2019 image was 94.50% and kappa statistics of 0.9024, indicating a very good accuracy of the LULC map. By using ArcGIS 10.8 software, land uses composition within the 1000-m buffer zones around the sampling stations was extracted from the LULC map. A 1000-m buffer scale is stronger than smaller scales in explaining land-use types and their water quality relations [30]. Percentages of these broad LULC types were used to examine the relationship between water quality parameters and LULC types.

2.3.2. Statistical Analyses

Descriptive statistics were used to explain the general characteristics of LULC and water quality parameters. Karl Pearson's correlation analysis was used to determine correlations between LULC patterns percentage and water quality parameters (WQPs) at statistical significance at a 5% level. Backward stepwise regression analysis was used to identify the relationship between the percentage of land usage composition within the 1000-m buffer zone and water quality properties. WQPs showing significant correlations with LULC types were considered for the backward stepwise regression analysis. In regression analysis, water quality parameters (BOD_5, temperature, EC, TDS, and TSS) were considered dependent variables, while LULC types (waterbody, built-up, vegetation, agricultural land, and barren land) were treated as independent variables. To identify the best combination of land uses for water quality estimation regression equations were compared with R^2 values (value closer to one indicates a greater accuracy of the model). The Statistical Packages for Social Science (IBM SPSS Statistics 20) for windows was used to perform all statistical analyses.

3. Results

3.1. Water Quality

The quality of water throughout the dry and wet seasons in 2010 and 2019 are presented in Tables 2 and 3, respectively, and in Figure 3. DO levels ranged from 3.6 mg/L to 4.4 mg/L in the dry season and 7.6 mg/L to 11.6 mg/L in the wet season in 2019. In the wet season, the highest DO was found at Kazir Bazar, while the lowest was found at Keane Bridge in dry season. The DO level increased in all stations during the wet season. In 2010, the DO level varied from 5.1 mg/L in the wet to 6.2 mg/L in the dry season. The Surma river's mean DO level in 2019 was higher than the DO level in 2010.

Table 2. Water quality during the dry and wet season in 2010.

Parameter		DO (mg/L)		BOD_5 (mg/L)		pH		Temp. (°C)		EC (µS/cm)		TDS (mg/L)		TSS (mg/L)		TS (mg/L)	
Season		Dry	Wet	Dry	Wet	Dry	Wet	Dry	Wet	Dry	Wet	Dry	Wet	Dry	Wet	Dry	Wet
Station No.	ST-1	6	4.1	1.2	1.1	7.4	7.4	20	29	280	100	400	300	140	100	540	400
	ST-2	6.2	5.1	1.3	1.2	7.4	7.5	20	30	290	120	500	310	100	90	600	400
	ST-3	6.3	6.2	1	1	7.4	5.6	21	30	300	160	400	430	110	110	510	540
Average		6.2	5.1	1.2	1.1	7.4	6.8	20.3	29.6	290	126.6	433.3	346.6	116.6	100	550	446.6
Mean		5.65		1.15		7.1		24.95		208.3		389.95		108.3		498.3	

Table 3. Water quality during the dry and wet season in 2019.

Parameter		DO (mg/L)		BOD_5 (mg/L)		pH		Temp. (°C)		EC (µS/cm)		TDS (mg/L)		TSS (mg/L)		TS (mg/L)	
Season		Dry	Wet	Dry	Wet	Dry	Wet	Dry	Wet	Dry	Wet	Dry	Wet	Dry	Wet	Dry	Wet
Station No.	ST-1	4.4	7.9	1	2.8	7.96	6.47	27.6	26.6	280.5	54.46	140.3	27.23	48.4	730	188.7	757.23
	ST-2	3.6	7.6	1	2	6.97	7.31	27.8	26.7	292.9	74.43	146.7	37.2	51.3	650	198	687.2
	ST-3	3.7	11.6	0.9	3.2	7.14	6.94	27.9	26.8	322.7	75.81	161.4	37.91	47.9	470	209.3	507.91
Average		3.9	9.03	0.97	2.67	7.36	6.91	27.77	26.7	298.7	68.23	149.47	34.11	49.2	616.67	198.67	650.78
Mean		6.47		1.82		7.13		27.23		183.47		91.79		332.93		424.72	

For BOD_5, in 2019, the recorded average concentration of BOD_5 for the Surma river water was 0.97 mg/L and 2.67 mg/L in the dry and wet seasons respectively. The seasonal comparison shows that during the wet season the BOD_5 level was higher than the dry season. The lowest and the highest were found at ST-3. In 2010, the average BOD_5 level was 1.2 mg/L (dry season) and 1.1 mg/L (wet season). The mean BOD_5 level slightly increased in 2019 as compared with 2010.

In 2019, the measured pH amongst different stations varied between 6.97 and 7.96 in the dry season, indicating almost neutral-to-slightly-alkaline water conditions. While in the wet season, pH level varied between 6.47 and 7.31. The average dry season pH was slightly alkaline, while the pH of the wet season was almost neutral. In 2010, Surma river water was almost neutral, and pH varied from 7.4 (dry season) to 6.8 (wet season). The mean value of pH was close to neutral in 2010 and 2019.

The average dry season water temperature was 20.3 °C in 2010 and 27.77 °C in 2019, while the average wet season water temperature was 29.6 °C in 2010 and 26.7 °C in 2019. Kazir Bazar had the highest dry season temperature in 2019 and Keane Bridge, as well as Kazir Bazar, had the highest wet season temperature in 2010. The recorded lowest wet season temperature was at ST-1 in the year 2010 and the lowest dry season temperature was at ST-1 and ST-2 in the year 2010. The mean temperature in 2019 is higher than the mean temperature in 2010. The temperature increased during the dry season from 2010 to 2019 but declined during the wet season. The maximum and minimum electrical conductivity (EC) of the Surma river was 322.7 µS/cm at Kazir Bazar in the dry season of 2019 and 54.46 µS/cm at Shahjalal Bridge in the wet season of 2019. During the dry season, EC was high. From 2010 to 2019, the mean EC level decreased.

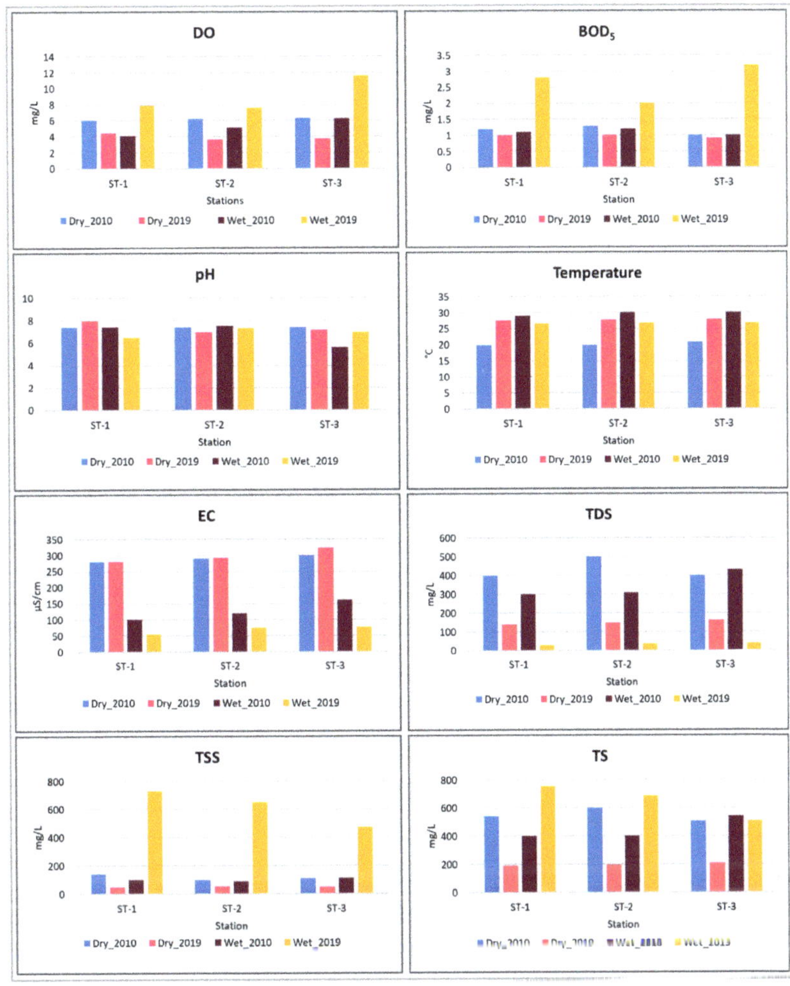

Figure 3. Seasonal and spatial variations of WQPs in 2010 and 2019.

The results of TDS showed that in 2019 the mean TDS concentrations were 149.47 mg/L (dry season) and 34.11 mg/L (wet season). In the dry and wet seasons, maximum TDS was found at Kazir Bazar (ST-3) and minimum TDS was found at ST-1. The mean TDS concentration in 2010 was significantly higher than the mean TDS concentration in 2019. The average maximum (616.67 mg/L) and minimum (49.9 mg/L) TSS concentrations were recorded during the wet and the dry season, respectively, in 2019. In 2019, the mean concentration of TSS was higher than the concentration of the year 2010. From 2010 to 2019, the TSS level increased, on the other hand, the TDS level decreases. In 2019, the recorded total solids (TS) in the dry season varied from 188.7 mg/L at Shahjalal Bridge (ST-1) to 209.9 mg/L at Kazir Bazar (ST-3). The TS ranged from 507.9 mg/L (minimum) at ST-3 to 757.23 mg/L (maximum) at ST-1. The seasonal comparison of the average TS level showed that the wet season's concentration was higher than the dry season's. In 2010, the average TSS concentration varied from 446.6 mg/L during the wet season to 550 mg/L during the dry season with a mean concentration of 498.3 mg/L that was greater than the mean concentration in 2019.

3.2. Descriptive Statistics of the Water Quality Parameters (WQPs)

The descriptive statistics of the WQPs of the Surma River from the year 2010 to 2019 are explained in Table 4.

Table 4. Descriptive statistics of the WQPs.

Water Quality Parameters	Minimum		Maximum		Mean		SD	
	2010	2019	2010	2019	2010	2019	2010	2019
DO (mg/L)	5.05	5.60	6.25	7.65	5.65	6.47	0.60	1.06
BOD_5 (mg/L)	1.00	1.50	1.25	2.05	1.13	1.82	0.13	0.28
pH	6.50	7.04	7.45	7.22	7.12	7.13	0.53	0.09
temperature (°C)	24.50	27.10	25.50	27.35	25.00	27.23	0.50	0.13
EC (μS/cm)	190.00	167.48	230.00	199.26	208.33	183.47	20.21	15.89
TDS (mg/L)	350.00	83.77	415.00	99.66	390.00	91.79	35.00	7.95
TSS (mg/L)	95.00	258.95	120.00	389.20	108.33	332.93	12.58	66.91
TS (mg/L)	470.00	358.61	525.00	472.97	498.33	424.72	27.54	59.24

3.3. LULC Change in the Buffer Zone

Land use change data, as the percentage of land of 1000-m buffer zone of monitoring stations, were derived from LULC change analysis from 2010 to 2019, and they were linked with water quality data. The results show that, in 2019, the central part was dominated by built-up area, while in 2010, built-up area was not clustered at a specific zone (Figure 4).

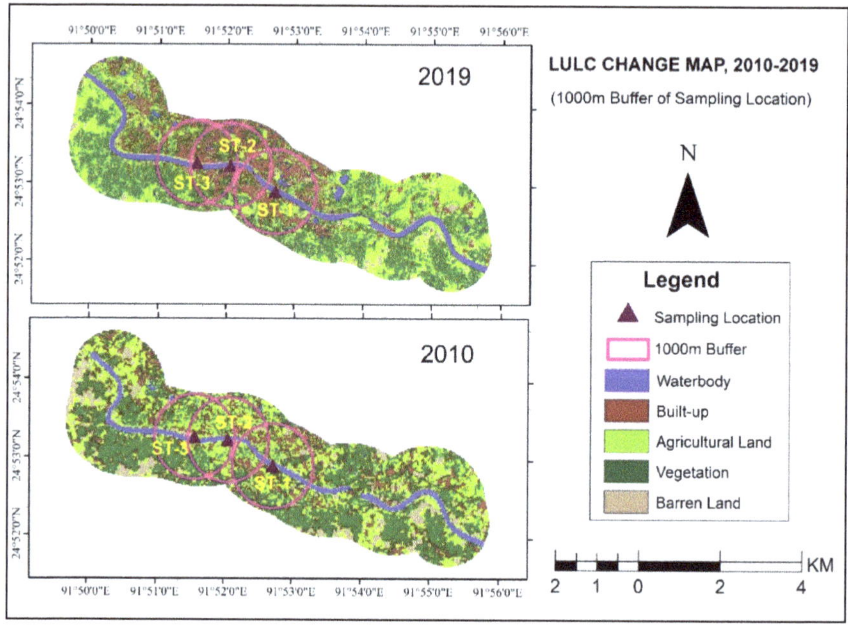

Figure 4. LULC change during the period 2010–2019.

The LULC change in the buffer zones showed that the agricultural land area decreased while the built-up area significantly increased from 2010 to 2019. During this period, the increase of barren land and decrease of vegetation area was also observed in the buffer zones. In 2019 the highest built-up area was found at the buffer zone of station ST-2.

The analysis illustrated that in 2010, at ST-1 (32.4%) and ST-2 (39.9%), agricultural land usage was dominant in the 1000-m buffer zones (Table 5). On the other hand, vegetation cover (40.6%) was the dominant land area in 2010 at ST-3. In 2019, the built-up area increased in all zones and become the dominant land-use type.

Table 5. LULC change within the 1000-m buffer zone of sampling stations.

LULC Type	Sampling Station					
	ST-1		ST-2		ST-3	
	2010 (%)	2019 (%)	2010 (%)	2019 (%)	2010 (%)	2019 (%)
Waterbody	6.3	7.5	5.4	6.9	5.8	7.1
Built-up	30.1	41.8	25.8	48.6	18.7	38.8
Agricultural Land	32.4	23.8	39.9	18.1	32.3	18.3
Vegetation	27.8	20.7	26.7	22.3	40.6	32.2
Barren Land	3.4	6.2	2.2	4.1	2.6	3.6

By comparing land usage distribution from 2010 to 2019, all zones had shown a notable decline in agricultural land area. For the built-up area, the ST-2 zone showed a maximum increase (22.8%), while the ST-1 zone showed a minimum increase (11.7%), and at the ST-3zone, 21.1% area increased.

3.4. Descriptive Statistics of the LULC Types

The descriptive statistics of LULC types within a 1000-m buffer zone at stations of the Surma River area from 2010 to 2019 are reported in Table 6.

Table 6. Descriptive statistics of the LULC types.

LULC (in %)	Minimum		Maximum		Mean		SD	
	2010	2019	2010	2019	2010	2019	2010	2019
waterbody	5.39	6.91	6.28	7.50	5.84	7.15	0.45	0.31
built-up	18.71	38.81	30.12	48.57	24.89	43.07	5.76	5.00
agricultural land	32.28	18.06	39.87	23.83	34.84	20.07	4.36	3.26
vegetation	26.72	20.67	40.62	32.26	31.73	25.10	7.72	6.26
barren land	2.17	3.58	3.36	6.18	2.70	4.62	0.61	1.37

In 2010, waterbody ranged from 5.39% to 6.28% with a mean value of 5.84% ± 0.45%. In 2019, the range and mean of waterbody are 6.91% to 7.5% and 7.15% ± 0.31% respectively. The built-up area, from 2010 to 2019, ranged from 18.71% to 30.12% and from 38.81% to 48.57%, with a mean value of 24.89% ± 5.76% and 43.07% ± 5.00%, respectively. Agricultural land use ranged from 32.28% to 39.87% in 2010, with a mean value of 34.84% ± 4.36%. On the other hand, in 2019, agricultural land use ranged from 18.06% to 23.83% with a mean value of 20.07% ± 3.26%.

Vegetation area in 2010 and 2019 ranged from 26.72% to 40.62% and from 20.67% to 32.26% with a mean value of 31.73% ± 7.72% and 25.1% ± 6.26%, respectively. Barren land ranged from 2.17% to 3.36% in 2010, with a mean value of 2.7% ± 0.61%. On the other hand, in 2019, barren land cover ranged from a minimum of 3.58% to a maximum of 6.18% with a mean value of 4.62% ± 1.37%.

3.5. Water Quality Relation with LULC

Correlation Analysis

A correlation analysis revealed that LULC patterns were correlated significantly with one or more parameters of water quality within the 1000-m buffer zone scale (Table 7). The analysis result reveals that only BOD_5, temperature, EC, TDS, and TSS had a strong and significant relationship at the 5% level of significance with the different LULC types. Other parameters viz. DO, pH, and TS had both positive and negative relationships with the LULC types but were not significant. BOD_5 was found to have a positive significant relationship with waterbody (0.82), which means that, if the waterbody should increase, the BOD_5 level will also increase at a significant level. At the same time, it was also found that temperature had a positive significant relationship with waterbody (0.82) but it had a negatively significant relationship with agricultural land (−0.90). This result reveals that, with the increase in agricultural land area, temperature will decrease at a significant level or vice-versa.

Table 7. Linear relationship (Pearson correlation, r) between WQPs and LULC types.

WQPs \ LULC Types	Waterbody	Built-Up	Agricultural Land	Vegetation	Barren Land
DO (mg/L)	0.38	0.12	−0.50	0.36	0.09
BOD_5 (mg/L)	**0.82**	0.72	−0.74	−0.42	0.66
pH	0.03	0.34	0.18	−0.76	0.15
temperature (°C)	**0.82**	0.79	**−0.90**	−0.32	0.63
EC (μS/cm)	−0.74	**−0.82**	0.47	**0.92**	**−0.84**
TDS (mg/L)	**−0.94**	**−0.93**	**0.92**	0.56	−0.79
TSS (mg/L)	**0.92**	**0.90**	**−0.83**	−0.64	**0.89**
TS (mg/L)	−0.61	−0.63	0.75	0.16	−0.25

NB: Bold letters indicates a significant relationship at the 5% level.

Electric conductivity (EC) had about a perfect positive significant relationship with vegetation (0.92) but a negative relationship with the land use types of built-up area (−0.82) and barren land (−0.84). Total dissolved solids (TDS) had also a significant relationship at the 5% level of significance, with three types of LULC types whereas total suspended solid (TSS) had four types of LULC types. Pearson's correlation result revealed that with the increase of waterbody and built-up area TSS value will increase significantly (a positive relationship) but the TDS value will decrease significantly (a negative relationship). A different result was also found for agricultural land; agricultural land had inverse relationships with the changes in TDS and TSS. The result reveals that, if agricultural land increases, TDS (0.92) will also increase significantly at the 5% level of significance, but TSS (−0.83) will decrease.

Vegetation cover was only correlated with the change of electric conductivity, which had about a perfect positive significant correlation.

3.6. Fulfillment of Distributional Assumption of Dependent Variables

The median of BOD_5 was found at 1.38, which was very similar to the average BOD_5 (1.48). The similarity between mean BOD_5 and median BOD_5 seems to follow the symmetrical distribution of BOD_5, which means that the dependent variable, BOD_5, fulfills the distributional assumption for the classical model.

The similarity between the mean (26.12) and median (26.30) temperatures indicates that temperature seems to follow the normal distribution (Figure 5). There were no observed outliers in temperature, thus we can proceed to the classical model with it. The variable EC seems to follow a bell-shaped distribution, because its second quartile (194.63) lay in the middle of the first (185.25) and third (203.56) quartile. With the fulfillment of the distributional assumption of EC, we can proceed to a classical model.

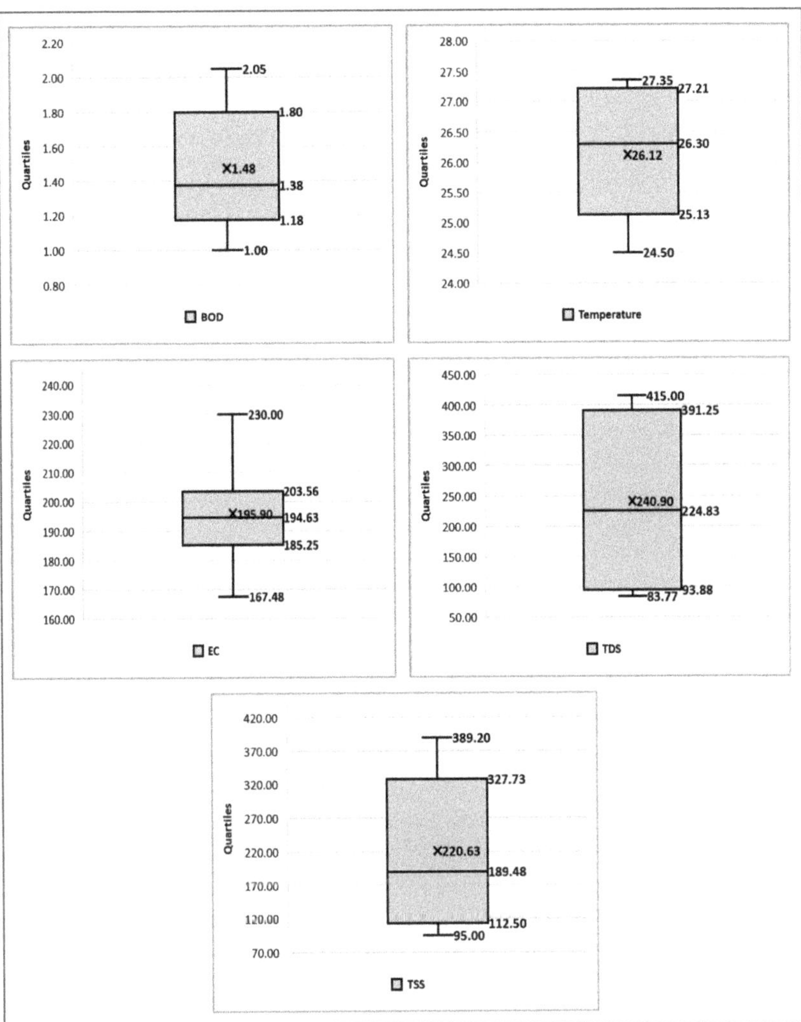

Figure 5. Box and whisker plots of dependent variables.

The median of TDS was found as 224.83 and lay in the middle of the first quartile (93.88) and third quartile (391.25). That's why the small departure of median TDS from mean TDS does not affect the symmetrical shape of TDS. Thus, we can proceed to a classical model for TDS.

The small departure of median TSS from the middle position of the box may have a small effect, an asymmetric shape, and also there was a difference between the mean TSS (220.63) and the median TSS (189.48). This means that TSS may follow a slightly skewed distribution. So, we can, but barely, proceed to the classical model with TSS.

3.7. Regression Analysis

Backward stepwise regression identified the relationship between WQP and LULC types that determine the combination of land uses for water quality estimation (Table 8). For the case of BOD_5, only the waterbody was used as a predictor, which was found significant in a Pearson's correlation analysis. Waterbody was not found as an especially strong predictor, as the adjusted r-squared value was 0.645. For temperature, waterbody and agricultural land were used as predictors (R^2 = 0.889). Similarly, for electric conductivity, built-up, vegetation, and barren land were used as predictors (R^2 = 0.833). For total dissolved solid, waterbody, built-up, and agricultural land were used as predictors (R^2 = 0.993). For TSS, waterbody, built-up, and barren land were used as predictors at R^2 of 0.922. From the regression analysis, it could be concluded that BOD_5 showed sensitivity to changing waterbody, whereas temperature was sensitive to changing waterbody and agricultural land. EC showed sensitivity on built-up, vegetation, and barren land. TDS showed sensitivity on the waterbody, built-up and agricultural land whereas, TSS for waterbody, built-up and barren land. In this study, different parameters of water quality viz. biological oxygen demand, electric conductivity, TDS, and TSS, tended to be most affected by built-up and waterbody land usage types.

Table 8. Linear regression models of LULC types on the WQPs.

Dependent Variable (WQPs)	Independent Variables (Land Usage Type)	Estimated Linear Regression Equations	Adjusted R^2
BOD_5	Waterbody	$BOD_5 = 0.737 + 0.594 \times Y\_2019 + 0.068 \times W$	0.645
temperature	waterbody, agricultural land	$Temp = 29.097 + 2.567 \times Y\_2019 - 0.546 \times W - 0.026 \times A$	0.889
EC	built-up, vegetation, barren land	$EC = 205.887 + 12.714 \times Y\_2019 - 0.802 \times Bu + 1.332 \times V - 7.369 \times Ba$	0.833
TDS	waterbody, built-up, agricultural land	$TDS = 726.425 - 184.029 \times Y\_2019 - 45.117 \times W - 2.988 \times Bu + 0.039 \times A$	0.993
TSS	waterbody, built-up, barren land	$TSS = 309.915 + 172.824 \times Y\_2019 - 68.363 \times W + 1.890 \times Bu + 55.799 \times Ba$	0.922

Y_2019 = year 2019 (dummy or indicator); Temp = temperature; W = waterbody, A = agricultural land; Bu = built-up; V = vegetation; Ba = barren land.

The equation for BOD_5 explains, that for a one-unit change of waterbody, BOD_5 would increase 0.068 times. For temperature, it would decrease, both for waterbody and agricultural land. In the same way, the equation for electric conductivity explains that, if the built-up area increases by one unit, EC will decrease 0.802 times, and for barren land, it will decrease 7.369 times. Yet, EC could increase 1.332 times if vegetation covers were increased by one unit. For TDS and TSS, the changes were the same, as is quantified and described in their equations.

4. Discussion

The water quality of any river is sturdily influenced by landscape characteristics, including land use land cover types and their spatial patterns [31]. LULC change mainly depends on how humans alter the natural landscapes and socioeconomic growth through space and time. Land covers involve the physical features of the Earth's surface that are occupied by vegetation, water, soil, and other characteristics of the Earth's surface created through human activities, whereas land used by human beings for habitats concerning economic activities is referred to as land cover [32].

River water qualities near the urban area changed due to several factors and LULC change was the most significant among them. LULC change has had a major impact on water quality. In the urbanization process, the built-up area increases rapidly. As a result, the quality of surrounding river water deteriorates. The rapid expansion of human settlements in the urbanized area discharge huge volumes of sewage water, as the point source of pollution, which includes a high level of nutrients and metals. Therefore, water physicochemical parameters such as pH, dissolved oxygen, biochemical oxygen demand, and chemical oxygen demand have significantly changed. The impact of land-use changes on water quality is usually studied by analyzing the relationships between land use and water quality parameters. Water quality differs according to location, time, weather, and pollution sources [33,34], and contamination is generally determined by studying the physical and chemical properties of the water bodies [35].

Findings from this study indicate that LULC types are considerably associated with one or more water quality parameters in the 1000-m buffer zone scale. It also reveals that only BOD, temperature, EC, TDS, and TSS had a strong and significant relationship with the different LULC types. In comparison, Kerala observed the water quality parameters of the Chalakudy river and compared them with diverse land use patterns over four seasons [36]. They found that urban land use was associated with poor water quality during the study period when there were changes in land use and land cover patterns [37]. Moreover, land use changes in the surrounding area of cities can modify the surface properties of watersheds that influence runoff quality and quantity. The impact of LULC changes on water quality involves analyzing the relationship between land use and water quality indicators [38]. In this research, among five types of land use and land cover, waterbody and built-up were the most significant variables in predicting water quality parameters. The Pearson correlation suggested that BOD is positively correlated with waterbody and built-up area but negatively correlated with agricultural land. A similar study by Tong and Chen [39] observed the water quality of a watershed in relation to land use change in Ohio State, USA, and their results indicate that BOD was positively correlated with residential and commercial lands but had only a non-significant correlation with agricultural land. Xiao et al. [40] conducted a multi-scale analysis of the relationship between urban river water qualities with landscape patterns in different seasons in Huzhou City, China, and their findings point out that, at a different scale, their relationships varied with the composition of land-use types—but, built-up land was most significant. These results suggest that with the development of different types of built-up land, water quality parameters change and exacerbate contamination.

Regression analysis of the present study showed that BOD was sensitive to changing waterbody, whereas temperature was sensitive to changing waterbody and agricultural land. Urban land is a mixture of different land uses types, such as residential, industrial, commercial, and other built-up areas. In contrast with other land use, wastewater is generated more in urban areas, and also urbanization increased coverage by impervious surfaces, which influences storm flow speed and runoff volumes [41]. Runoff and the huge volume of storm and drainage water are mainly responsible for this relationship between BOD and waterbody sensitivity because a higher BOD value indicates that a greater amount of organic matter is present; thus, storm flows and drainage water deliver more pollutants to the surrounding urban catchment, especially in a river. Due to the huge volume of drainage and stormwater, the pollutant quantity increased for this reason in the wet seasons, while BOD was high at a few stations.

The present study's Pearson's correlation results showed that, with the increase of waterbody and built-up, TSS will increase significantly (a positive relationship) and backward stepwise regression identified a relationship between WQP and LULC types, indicating that TDS were sensitive to waterbody, built-up, and agricultural land, whereas TSS was found sensitive for waterbody, built-up and barren land. Wang and Zhang [42] studied the relationship between landscape types and water quality index (WQI), in a multi-scale analysis in the Ebinur Lake oasis; their findings revealed that, for different buffers, both positive and negative relationships exist between certain land use and land cover (LULC) types and the water quality index, but there was considerable correlation between water quality index and landscape index. Li et al. [43] found a relationship between land use/cover and water quality using correlation and regression analyses in the Liao River basin, China, also indicating that BOD_5, COD, sediment, and hardness were considerably associated with land use.

Although, in reviewing the literature, we found similar studies from different parts of the world, but, considering that the present work is the first such kind of investigation for a data-scarce region such as our study area, these results hold a lot of merit to both scientific communities as well as policy makers. This scientific evidence will lay a foundation for designing more robust adaptation and mitigation measures for water resource management in a timely manner.

5. Conclusions

The study revealed the concentrations of several physicochemical parameters of river water in relation to LULC changes. However, the result of the present investigation indicates that LULC changes and seasonal variations (influences the concentration) have a significant impact on water quality parameters. The results of the LULC change analysis indicate built-up, waterbody, and barren land increased and agricultural land and vegetation decreased. Built-up area is dominant in LULC types and the change of LULC pattern within 1000-m buffer zones had a significant impact on the water quality parameters of the Surma river. LULC information, in relation to the surrounding river water quality in urban areas, is very important for planning, monitoring, and management of river water because, in urbanized and densely populated cities, river water is also used for drinking purposes, after treatment, and also for recreational purposes. LULC change causes severe environmental problems worldwide and poses a threat to water quality. Spatiotemporal information about LULC change patterns with water quality helps in finding a solution to this problem. The study provides useful tools for future study, which, combined with LULC change and its relationship with different water quality parameters, can help decision-makers in formulating of rules and guidelines about sustainable land use, especially in city areas, and aid in minimizing negative impacts on water quality.

Author Contributions: Conceptualization: A.K., Z.A., M.M.U., Z.X., P.K.; methodology: A.K., Z.A.; formal analysis: A.K., Z.A.; writing—original draft preparation: A.K., Z.A., M.M.U., Z.X., P.K.; writing—review and editing: A.K., Z.A., M.M.U., Z.X., P.K.; funding acquisition: P.K. All authors have read and agreed to the published version of the manuscript.

Funding: Research work was supported by the Ministry of Science and Technology, Bangladesh, under the National Science and Technology (NST) Fellowship (FY 2018-19). Also, the publication is supported by the Asia-Pacific Network for Global Change Research (APN) under Collaborative Regional Research Programme (CRRP) with project reference number CRRP2019-01MY-Kumar.

Institutional Review Board Statement: Not applicable.

Informed Consent Statement: Not applicable.

Data Availability Statement: Not applicable.

Conflicts of Interest: The authors declare no conflict of interest.

References

1. Haygarth, P.M.; Jarvis, S.C. (Eds.) *Agriculture, Hydrology, and Water Quality*; CABI Publishing: Wallingford, UK, 2002.
2. Zhou, T.; Wu, J.; Peng, S. Assessing the effects of landscape pattern on river water quality at multiple scales: A case study of the Dongjiang River watershed, China. *Ecol. Indic.* **2012**, *23*, 166–175. [CrossRef]
3. Naher, T.; Chowdhury, M.A.I. Assessment and Correlation Analysis of Water Quality Parameters: A Case Study of Surma River at Sylhet Division, Bangladesh. *Int. J. Eng. Trends Technol.* **2017**, *53*, 126–136. [CrossRef]
4. UNEP-IETC. *Planning and Management of Lakes and Reservoirs: An Integrated Approach to Eutrophication*; UNEP-IETC: Osaka, Japan, 1997; Volume 11.
5. Bhat, S.; Jacobs, J.M.; Hatfield, K.; Prenger, J. Relationships between stream water chemistry and military land use in forested watersheds in Fort Benning, Georgia. *Ecol. Indic.* **2006**, *6*, 458–466. [CrossRef]
6. Hopkins, R.L. Use of landscape pattern metrics and multiscale data in aquatic species distribution models: A case study of a freshwater mussel. *Landsc. Ecol.* **2009**, *24*, 943–955. [CrossRef]

7. Bhaduri, B.; Harbor, J.O.N.; Engel, B.; Grove, M. Assessing watershed-scale, long-term hydrologic impacts of land-use change using a GIS-NPS model. *Environ. Manag.* **2000**, *26*, 643–658. [CrossRef]
8. Kumar, P. Numerical quantification of current status quo and future prediction of water quality in eight Asian megacities: Challenges and opportunities for sustainable water management. *Environ. Monit. Assess.* **2019**, *191*, 319. [CrossRef]
9. Zhu, C. Land Use/Land Cover Change and Its Hydrological Impacts from 1984 to 2010 in the Little River Watershed, Tennessee. 2011. Available online: https://trace.tennessee.edu/cgi/viewcontent.cgi?article=2228&context=utk_gradthes (accessed on 5 February 2020).
10. Hossain, M.S. Impact of land use change on stream water quality: A review of modelling approaches. *J. Res. Eng. Appl. Sci.* **2017**, *2*.
11. Griffith, J.A. Geographic techniques and recent applications of remote sensing to landscape-water quality studies. *Water Air Soil Pollut.* **2002**, *138*, 181–197. [CrossRef]
12. Li, K.; Chi, G.; Wang, L.; Xie, Y.; Wang, X.; Fan, Z. Identifying the critical riparian buffer zone with the strongest linkage between landscape characteristics and surface water quality. *Ecol. Indic.* **2018**, *93*, 741–752. [CrossRef]
13. Goldstein, R.M.; Carlisle, D.M.; Meador, M.R.; Short, T.M. Can basin land use effects on physical characteristics of streams be determined at broad geographic scales? *Environ. Monit. Assess.* **2007**, *130*, 495–510. [CrossRef]
14. Pegram, G.C.; Görgens, A.H.M. *A Guide to Non-Point Source Assessment: To Support Water Quality Management of Surface Water Resources in South Africa*; Water Research Commission: Cape Town, South Africa, 2001.
15. Uriarte, M.; Yackulic, C.B.; Lim, Y.; Arce-Nazario, J.A. Influence of land use on water quality in a tropical landscape: A multi-scale analysis. *Landsc. Ecol.* **2011**, *26*, 1151. [CrossRef]
16. Hassan, Z.U.; Shah, J.A.; Kanth, T.A.; Pandit, A.K. Influence of land use/land cover on the water chemistry of Wular Lake in Kashmir Himalaya (India). *Ecol. Process.* **2015**, *4*, 9. [CrossRef]
17. McMillan, S.K.; Tuttle, A.K.; Jennings, G.D.; Gardner, A. Influence of restoration age and riparian vegetation on reach-scale nutrient retention in restored urban streams. *JAWRA J. Am. Water Resour. Assoc.* **2014**, *50*, 626–638. [CrossRef]
18. Giao, N.T.; Cong, N.V.; Nhien, H.T.H. Using Remote Sensing and Multivariate Statistics in Analyzing the Relationship between Land Use Pattern and Water Quality in Tien Giang Province, Vietnam. *Water* **2021**, *13*, 1093. [CrossRef]
19. Li, S.; Peng, S.; Jin, B.; Zhou, J.; Li, Y. Multi-scale relationship between land use/land cover types and water quality in different pollution source areas in Fuxian Lake Basin. *PeerJ* **2019**, *7*, e7283. [CrossRef] [PubMed]
20. Gu, Q.; Hu, H.; Ma, L.; Sheng, L.; Yang, S.; Zhang, X.; Zhang, M.; Zheng, K.; Chen, L. Characterizing the spatial variations of the relationship between land use and surface water quality using self-organizing map approach. *Ecol. Indic.* **2019**, *102*, 633–643. [CrossRef]
21. Chen, D.; Elhadj, A.; Xu, H.; Xu, X.; Qiao, Z. A Study on the Relationship between Land Use Change and Water Quality of the Mitidja Watershed in Algeria Based on GIS and RS. *Sustainability* **2020**, *12*, 3510. [CrossRef]
22. Ribeiro, K.H.; Favaretto, N.; Dieckow, J.; Souza, L.C.D.P.; Minella, J.P.G.; Almeida, L.D.; Ramos, M.R. Quality of surface water related to land use: A case study in a catchment with small farms and intensive vegetable crop production in southern Brazil. *Rev. De Ciência Do Solo* **2014**, *38*, 656–668. [CrossRef]
23. Xiao, H.; Ji, W. Relating landscape characteristics to non-point source pollution in mine waste-located watersheds using geospatial techniques. *J. Environ. Manag.* **2007**, *82*, 111–119. [CrossRef] [PubMed]
24. Yang, X. An assessment of landscape characteristics affecting estuarine nitrogen loading in an urban watershed. *J. Environ. Manag.* **2012**, *94*, 50–60. [CrossRef]
25. Mitchell, M.G.; Bennett, E.M.; Gonzalez, A. Linking landscape connectivity and ecosystem service provision: Current knowledge and research gaps. *Ecosystems* **2013**, *16*, 894–908. [CrossRef]
26. Shen, Z.; Hou, X.; Li, W.; Aini, G.; Chen, L.; Gong, Y. Impact of landscape pattern at multiple spatial scales on water quality: A case study in a typical urbanised watershed in China. *Ecol. Indic.* **2015**, *48*, 417–427. [CrossRef]
27. Chowdhury, M.H. Surma River. 2015. Available online: http://en.banglapedia.org/index.php?title=Surma_River (accessed on 3 January 2020).
28. American Public Health Association (APHA). *Standard Methods for the Examination of Water and Wastewater*; APHA: Washington DC, USA, 2005.
29. Eastman, J.R. *IDRISI Taiga Guide to GIS and Image Processing*; Clark Labs Clark University: Worcester, MA, USA, 2009.
30. Zhang, Z.; Zhang, F.; Du, J.; Chen, D.; Zhang, W. Impacts of land use at multiple buffer scales on seasonal water quality in a reticular river network area. *PLoS ONE* **2021**, *16*, e0244606. [CrossRef] [PubMed]
31. Bateni, F.; Fakheran, S.; Soffianian, A. Assessment of land cover changes & water quality changes in the Zayandehroud River Basin between 1997–2008. *Environ. Monit. Assess.* **2013**, *185*, 10511–10519.
32. Rawat, J.S.; Kumar, M. Monitoring land use/cover change using remote sensing and GIS techniques: A case study of Hawalbagh block, district Almora, Uttarakhand, India. *Egypt. J. Remote Sens. Space Sci.* **2015**, *18*, 77–84. [CrossRef]
33. Giri, S.; Qiu, Z. Understanding the relationship of land uses and water quality in twenty first century: A review. *J. Environ. Manag.* **2016**, *173*, 41–48. [CrossRef]
34. Camara, M.; Jamil, N.R.; Abdullah, A.F.B. Impact of land uses on water quality in Malaysia: A review. *Ecol. Process.* **2019**, *8*, 10. [CrossRef]

35. Duran, M.; Suicmez, M. Utilization of both benthic macroinvertebrates and physicochemical parameters for evaluating water quality of the stream Cekerek (Tokat, Turkey). *J. Environ. Biol.* **2007**, *28*, 231–236.
36. Chattopadhyay, S.; Rani, L.A.; Sangeetha, P.V. Water quality variations as linked to land use pattern: A case study in Chalakudy river basin, Kerala. *Curr. Sci.* **2005**, *89*, 2163–2169.
37. Chauhan, A.; Verma, S.C. Impact of Agriculture, Urban and Forest Land Use on Physico-Chemical Properties of Water-A Review. *Int. J. Curr. Microbiol. Appl. Sci.* **2015**, *4*, 18–22.
38. Tu, J. Spatial and temporal relationships between water quality and land use in northern Georgia, USA. *J. Integr. Environ. Sci.* **2011**, *8*, 151–170. [CrossRef]
39. Tong, S.T.; Chen, W. Modeling the relationship between land use and surface water quality. *J. Environ. Manag.* **2002**, *66*, 377–393. [CrossRef] [PubMed]
40. Xiao, R.; Wang, G.; Zhang, Q.; Zhang, Z. Multi-scale analysis of relationship between landscape pattern and urban river water quality in different seasons. *Sci. Rep.* **2016**, *6*, 25250. [CrossRef] [PubMed]
41. Waters, E.R.; Morse, J.L.; Bettez, N.D.; Groffman, P.M. Differential carbon and nitrogen controls of denitrification in riparian zones and streams along an urban to exurban gradient. *J. Environ. Qual.* **2014**, *43*, 955–963. [CrossRef] [PubMed]
42. Wang, X.; Zhang, F. Multi-scale analysis of the relationship between landscape patterns and a water quality index (WQI) based on a stepwise linear regression (SLR) and geographically weighted regression (GWR) in the Ebinur Lake oasis. *Environ. Sci. Pollut. Res.* **2018**, *25*, 7033–7048. [CrossRef]
43. Li, Y.L.; Liu, K.; Li, L.; Xu, Z.X. Relationship of land use/cover on water quality in the Liao River basin, China. *Procedia Environ. Sci.* **2012**, *13*, 1484–1493. [CrossRef]

Article

Analysis of Factors Influencing the Trophic State of Drinking Water Reservoirs in Taiwan

Cheng-Wei Hung [1] and Lin-Han Chiang Hsieh [2,*]

[1] Graduate Institute of Environmental Engineering, National Taiwan University, Taipei 106, Taiwan; hungjason111210@gamil.com
[2] Department of Environmental Engineering, College of Engineering, Chung Yuan Christian University, Zhongli 320, Taiwan
* Correspondence: chianghsieh@cycu.edu.tw; Tel.: +886-3-265-4908

Abstract: Eutrophication is an environmental pollution problem that occurs in natural water bodies. Regression analyses with interaction terms are carried out to identify the factors influencing the Shimen, Mingde, and Fongshan Reservoirs in Taiwan. The results indicate that the main factor influencing these reservoirs is total phosphorus. In the Shimen and Mingde Reservoirs, the influence of total phosphorus, when interacting with other factors, on water quality trophic state is more serious than that of total phosphorus per se. This implies that the actual influence of total phosphorus on the eutrophic condition could be underestimated. Furthermore, there was no deterministic causality between climate and water quality variables. In addition, time lagged effects, or the influence of their interaction with other variables, were considered separately in this study to further determine the actual relationships between water trophic state and influencing factors. The influencing patterns for three reservoirs are different, because the type, size, and background environment of each reservoir are different. This is as expected, since it is difficult to predict eutrophication in reservoirs with a universal index or equation. However, the multiple linear regression model used in this study could be a suitable quick-to-use, case-by-case model option for this problem.

Keywords: eutrophication; variable interactions; multiple linear regression; reservoir

Citation: Hung, C.-W.; Chiang Hsieh, L.-H. Analysis of Factors Influencing the Trophic State of Drinking Water Reservoirs in Taiwan. *Water* **2021**, *13*, 3228. https://doi.org/10.3390/w13223228

Academic Editors: Pankaj Kumar and Ram Avtar

Received: 13 October 2021
Accepted: 11 November 2021
Published: 14 November 2021

1. Introduction

Constructing reservoirs is one of the most effective ways of storing water in Taiwan. Surface runoff may be caught during periods of high flow and provide water for people's livelihood, industry, and agriculture during periods of water shortage.

Eutrophication, the most challenging water pollution problem in water bodies, will eventually become an issue in many reservoirs [1]. Eutrophication negatively affects the water quality, safety, ecological integrity, and sustainability of global water resources [2–4]. It has long been believed that excessive phosphorus is the main reason for eutrophication [5]. However, population density, urbanization, and agricultural activities are also factors that influence the water quality of freshwater systems [6–10]. Since the 1940s, a substantial population increase, land-use intensification, and the use of agricultural fertilizers from developed countries [11], as well as the use of detergents containing phosphate compounds since the 1950s, have accelerated the eutrophication of waterbodies [12].

Eutrophication influences the water volume and quality in reservoirs. Regarding water volume, algae distributed on the water surface causes water hypoxia and a decrease in water transparency [13,14], leading to a substantial death of aquatic organisms, which are then deposited on the bottom of the reservoir, which, in turn, reduces the reservoir's capacity over time. Regarding water quality, the proliferation of algae causes algal blooms and releases algal poison, which both influences water quality conditions such as dissolved oxygen, transparency, odor, and pH value, and it also causes problems during the filtration of drinking water, increasing health risks to people [15,16]. Eutrophication also causes the depletion of dissolved oxygen in water, which may have potentially harmful influences on the habitats of fish and macroinvertebrates [17,18]. In addition, eutrophication results in the release of nutrients such as phosphorus and ammonium into the water [19–21], potentially producing toxic heavy metal ions [22].

At present, reservoirs in Taiwan are facing a crisis of deteriorating water quality. According to the Environmental Water Quality Monitoring Annual Report of the Environmental Protection Agency, among the 26 reservoirs on the main island of Taiwan, 7 reservoirs are in a eutrophic state, 18 are in a mesotrophic state, and only 1 is in an oligotrophic state.

However, the Carlson trophic state index (CTSI) for assessing the water quality of reservoirs is not completely suitable in Taiwan. Taiwan is affected by frequent heavy rainfall, and particularly after typhoons and heavy rains, large amounts of sand, soil, and rock flow into the reservoirs, leading to a large increase in the concentration of suspended solids and a decrease in water transparency. Thus, it is quite likely that a water body with a high CTSI value, but without a large amount of algae distributed on the water surface, may be identified as eutrophic.

This study aims to investigate the correlation between weather and water quality factors and their degree of influence on the trophic state of reservoirs both as single variables and as interactions of variables. We also discuss the suitability of CTSI for assessing water quality in reservoirs in Taiwan. We used weather and water quality data from 2017 to 2019 from three main reservoirs in Taiwan: Shimen Reservoir, Mingde Reservoir, and Fongshan Reservoir. Chlorophyll a was used as an indicator to illustrate the degree of eutrophication, and data were analyzed using multiple linear regressions (MLR) including time lags and variable interactions.

2. Materials and Methods

2.1. Characteristics of Reservoir

The Shimen Reservoir is a stable source of water supply in northern Taiwan and is the third largest reservoir in Taiwan (Figure 1). It is a multi-objective water conservancy project that combines benefits such as irrigation, power generation, water supply, flood control, sightseeing, and recreation. The Shimen Reservoir has a catchment area of 763 km^2, a full water level of 8 km^2, a total storage capacity of 309 million m^3, and an effective storage capacity of 197 million m^3 [23]. The irrigation area of the Shimen Reservoir includes three counties: Hsinchu, Taoyuan, and New Taipei City. It provides for a daily consumption of 800,000 m^3 of livelihood water and also the Shimen Power Plant with 230 million kWh of power generation per year [24].

Figure 1. Location of Shimen Reservoir (including water catchment area).

The Mingde Reservoir provides more water for consumption in the Miaoli County, where there are more mountains and few fields (Figure 2). The Mingde Reservoir has a catchment area of 61 km^2, a full water level of 1.7 km^2, a total storage capacity of 17.7 million m^3, and an effective storage capacity of 12.2 million m^3. The irrigation area of the Mingde Reservoir is 13 km^2, and it provides 27,000 m^3 of water daily [25].

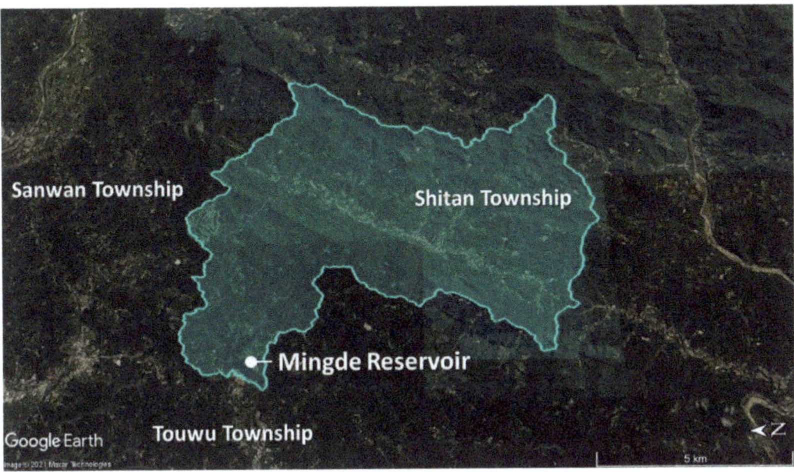

Figure 2. Location of Mingde Reservoir (including water catchment area).

The Fongshan Reservoir is an off-site reservoir located in Kaohsiung City and provides Kaohsiung with a large additional water supply because of the high population concentration and the rapid development of industry and commerce, which increases water consumption in the area (Figure 3). It has a catchment area of 2.75 km^2, a full water level of 0.75 km^2, a total storage capacity of 9.2 million m^3, and an effective storage capacity of 8.5 million m^3 [26]. The Fongshan Reservoir supplies 1.6 million tons of water daily, of which 350,000 tons, 22%, caters for industrial consumption [27].

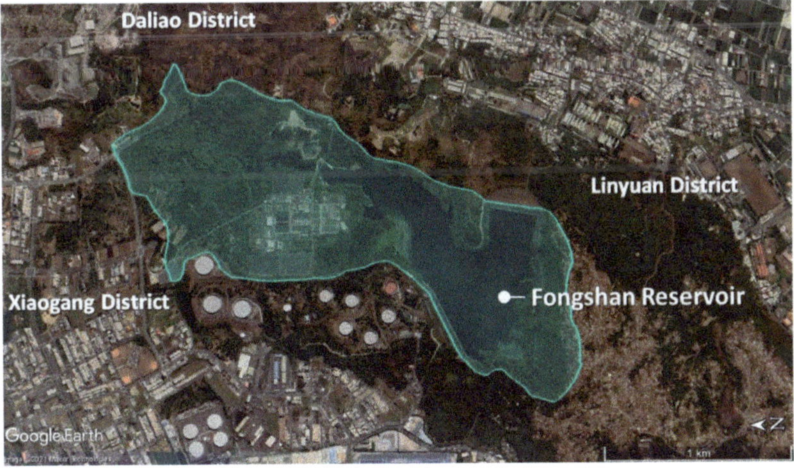

Figure 3. Location of Fongshan Reservoir (off-stream).

2.2. Dataset

The Taiwan Environmental Protection Administration (EPA) changed reservoir water quality monitoring from quarterly monitoring in previous years to monthly monitoring from January 2017. In this study, we used monthly weather and water quality data from the Shimen, Mingde, and Fongshan Reservoirs from January 2017 to December 2019.

Weather data included the daily statistics of rainfall (mm) and inflow (mm) data in the catchment area from 2017 to 2019 and was downloaded from the disaster prevention information service website of the Water Resources Agency (WRA). The data collection period corresponds to that of the EPA water quality monitoring data from each reservoir. In addition, monthly water temperature (WT) data from 2017 to 2019 were collected from the national water quality monitoring information website of the EPA [28].

Water quality data were collected from the monthly statistics on the national water quality monitoring information website of the EPA and included chlorophyll a (Chl-a), dissolved oxygen (DO), transparency (SD), total phosphorus (TP), pH, conductivity, suspended solids (SS), chemical oxygen demand (COD), and ammonia nitrogen (AN) sampled from 2017 to 2019 [28].

In order to investigate whether the different seasons affect the degree of influence, March to May were designated as Season 1 (S1), June to August were designated as Season 2 (S2), September to November were designated as Season 3 (S3), and December to February were designated as Season 4 (S4). S4 was used as a control, and S1 to S3 were analyzed to identify the degree of influence that each variable has on water quality in the different seasons.

2.3. Methodology

2.3.1. Regression Analysis

We used multiple linear regression (MLR) using regression analysis (Equation (1)). In order to identify how much each factor influences the eutrophication of the reservoirs, we used rapidly adjusting variables in regression models to analyze weather and water quality factors.

$$Y_i = \beta_0 + \beta_1 X_{1i} + \beta_2 X_{2i} + \cdots + \beta_k X_{ki} + \varepsilon_i \quad (1)$$

where Y_i is the i-th observation value in the dependent variable, which represents the concentration of chlorophyll a in this study. The independent variables X_{1i} to X_{ki} are weather (rainfall, inflow, and water temperature) and water quality (Chl-a, DO, etc.) factors. β_0 is an intercept term, β_1 to β_k are the slope terms, and also the unknown coefficients corresponding to the independent variables X_{1i} to X_{ki}. ε_i is a random error term.

2.3.2. Time-Lag

The reason of applying time-lag variables in this study is that the current weather or water quality factors do not necessarily have an immediate influence on the Chl-a concentration. These time lag situations might be one week, one month, or even more, so it is not suitable to use the monitoring current data in the regression analysis.

Basic analysis, Lag 1, and Lag 2 data in the regression model were selected using the following steps: Firstly, we select the significant data of the three types of data. Secondly, if more than one of the three types of data were significant, we selected the data collected closest to the monitoring date, i.e., basic analysis data take precedence over Lag 1 data, which takes precedence over Lag 2 data. Thirdly, if there was no significance in the three types of data, basic analysis data were selected as a representative term.

2.3.3. Interaction Terms

We also test whether the influence of weather and water quality factors on the water trophic state is affected by their interaction. The traditional ordinary least square (OLS) formula can be illustrated as shown in Equation (2):

$$Y = \beta_1 X_1 + \cdots + \beta_4 X_4 \quad (2)$$

Regarding the interaction of factors, we need to consider whether factors are correlated. We tested the pair-by-pair interaction of factors that are theoretically related using an MLR equation and then analyzed whether the correlation was still significant after the interaction. If it was significant, the two factors were grouped to form an interaction term that was added in the equation, and the correlation was analyzed as shown in Equations (3) and (4). $\beta_5 X_1 X_2$ and $\beta_6 X_3 X_4$ are the interaction terms added on the basis of an MLR.

$$Y_1 = \beta_1 X_1 + \cdots + \beta_4 X_4 + \beta_5 X_1 X_2 \quad (3)$$

$$Y_2 = \beta_1 X_1 + \cdots + \beta_4 X_4 + \beta_6 X_3 X_4 \quad (4)$$

If the individual analysis results of interaction terms $\beta_5 X_1 X_2$ and $\beta_6 X_3 X_4$ in the regressions of Y_1 and Y_2 were both significant, then the two factors were analyzed in the same regression formula for Y_3, as shown in Equation (5).

$$Y_3 = \beta_1 X_1 + \cdots + \beta_4 X_4 + \beta_5 X_1 X_2 + \beta_6 X_3 X_4 \quad (5)$$

If the regression results of $\beta_5 X_1 X_2$ and $\beta_6 X_3 X_4$ in Y_3 were both significantly correlated, the combination was retained, and the method was repeated on other interaction terms. At most, two interaction groups were included in the regression formula of each reservoir, and the factors in the two groups did not overlap with each other.

We used three conditions for selecting two groups of interaction terms with simultaneous significant correlations: Firstly, we considered TP and AN that represent the nutrient factors in the two interaction groups. Secondly, we considered WT, rainfall, and inflow that are representative of the weather factors, and WT had priority over rainfall, and rainfall had priority over inflow. Thirdly, we considered the R^2 value that could be explained by applying each group to the regression formula as a final selection step.

2.3.4. Interrelationship of Interaction Terms

Equation (2) was simplified by reducing variables and adding an interaction term, as shown in Equation (6).

$$Y = \beta_1 X_1 + \beta_2 X_2 + \beta_3 X_1 X_2 \quad (6)$$

Equation (6) was rewritten as Equation (7), in which other variables are fixed, X_1 increases by 1 unit, and X_2 does not increase, and the dependent variable becomes Y_1.

$$Y_1 = \beta_1 (X_1 + 1) + \beta_2 X_2 + \beta_3 (X_1 + 1) X_2 \quad (7)$$

Equation (7) minus Equation (6) provides Equation (8), which represents a situation in which other variables are fixed, X_1 increases by 1 unit and X_2 does not increase, and the unit amount ΔY_1 is the dependent variable that will increase.

$$\Delta Y_1 = Y_1 - Y = \beta_1 + \beta_3 X_2 \quad (8)$$

Similarly, if other variables are fixed but X_1 does not increase and X_2 increases by 1 unit, the dependent variable becomes Y_2 (Equation (9)). Equation (9) minus Equation (6) provides Equation (10), which represents the unit amount ΔY_2—the dependent variable that will increase.

$$Y_2 = \beta_2 + \beta_3 X_1 \quad (9)$$

$$\Delta Y_2 = Y_2 - Y = \beta_2 + \beta_3 X_1 \quad (10)$$

When both X_1 and X_2 increase by 1 unit, and other variables are fixed, Equation (11) is obtained. Equation (11) minus Equation (6) provides Equation (12), which represents the unit amount ΔY_3—the dependent variable that will increase.

$$Y_3 = \beta_1 (X_1 + 1) + \beta_2 (X_2 + 1) + \beta_3 (X_1 + 1)(X_2 + 1) \quad (11)$$

$$\Delta Y_3 = Y_3 - Y = \beta_1 + \beta_2 + \beta_3 + \beta_3(X_1 + X_2) \quad (12)$$

Equations (8), (10), and (12) were integrated (as shown in Table 1), where ΔX_1 and ΔX_2 are the unit amounts by which the independent variables, X_1 and X_2, are increased. Table 1 shows that when X_1 and X_2 are increased by 1 unit separately or simultaneously, the unit amount ΔY, which is the dependent variable, will increase.

Table 1. Interrelationship of interaction terms.

ΔX_1	ΔX_2	ΔY
1	0	$\beta_1 + \beta_3 X_2$
0	1	$\beta_2 + \beta_3 X_1$
1	1	$\beta_1 + \beta_2 + \beta_3 + \beta_3(X_1 + X_2)$

2.4. Model

In this study, Chl-a is set as the dependent variable, and the eleven weather and water quality factors, WT, DO, SD, TP, pH, conductivity, SS, COD, AN, rainfall, and inflow, are set as independent variables. STATA (version 13) is used in this study to perform correlation analysis and regression analysis.

The research model of this study is based on MLR. Using the result of a Hausman Test, we chose to include a random effects model in the MLR model to avoid or reduce ignoring differences between the data (which results in the omission of variables and leads to estimation errors) and also to reduce the occurrence of collinearity problems between variables.

The MLR model built in this study includes the random effects model and performs a time-lag analysis as well as adding specific interaction terms separately based on the difference in the reservoir data.

3. Results and Discussion

3.1. Descriptive Statistics

This study lists the descriptive statistics of the Shimen, Mingde, and Fongshan Reservoirs (Table 2). The minimums of TP, SS, COD, and AN were below the detection limit (ND).

Table 2. Descriptive statistics from the Shimen, Mingde, and Fongshan Reservoir data.

Factors		Chl-a	WT	DO	SD	TP	pH	Conductivity	SS	COD	AN	Rainfall	Inflow
Unit		μg/L	°C	mg/L	m	mg/L	-	μmho/cm	mg/L	mg/L	mg/L	mm	cms
Mean	SR	4.078	23.741	8.906	1.909	0.021	8.396	210.282	3.808	3.818	0.021	5.642	302.692
	MR	20.395	25.681	10.108	1.145	0.023	8.714	242.361	6.753	9.218	0.040	4.575	15.509
	FR	49.297	27.540	7.235	0.704	0.810	7.983	646.488	17.016	16.822	1.052	1.653	23.329
Max	SR	21.900	31.200	11.700	4.400	0.059	9.360	302.000	87.000	10.600	0.130	27.900	1114.350
	MR	54.400	32.700	16.500	2.000	0.033	9.720	368.000	28.000	15.400	0.120	101.100	150.320
	FR	258.000	32.100	14.500	1.300	1.710	8.780	1130.000	46.000	47.600	6.850	35.500	32.700
Min	SR	0.500	15.000	5.600	0.400	ND	7.000	150.000	ND	ND	ND	0.000	47.780
	MR	2.000	11.700	5.400	0.400	0.007	7.380	176.000	2.400	ND	ND	0.000	0.030
	FR	5.700	19.100	1.700	0.300	0.091	4.460	343.000	5.200	4.400	ND	0.000	9.800
Std. Dev.	SR	2.878	4.413	1.085	0.768	0.011	0.584	26.936	6.346	3.421	0.019	7.226	260.253
	MR	12.267	5.012	2.160	0.324	0.006	0.499	37.782	3.883	2.967	0.029	16.893	27.482
	FR	44.075	3.093	2.699	0.218	0.440	0.324	177.133	8.298	8.901	1.493	6.328	6.648

SR: Shimen Reservoir, MR: Mingde Reservoir, FR: Fongshan Reservoir, ND: The monitoring data is smaller than the detection limit

The average values of Chl-a, TP, Conductivity, SS, COD, and AN in the Fongshan Reservoir are several times more than the average values of the other two reservoirs. The concentration of Chl-a is 12 times that of the Shimen Reservoir and 2.4 times that of the Mingde Reservoir. The concentration of TP is 40 times that of the Shimen and Mingde Reservoirs. The concentration of AN is 52.5 times that of the Shimen Reservoir and 56.2 times that of the Mingde Reservoir.

The maximum Chl-a, TP, Conductivity, COD, and AN were recorded in the Fongshan Reservoir, the maximum SS was recorded in the Shimen Reservoir, and the maximum pH was recorded in the Mingde Reservoir. There were large variations in the daily rainfall and inflow data. It is possible that there was no rain on the day monitoring was conducted but heavy rainfalls the next day; therefore, we do not discuss the maximum and minimum rainfall and inflow.

The minimum DO of the Fongshan Reservoir was about three times lower than those of the Shimen and Mingde Reservoirs. However, its minimum TP concentration was still much larger than that of the Shimen and Mingde Reservoirs. The Fongshan Reservoir was the only reservoir with a pH of less than 7. The minimum conductivity of the Fongshan Reservoir was more than two times that of the Shimen and Mingde Reservoirs. The minimum SS of the Fongshan Reservoir was about two times that of the Mingde Reservoir.

3.2. Correlation Analysis

Results from the correlation analysis of the Shimen, Mingde, and Fongshan Reservoirs are show in Tables 3 and 4. The absolute value of the correlation coefficient was above 0.6, which can be regarded as a high correlation between the two factors.

Table 3. Correlation coefficients between factors from correlation analyses (1/2).

		Chl-a	WT	DO	SD	TP	pH	Conductivity	SS	COD	AN	Rainfall	Inflow
Chl-a	SR	1.000											
	MR	1.000											
	FR	1.000											
WT	SR	0.266	1.000										
	MR	0.208	1.000										
	FR	0.241	1.000										
DO	SR	0.260	−0.126	1.000									
	MR	−0.211	0.055	1.000									
	FR	−0.022	0.355	1.000									
SD	SR	0.053	−0.005	0.093	1.000								
	MR	−0.156	0.067	0.177	1.000								
	FR	−0.438	−0.388	−0.231	1.000								
TP	SR	0.369	−0.011	0.180	−0.141	1.000							
	MR	0.019	−0.034	−0.180	−0.126	1.000							
	FR	0.321	−0.282	−0.488	−0.025	1.000							
pH	SR	0.308	0.700	0.297	0.079	0.114	1.000						
	MR	0.011	0.591	0.604	0.145	−0.008	1.000						
	FR	0.042	0.549	0.700	−0.212	−0.549	1.000						

Table 4. Correlation coefficients between factors from correlation analyses (2/2).

		Chl-a	WT	DO	SD	TP	pH	Conductivity	SS	COD	AN	Rainfall	Inflow
Conductivity	SR	−0.369	−0.200	−0.288	−0.237	0.004	−0.413	1.000					
	MR	−0.191	−0.644	−0.296	−0.233	−0.049	−0.649	1.000					
	FR	0.206	−0.210	−0.495	0.021	0.854	−0.543	1.000					
SS	SR	−0.029	−0.118	−0.105	−0.350	0.317	−0.160	0.267	1.000				
	MR	0.108	−0.154	−0.137	−0.575	0.088	−0.250	0.372	1.000				
	FR	0.352	0.261	0.101	−0.600	−0.051	0.146	−0.138	1.000				
COD	SR	−0.039	0.315	0.075	0.019	−0.108	0.427	−0.076	−0.029	1.000			
	MR	0.440	0.257	−0.139	−0.131	−0.079	0.077	−0.049	0.171	1.000			
	FR	0.602	0.089	−0.067	−0.488	0.599	−0.030	0.454	0.473	1.000			
AN	SR	0.193	0.091	0.115	−0.009	0.078	0.220	−0.110	−0.011	0.197	1.000		
	MR	0.224	0.036	−0.186	−0.044	0.110	−0.099	0.108	0.221	0.074	1.000		
	FR	0.253	−0.180	−0.402	−0.140	0.827	−0.479	0.696	−0.007	0.469	1.000		
Rainfall	SR	−0.078	0.121	−0.455	−0.037	0.107	−0.003	0.082	0.171	0.054	−0.101	1.000	
	MR	−0.179	0.014	0.064	0.044	0.120	0.110	0.093	0.068	0.007	−0.087	1.000	
	FR	−0.045	0.071	−0.233	0.160	0.091	−0.231	−0.051	0.049	−0.005	0.026	1.000	
Inflow	SR	0.332	0.096	0.086	0.003	0.356	0.275	−0.354	0.027	0.092	0.220	0.262	1.000
	MR	0.055	0.115	0.186	0.085	0.157	0.243	−0.177	−0.006	0.013	−0.038	0.826	1.000
	FR	0.096	−0.027	−0.034	−0.014	0.226	−0.101	0.103	0.023	0.310	0.253	0.142	1.000

In the Shimen Reservoir, WT and pH were highly correlated with a correlation coefficient of 0.700 (Table 3).

In the Mingde Reservoir, the pairs of WT and conductivity, DO and pH, pH and conductivity, and rainfall and inflow were highly correlated with correlation coefficients of −0.644, 0.634, −0.649, and 0.826 respectively (Tables 3 and 4).

In the Fongshan Reservoir, the pairs of Chl-a and COD, DO and pH, SD and SS, TP and Conductivity, TP and AN, and conductivity and AN were highly correlated with correlation coefficients of 0.602, 0.700, −0.600, 0.854, 0.827, and 0.696, respectively (Tables 3 and 4).

The correlation coefficient of rainfall and inflow in the Mingde Reservoir as well as the correlation coefficient of TP with conductivity and TP with AN in the Fongshan Reservoir exceeded 0.8. However, when performing an MLR analysis, a high correlation coefficient between the independent variables causes the problem of collinearity in the regression results. This problem means that the lower the correlation between the independent variables, the more it reflects the relationship with dependent variables.

3.3. Analysis Result

Tables 5–7 show the regression analysis result after including time-lag and interaction terms from the Shimen, Mingde, and Fongshan Reservoir data, respectively. This study divides independent variables into "Basic" analysis without time lag, "Lag 1" data with one month lagged, and "Lag 2" data with two months lagged. For example, there is a time lag between an increase in water temperature, the growth of algae, the flow time of rainfall runoff to the reservoir, etc.

Table 5. Results from regression analyses of the Shimen Reservoir data.

	Coefficient	**SE**	***p*-Value**
Intercept	8.408	4.119	0.041 *
WT	0.291	0.069	0.000 *
DO	0.511	0.312	0.102
SD	0.028	0.125	0.822
TP	−26.874	30.033	0.371
pH	−1.078	0.199	0.000 *
Conductivity	−0.038	0.005	0.000 *
SS—Lag 1	0.033	0.013	0.014 *
COD	0.478	0.123	0.000 *
AN	−54.684	22.944	0.017 *
Rainfall—Lag 1	0.047	0.020	0.015 *
Inflow	0.001	0.001	0.075
WT × COD	−0.025	0.005	0.000 *
TP × AN	3817.198	788.920	0.000 *
S1	1.601	0.756	0.034 *
S2	0.656	0.524	0.211
S3	1.605	0.614	0.009 *
R^2	0.517		
Obs.	210		

* $p < 0.05$.

According to the three conditions listed in Section 2.3.3, 'WT × COD' and 'TP × AN' were selected as interaction terms for the Shimen Reservoir regression model; 'WT × pH' and 'DO × TP' were the interaction terms for the Mingde Reservoir regression model. Since there was no significant interaction term for the Fongshan Reservoir, we used the result of the time-lag analysis as the final analysis result. The coefficient and correlation of WT, TP, COD, and AN in the Shimen and Mingde Reservoirs should be considered using interaction terms.

Table 6. Results from regression analyses of the Mingde Reservoir data.

	Coefficient	SE	*p*-Value
Intercept	−474.185	36.471	0.000 *
WT	17.831	0.536	0.000 *
DO	−3.540	0.830	0.000 *
SD—Lag 2	1.692	0.721	0.019 *
TP	−1697.474	178.954	0.000 *
pH	60.517	3.713	0.000 *
Conductivity—Lag 1	0.073	0.050	0.145
SS—Lag 2	0.047	0.155	0.760
COD	1.502	0.127	0.000 *
AN	76.633	16.141	0.000 *
Rainfall	−0.661	0.036	0.000 *
Inflow	0.411	0.021	0.000 *
WT × pH	−2.256	0.079	0.000 *
DO × TP	174.674	17.448	0.000 *
S1	2.862	2.468	0.246
S2	23.801	0.672	0.000 *
S3	22.537	2.738	0.000 *
R^2	0.701		
Obs.	102		

* $p < 0.05$.

Table 7. Results from regression analyses of the Fongshan Reservoir data.

	Coefficient	SE	*p*-Value
Intercept	417.270	62.137	0.000 *
WT	5.237	1.654	0.002 *
DO—Lag 1	−0.139	1.294	0.915
SD	−40.372	12.974	0.002 *
TP—Lag 2	36.883	5.015	0.000 *
pH—Lag 1	−55.004	11.780	0.000 *
Conductivity	−0.117	0.042	0.006 *
SS	−0.440	0.094	0.000 *
COD	2.387	0.704	0.001 *
AN—Lag 1	−6.933	4.499	0.123
Rainfall	−0.959	0.982	0.329
Inflow—Lag 1	0.321	0.317	0.307
S1	6.626	13.378	0.620
S2	−32.329	7.933	0.000 *
S3	−15.980	10.541	0.130
R^2	0.605		
Obs.	101		

* $p < 0.05$.

For example, when using the interaction term 'TP × AN' for the Shimen Reservoir in Equation (6), the concentration of Chl-a is designated as the dependent variable Y, the concentration of TP is designated as the independent variable X_1, and the concentration of AN is designated as the independent variable X_2. β_1, β_2, and β_3 are the coefficients of TP, AN, and the 'TP × AN' interaction term, respectively.

Assuming the value of TP is 0.02 mg/L and AN is 0.03 mg/L, then 'TP × AN' is 0.0006 after multiplying the two. Inserting the above values and the coefficients '−26.874', '−54.684', and '3817.198' into Equation (6) results in a concentration of 0.112 (mg/L) Chl-a when other variables are fixed and the concentration of TP and AN are 0.02 and 0.03 mg/L (Equation (13)).

$$[(-26.874 \times 0.02)] + [(-54.684) \times 0.03] + (3817.198 \times 0.0006) = 0.112 \tag{13}$$

When TP and AN both increase by 1 unit, the concentration of TP becomes 1.02 mg/L and AN becomes 1.03 mg/L, and, thus, the value of 'TP × AN' will become 1.0506. Then, the concentration of Chl-a is calculated as 3926.5 mg/L. This means that the unit value of Chl-a increases when other variables are fixed and TP and AN are both increased by 1 unit (Equation (14)).

$$(-26.874) + (-54.684) + 3817.198 + 3817.198 \times (0.02 + 0.03) = 3926.5 \quad (14)$$

These results show that when evaluating the water quality trophic state, time-lag and the additive relationship of interactions between factors should also be taken into account to evaluate more accurately the potential for water eutrophication.

Note that the final model for each reservoir is different. This reflects the fact that the eutrophication process in reservoirs is a complex of many different factors, and the process is highly case-by-case. As shown in the introduction, influencing factors sometimes work in opposite directions in different cases from different studies. The result of this study further confirms that even reservoirs in Taiwan show very different patterns when it comes to the factors influencing the trophic state.

3.4. Standardization Coefficient

The regression coefficient of each factor was multiplied by the standard deviation of its data to form a 'standardized coefficient'. The basic amounts and units of each factor were different, and it is difficult to predict the dependent variable values based on only the regression coefficient value of each factor. The relative importance of each factor can be compared by standardizing the coefficients, making it is easier to intuitively understand the degree of influence of each independent variable on the dependent variable.

Tables 8–10 show the standardization coefficient of the Shimen, Mingde, and Fongshan Reservoirs. These results show the degree of influence of each factor on Chl-a while taking into account time-lag and variable interactions.

Table 8. Standardization coefficients of variables measured at the Shimen Reservoir.

	Coef.	SD	Std. Coef. **	p-Value
WT	0.291	4.413	1.284	0.000 *
DO	0.511	1.085	0.554	0.102
SD	0.028	0.768	0.022	0.822
TP	−26.874	0.011	−0.296	0.371
pH	−1.078	0.584	−0.630	0.000 *
Conductivity	−0.038	26.936	−1.024	0.000 *
SS—Lag 1	0.033	6.346	0.209	0.014 *
COD	0.478	3.421	1.635	0.000 *
AN	−54.684	0.019	−1.039	0.017 *
Rainfall—Lag 1	0.047	7.226	0.340	0.015 *
Inflow	0.001	260.253	0.260	0.075

* $p < 0.05$. ** 'Std. Coef.': Is the factor regression coefficient multiplied by the standard deviation. All other factors are fixed, the degree of influence of each increase by one standard deviation of the factor affects the concentration of Chl-a.

At the Shimen Reservoir (Table 8), the influence of interactions must be considered when analyzing WT, TP, COD, and AN, so these variables will not be discussed separately. Other highly influential factors are conductivity (standardization coefficient = −1.024), secondly pH (standardization coefficient = −0.630), and lastly SS with a lag of one month (standardization coefficient = 0.209).

At the Mingde Reservoir (Table 9), the influence of interactions must be considered when analyzing WT, DO, TP, and pH, so these variables will not be discussed separately. Other factors with a high influence were inflow (standardization coefficient = 11.295), followed by rainfall (standardization coefficient = −11.166), and lastly SD with two months lag (standardization coefficient = 0.548).

Table 9. Standardization coefficients of variables measured at the Mingde Reservoir.

	Coef.	Std. Dev.	Std. Coef. **	p-Value
WT	17.831	5.012	89.369	0.000 *
DO	−3.540	2.16	−7.646	0.000 *
SD–Lag 2	1.692	0.324	0.548	0.019 *
TP	−1697.474	0.006	−10.185	0.000 *
pH	60.517	0.499	30.198	0.000 *
Conductivity—Lag 1	0.073	37.782	2.758	0.145
SS—Lag 1	0.047	3.883	0.183	0.760
COD	1.502	2.967	4.456	0.000 *
AN	76.633	0.029	2.222	0.000 *
Rainfall—Lag 1	−0.661	16.893	−11.166	0.000 *
Inflow	0.411	27.482	11.295	0.000 *

* $p < 0.05$. ** 'Std. Coef.': Is the factor regression coefficient multiplied by the standard deviation. All other factors are fixed, the degree of influence of each increase by one standard deviation of the factor affects the concentration of Chl-a.

Table 10. Standardization coefficients of variables measured at the Fongshan Reservoir.

	Coef.	Std. Dev.	Std. Coef. **	p-Value
WT	5.237	3.093	16.198	0.002 *
DO—Lag 2	−0.139	2.699	−0.375	0.915
SD	−40.372	0.218	−8.801	0.002 *
TP—Lag 2	36.883	0.440	16.229	0.000 *
pH—Lag 1	−55.004	0.324	−17.821	0.000 *
Conductivity	−0.117	177.133	−20.725	0.006 *
SS	−0.440	8.298	−3.651	0.000 *
COD	2.387	8.901	21.247	0.001 *
AN—Lag 1	−6.933	1.493	−10.351	0.123
Rainfall	−0.959	6.328	−6.069	0.329
Inflow—Lag 1	−0.324	6.648	−2.154	0.307

* $p < 0.05$. ** 'Std. Coef.': Is the factor regression coefficient multiplied by the standard deviation. All other factors are fixed, the degree of influence of each increase by one standard deviation of the factor affects the concentration of Chl-a.

Data from the Fongshan Reservoir (Table 10) showed that COD had the greatest influence on Chl-a (standardization coefficient = 21.247), which was followed by conductivity (standardization coefficient = −20.725) and lastly SS (standardization coefficient = −3.651).

To summarize, at the Shimen Reservoir, Chl-a was significantly and immediately affected by WT, pH, Conductivity, COD, and AN, and significantly, but not immediately, affected by SS and rainfall. DO, SD, TP, and inflow did not significantly affect Chl-a at the Shimen Reservoir. Chl-a at the Mingde Reservoir was significantly and immediately affected by WT, DO, TP, pH, COD, AN, rainfall, and inflow, while the effect of SD was significant but not immediate. Conductivity and SS did not significantly affect Chl-a at the Mingde Reservoir. WT, SD, Conductivity, SS, and COD significantly and immediately affected Chl-a at the Fongshan Reservoir; TP and pH also significantly affected Chl-a, but not immediately; while the effects of DO, AN, rainfall, and inflow were not significant.

3.5. Correlation of Factors

Table 11 illustrates the correlation results of each weather and water quality factor. Table 11 clearly indicates that the correlation of WT in the Shimen and Mingde Reservoirs needs to be considered within an interaction and is significantly positive in the Fongshan Reservoir. The correlation of DO in the Mingde Reservoir needs to be considered within an interaction, and it is not significantly correlated in the Shimen and Fongshan Reservoirs. The correlation of SD is significantly positive in the Mingde Reservoir and significantly negative in the Fongshan Reservoir, but it is not significantly correlated in the Shimen Reservoir. The correlation of TP in the Shimen and Mingde Reservoirs needs to be considered within an interaction and is significantly positive in the Fongshan Reservoir. The correlation of pH in the Mingde Reservoir needs to be considered within an interaction and is significantly negative in both the Shimen and Fongshan Reservoirs. The correlations of conductivity in the Shimen and Fongshan Reservoirs are significantly negative but not significant in the Mingde Reservoir. The correlation of SS is significantly positive in the Shimen Reservoir and significantly negative in the Fongshan Reservoir, but there is no significant correlation in the Mingde Reservoir. The correlation of COD in the Shimen Reservoir needs to be considered within an interaction, but it is significantly positive in both the Mingde and Fongshan Reservoirs. The correlation of AN in the Shimen Reservoir needs to be considered within an interaction, and it is significantly positive in the Mingde Reservoir but not significantly correlated in the Fongshan Reservoir. The correlation of rainfall is significantly positive in the Shimen Reservoir and significantly negative in the Mingde Reservoir, but there is no significant correlation in the Fongshan Reservoir. The correlation of inflow is significantly positive in the Mingde Reservoir, but there is no significant correlation in the Shimen and Fongshan Reservoirs.

Table 11. Correlation of factors from the Shimen, Mingde, and Fongshan Reservoirs.

	Shimen	**Mingde**	**Fongshan**
WT	I	I	+
DO	X	I	X
SD	X	+	−
TP	I	I	+
pH	−	I	−
Conductivity	−	X	−
SS	+	X	−
COD	I	+	+
AN	I	+	X
Rainfall	+	−	X
Inflow	X	+	X
WT × COD	−	N	N
TP × AN	+	N	N
WT × pH	N	−	N
DO × TP	N	+	N

「 I 」: The correlation of factors needs to take into account interaction relationships. 「 X 」: Factors were not significantly correlated. 「 + 」: Factors have a significant positive correlation. 「 − 」: Factors have a significant negative correlation. 「 N 」: The variable is not relevant for the reservoir.

There was no deterministic causality between climate and water quality variables. For example, the pH in the Fongshan Reservoir is negatively correlated with Chl-a, but Zang (2011) shows that Chl-a is positively correlated with pH and DO [29]. The same case as Blumberg (1990) finds that WT is negatively correlated with DO [30], but Chen (2007) shows that WT is positively correlated with DO [31]. In another, case Watson (2016) shows that Chl-a is negatively correlated with DO [32], but Zang (2011) shows a positive correlation [29].

The interaction combinations for the Shimen Reservoir are 'WT × COD' and 'TP × AN', and for the Mingde Reservoir, they are 'WT × pH' and 'DO × TP'. There were no significant correlation interactions for the Fongshan Reservoir. Note that temperature and total phosphorus are the only two factors that have a positive influence among all three reservoirs. However, the effect of temperature negatively interacts with COD in Shimen and with pH in Mingde, making the effect of temperature on the trophic state actually more minor than expected. On the other hand, the interaction terms related to the total phosphorus in Shimen and Mingde are magnifying the effect. This result further supports that total phosphorus is the main factor for the trophic state.

To summarize, these results indicate that the influencing factors of the trophic state in reservoirs defer from case to case; thus, it is difficult to find a one-size-fits-all equation to be perfectly suitable in all cases.

4. Conclusions

The main factor influencing the three reservoirs is total phosphorus. At the Shimen and Mingde Reservoirs, in particular, the interactive effect of TP with other factors on the water quality trophic state was greater than that of TP alone, indicating that more attention should be paid to the interaction effect between the influencing factors. However, there is no significant interaction effect found to further aggravate the trophic state between weather and water quality factors. In the case of these three reservoirs in Taiwan, an additional deterioration of eutrophication from the climate-change-related interaction effect is not a concern.

The analysis of characteristics influenced by time lags and the analysis of the interactions between factors provide a deeper understanding of the correlation between each factor and the degree to which they influence the water quality trophic state. Furthermore, the length of the time lag and the significant combinations of influencing factors vary from reservoir to reservoir, indicating that the patterns of eutrophication might differ according to different reservoir conditions. These results imply that factors influencing the tropic state in a reservoir might vary by reservoir type, geological and meteorological conditions, as well as other potential factors. In other words, forming a model that describes the tropic state for a reservoir is highly case sensitive. The perfect solution of a one-size-fits-all model might not exist. Researchers should carefully review all possible factors before finalizing a model.

In this study, the R^2 values of the MLR model developed for the three reservoirs were all above 0.5, indicating that the regression model for each reservoir explains more than half of the cause of the water quality trophic state. The results indicate that the regression model developed during this study and the methods used are both feasible for assessing the water quality trophic state.

Author Contributions: Conceptualization, L.-H.C.H.; methodology, C.-W.H. and L.-H.C.H.; software, C.-W.H.; validation, L.-H.C.H.; writing—original draft preparation, C.-W.H.; writing—review and editing, L.-H.C.H.; visualization, C.-W.H. All authors have read and agreed to the published version of the manuscript.

Funding: Ministry of Science and Technology, Taiwan: 109-2221-E-033-004-MY2.

Data Availability Statement: Environmental Protection Administration, Environmental Water Quality Information (https://wq.epa.gov.tw/EWQP/en/Default.aspx, accessed on 31 August 2021); Water Resources Agency, Disaster Prevention Information Service (https://fhy.wra.gov.tw/fhy/Monitor/Reservoir, accessed on 31 August 2021).

Acknowledgments: The authors would like to thank Uni-edit (www.uni-edit.net, 31 August 2021) for editing and proofreading this manuscript.

Conflicts of Interest: The authors declare no conflict of interest.

References

1. Smith, V.H.; Schindler, D.W. Eutrophication Science: Where do we go from here? *Trends Ecol. Evol.* **2009**, *24*, 201–207. [CrossRef] [PubMed]
2. World Heath Organization (WHO). *Guidelines for Drinking Water Quality*, 2nd ed.; World Health Organization: Geneva, Switzerland, 1998; Volume 1.
3. Paerl, H.W.; Fulton, R.S. Ecology of harmful cyanobacteria. In *Ecology of Harmful Marine Algae*; Graneli, E., Turner, J., Eds.; Springer: Berlin/Heidelberg, Germany, 2006; pp. 95–107.
4. National Research Council (NRC). *Clean Coastal Waters: Understanding and Reducing the Effects of Nutrient Pollution*; National Academy Press: Washington, DC, USA, 2000.
5. Schindler, D.W. Evolution of Phosphorus Limitation in Lakes. *Science* **1977**, *195*, 260–262. [CrossRef] [PubMed]
6. Keatley, B.E.; Bennett, E.M.; MacDonald, G.K.; Taranu, Z.E.; Gregory Eaves, I. Land-Use Legacies Are Important Determinants of Lake Eutrophication in the Anthropocene. *PLoS ONE* **2011**, *6*, e15913. [CrossRef] [PubMed]
7. Carpenter, S.R.; Caraco, N.F.; Correll, D.L.; Howarth, R.W.; Sharpley, A.N.; Smith, V.H. Nonpoint Pollution of Surface Waters with Phosphorus and Nitrogen. *Ecol. Appl.* **1998**, *8*, 559–568. [CrossRef]

8. Sharpley, A.N.; Chapra, S.C.; Wedepohl, R.; Sims, J.T.; Daniel, T.C.; Reddy, K.R. Managing Agricultural Phosphorus for Protection of Surface Waters. Issues and Options. *J. Environ. Qual.* **1994**, *23*, 437–451. [CrossRef]
9. Smith, V.H.; Tilman, G.D.; Nekola, J.C. Eutrophication: Impacts of Excess Nutrient Inputs on Freshwater, Marine, and Terrestrial Ecosystems. *Environ. Pollut.* **1999**, *100*, 179–196. [CrossRef]
10. Caraco, N.F. Influence of Human Populations on P transfers to Aquatic Ecosystems: A Regional Scale Study Using Large Rivers. In *Phosphorus in the Global Environment*; Tiessen, H., Ed.; John Wiley: New York, NY, USA, 1995; pp. 235–244.
11. Chloupek, O.; Hrstkova, P.; Schweigert, P. Yield and its Stability, Crop Diversity, Adaptability and Response to Climate Change, Weather and Fertilization over 75 years in the Czech Republic in Comparison to some European Countries. *Field Crop. Res.* **2004**, *85*, 167–190. [CrossRef]
12. Kroes, H.W. Replacement of Phosphates in Detergents. *Aquat. Ecol.* **1980**, *14*, 90–93. [CrossRef]
13. Bonsdorff, E.; Blomqvist, E.M.; Mattila, J.; Norkko, A. Coastal eutrophication: Causes, consequences and perspectives in the Archipelago areas of the northern Baltic Sea. *Estuar. Coast. Shelf Sci.* **1997**, *44*, 63–72. [CrossRef]
14. Environmental Protection Administration, EPA. What is Hypoxia and What Causes It? Available online: https://www.epa.gov/ms-htf/hypoxia-101 (accessed on 20 September 2020).
15. Carmichael, W.W. Health Effects of Toxin-Producing Cyanobacteria: "The CyanoHABs". *Hum. Ecol. Risk Assess. Int. J.* **2001**, *7*, 1393–1407. [CrossRef]
16. Smith, V.H. Cultural Eutrophication of Inland, Estuarine, and Coastal Waters. In *Successes, Limitations and Frontiers in Ecosystem Science*; Springer: New York, NY, USA, 1998; pp. 7–49.
17. Regier, H.A.; Holmes, J.A.; Pauly, D. Influence of Temperature Changes on Aquatic Ecosystems—An Interpretation of Empirical-Data. *Trans. Am. Fish. Soc.* **1990**, *119*, 374–389. [CrossRef]
18. Brinkhurst, R.O. *The Benthos of Lakes*; MacMillan Press: London, UK, 1974.
19. Mortimer, C.H. The Exchange of Dissolved Substances between Mud and Water in Lakes. *J. Ecol.* **1941**, *29*, 280–329. [CrossRef]
20. Nurnberg, G.K. The Prediction of Internal Phosphorus Load in Lakes with Anoxic Hypolimnia. *Limnol. Oceanogr.* **1984**, *29*, 111–124. [CrossRef]
21. Wetzel, R.G. *Limnology: Lake and River Ecosystems*, 3rd ed.; Academic Press: New York, NY, USA, 2001.
22. Davison, W. Supply of Iron and Manganese to an Anoxic Lake Basin. *Nature* **1981**, *290*, 241–243. [CrossRef]
23. Water Resources Agency. Shimen Reservoir Theme Network. Available online: https://www.wranb.gov.tw/3517/4697/4698/22701/ (accessed on 4 November 2020).
24. Water Resources Agency Ministry of Economic Affairs, Northern Region Water Resources Office. Shimen Reservoir Introduction. 2019. Available online: https://www.wranb.gov.tw/ (accessed on 4 November 2020).
25. Agency, Water Resources. Introduction of Mingde Reservoir. Available online: https://www.wra.gov.tw/News_Content.aspx?n=3254&s=19374 (accessed on 4 November 2020).
26. Chen, S.C.; Kuo, J.T. Simplified Eutrophic Modeling in Fonsan Reservoir. *J. Chin. Soil Water Conserv.* **1992**, *23*, 41–56.
27. Seventh Branch, Taiwan Water Corporation. Branch Introduction. Available online: https://www.water.gov.tw/dist7/Contents?nodeId=6787 (accessed on 7 October 2020).
28. Environmental Protection Administration, EPA. Environmental Water Quality Monitoring Annual Report. 2019. Available online: https://wq.epa.gov.tw/EWQP/zh/ConService/DownLoad/AnnReport.aspx (accessed on 30 August 2020).
29. Zang, C.J.; Huang, S.L.; Wu, M.; Du, S.L.; Scholz, M.; Gao, F.; Lin, C.; Guo, Y.; Dong, Y. Comparison of Relationships Between pH, Dissolved Oxygen and Chlorophyll a for Aquaculture and Non-aquaculture Waters. *Water Air Soil Pollut.* **2011**, *219*, 157–174. [CrossRef]
30. Blumberg, A.F.; Ditoro, D.M. Effects of Climate Warming on Dissolved-Oxygen Concentrations in Lake Erie. *Trans. Am. Fish. Soc.* **1990**, *119*, 210–223. [CrossRef]
31. Chen, H.L.; Tsai, D.W. Colinearity Analysis and Model Selection of Water Quality Factors Related to Eutrophication. *J. Soil Water Conserv.* **2007**, *39*, 229–246.
32. Watson, S.B.; Miller, C.; Arhonditsis, G.; Boyer, G.L.; Carmichael, W.; Charlton, M.N.; Confesor, R.; Depew, D.C.; Hook, T.O.; Ludsin, S.A.; et al. The re-eutrophication of Lake Erie: Harmful algal blooms and hypoxia. *Harmful Algae* **2016**, *56*, 44–66. [CrossRef] [PubMed]

 water

MDPI

Article

Assessment of the Ecological Risk from Heavy Metals in the Surface Sediment of River Surma, Bangladesh: Coupled Approach of Monte Carlo Simulation and Multi-Component Statistical Analysis

Arup Acharjee [1], Zia Ahmed [1], Pankaj Kumar [2,*], Rafiul Alam [3], M. Safiur Rahman [4] and Jesus Simal-Gandara [5]

1 Department of Geography and Environment, Shahjalal University of Science and Technology, Sylhet 3114, Bangladesh; aruporgho@gmail.com (A.A.); ziaahmed-gee@sust.edu (Z.A.)
2 Institute for Global Environmental Strategies, Hayama, Kanagawa 240-0115, Japan
3 BRAC James P Grant School of Public Health, BRAC University, Dhaka 1212, Bangladesh; rafiulalamctg@gmail.com
4 Water Quality Research Laboratory, Chemistry Division, Atomic Energy Centre (AECD), Bangladesh Atomic Energy Commission, Dhaka 1207, Bangladesh; safiur.rahman@dal.ca
5 Department of Analytical Chemistry and Food Science, Faculty of Science, Universidade de Vigo, E-32004 Ourense, Spain; jsimal@uvigo.es
* Correspondence: kumar@iges.or.jp

Citation: Acharjee, A.; Ahmed, Z.; Kumar, P.; Alam, R.; Rahman, M.S.; Simal-Gandara, J. Assessment of the Ecological Risk from Heavy Metals in the Surface Sediment of River Surma, Bangladesh: Coupled Approach of Monte Carlo Simulation and Multi-Component Statistical Analysis. *Water* **2022**, *14*, 180. https://doi.org/10.3390/w14020180

Academic Editors: Chin H. Wu and George Arhonditsis

Received: 23 November 2021
Accepted: 3 January 2022
Published: 10 January 2022

Publisher's Note: MDPI stays neutral with regard to jurisdictional claims in published maps and institutional affiliations.

Abstract: River sediment can be used to measure the pollution level in natural water, as it serves as one of the vital environmental indicators. This study aims to assess heavy metal pollution namely Copper (Cu), Iron (Fe), Manganese (Mn), Zinc (Zn), Nickel (Ni), Lead (Pb), and Cadmium (Cd) in Surma River. Further, it compares potential ecological risk index values using Hakanson Risk Index (RI) and Monte Carlo Simulation (MCS) approach to evaluate the environmental risks caused by these heavy metals. in the study area. With obtained results, enrichment of individual heavy metals in the study area was found in the order of Ni > Pb > Cd > Mn > Cu > Zn. Also, variance in MCS index contributed by studied metals was in the order of Cd > Pb > Ni > Zn > Cu. None of the heavy metals, except Ni, showed moderate contamination of the sediment. Risk index values from RI and MCS provide valuable insights in the contamination profile of the river, indicating the studied river is currently under low ecological risk for the studied heavy metals. This study can be utilized to assess the susceptibility of the river sediment to heavy metal pollution near an urban core, and to have a better understanding of the contamination profile of a river.

Keywords: heavy metals; ecological risk; Surma River; Monte Carlo simulation; multivariate analysis; Hakanson risk index

1. Introduction

In developing countries, heavy metal contamination in river water and sediment is a matter of concern [1]. The general ways these heavy metals reach river bodies are via weathering, erosion of rocks, and an array of anthropogenic sources. The sources of contamination are found to be, generally, occurring from industrial and agricultural activities, surface runoff, and sewage disposal [2]. The sources of the contamination can be either point or non-point in nature [3]. River sediment can be used to measure the pollution levels in natural waters, as it serves as one of the vital environmental indicators [4]. Though soil pollution occurs by a diverse variety of heavy metals, some of them (Cu, Ni, Cd, Zn, Cr, and Pb) are more significant because of their distinct toxicity [5]. Iron and zinc have been reported to be biologically important for human beings and their diet and medicinal preparations, but in contrast to these metals, Hg, Cd, and Pb have no biological significance to humans of any sort, and ingestion of them can be harmful, owing to the high toxicity [6]. The level of harm done to riverine ecosystems by the waste discharges from anthropogenic and industrial sources can be measured by conducting a thorough inspection of pollution attributable to the heavy metals in the river sediment [7]. Disposed urban wastes, untreated industry effluents, and agrochemicals in the most adjacent water bodies are the most contributing factors to the heavy metal pollution in Bangladesh [4]. Heavy metals are dangerous over critical limits, despite being essential micronutrients for floral and faunal lifeforms; examples of such metals include Fe, Mn, Co, and Zn [8]. A myriad of diseases is caused by exposure to heavy metals, as well as other physiological complications, including inhibition of development, renal failure, genetic mutation, and a disruptive effect on intelligence and behavior [9].

Sediments are unavoidable constituent elements in a riverine environment, where they provide living organisms with sustenance, as well as work as a natural sink for hazardous chemicals [10]. However, the accumulated hazardous chemicals in the sediment continue to pose a threat to ecological and biological entities, even though the contaminants are seized from being released from different sources [11]. Risk assessment methods should be applied for a correct understanding of heavy metal contamination, its management, and pollution monitoring [12]. However, risk assessment is a complex process that intrinsically allows a degree of uncertainty [13]. The uncertainty can be attributed to these factors: lack of accurate understanding; data scarcity; and variability, which is a common feature of the environmental domain and dynamics [14–17]. Hakanson's Risk Index naturally aims at achieving a definite estimation of risk by integrating average and worst-case point values of risk [18,19].

There has been little scientific investigation on heavy metal contamination in the bottom sediment of the important rivers of Bangladesh, whereas more concentration has been given on river water quality. In Bangladesh, the Surma River forms the important Surma–Meghna river system, which is the longest river system in the country. Sylhet, on the edge of River Surma, is a north-eastern city of Bangladesh. Excessive production of waste materials is a general outcome of population growth in a city. On a typical day, the city produces approximately 215 tons of waste product [20]. Generally, industrial effluents and municipal wastewaters are enriched with high levels of heavy metals, such as As, Cd, Cr, Cu, Fe, Hg, Mn, Ni, Pb, and Zn [21]. The study river is a recipient of an excessive amount of domestic waste, and industrial effluents through municipal sewage outlets. Non-point sources include urban runoff and agricultural runoff supposedly carrying heavy metals into the river. To the best of our knowledge, this research work involving the assessment of ecological risk with a view of eradicating the uncertainty principle is the first scientific assessment of ecological risk in river-bed sediment in Bangladesh through the coupled application of Monte Carlo Simulation (MCS) and Hakanson Risk Index (RI). In addition, the findings of the study will provide a significant contribution to the formulation of policies related to river pollution, and will help in taking apposite initiatives for the management of domestic sewage disposal from the urban complex settlements. The principal objectives of this study are to assess the contamination of the bottom sediment using multiple pollution indicators, and to estimate ecological risks due to the heavy metals using the concerted approach of traditional ecological risk index and the relatively new Monte Carlo Simulation technique.

2. Materials and Methods

2.1. Study Area

The study was conducted on the Surma River, which forms the longest river system, the Surma–Meghna river system (669 km), in Bangladesh, and flows through the north-eastern city of Sylhet. Sylhet City is located at 24°53′ N latitude and 91°53′ E longitude, with an estimated population of 0.6 million, and a population growth rate of 4% per annum [20], in contrast with the annual growth rate of 2.01% in Bangladesh [22]. The study river originates from the Shillong Hills in Meghalaya, India. Our study river starts from its source, which is the slopes of the Naga–Manipur catchment area, and is known as River Barak.

The river gets divided into two distinct branches at Cachar in Assam District in India. The northern branch is known as Surma, which enters Bangladesh through the Sylhet District, and the southern branch is Kushiara. Both of these distributary rivers meet at Madna at the lower segment of the river courses [23]. The river segment covering the metropolitan encroachment limits (from Tukerbazar Ghat to Kushi Ghat) was selected as the study area, owing to the increasing major industrial activity, agricultural activity, and urban land use in this region. The study area is, as shown in Figure 1, located at the Sylhet Metropolitan stretch, which is approximately 15 km within the latitudes 24°54′36.81″ N 24°52′29.64″ N, and longitudes 91°49′23.9988″ E 91°54′10.0008″ E. The geographical coordinates of the sampling locations are given in the Table S1 in the supplementary file. A pilot survey was conducted in the area before sample collection. A total of 15 sampling locations were selected, and are shown in Figure 1.

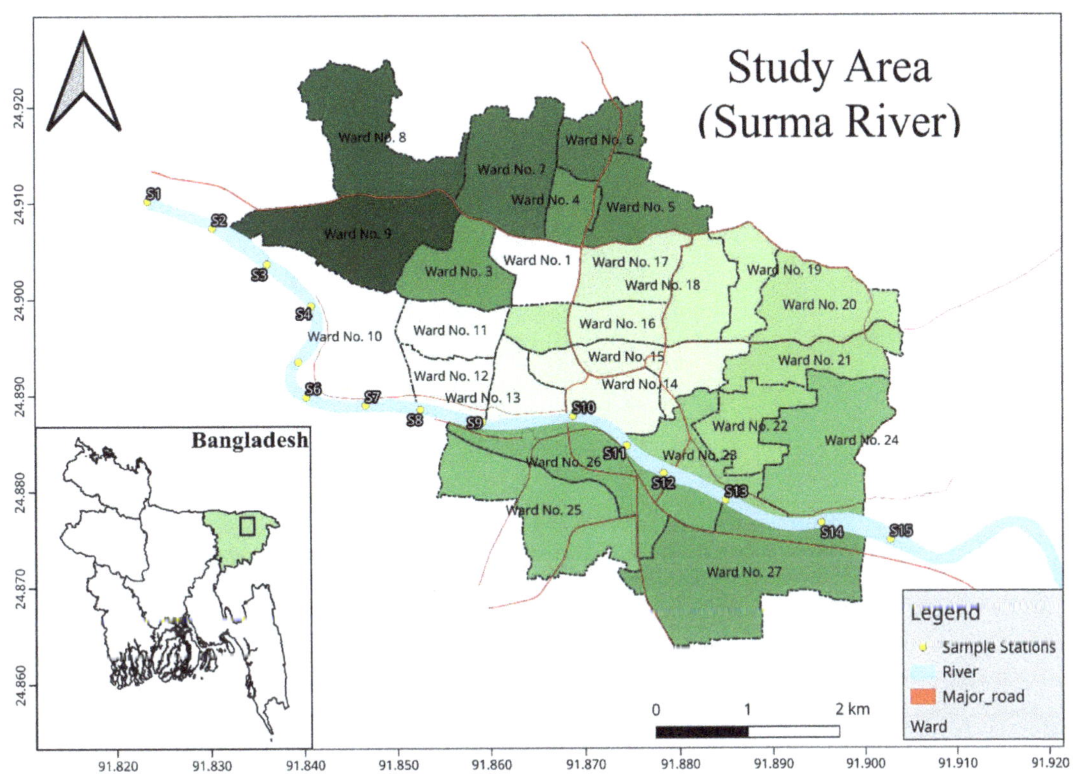

Figure 1. Study area and sampling stations.

2.2. Sample Collection and Preparation

Fifteen sediment samples were collected from the selected sampling locations. Sampling locations were selected based on locational interest, such as industrial sewage outlets and municipality sewage outlets. Sediment samples were collected from a depth of 0–30 cm with a 1.5 m long PVC corer (RFL Group, Dhaka, Bangladesh) with 10 cm diameter, manually attached with a galvanized iron pipe (Simex Bangladesh, Dhaka, Bangladesh) and transferred in polyethylene bags immediately. All geographical coordinates were taken with a handheld GPS device (Garmin eTrex 32x, American multinational technology company, Olathe, KS, USA). Before the sampling procedure, the polyethylene bags were cleansed with a diluted 10% nitric acid solution and distilled water [24,25]. Samples were brought to the Soil Resource Development Institute (SRDI), Sylhet, Bangladesh. The sediment samples were air-dried in a dry, dust-free room at room temperature. The samples were grounded after discarding the plant roots and inorganic debris, and sieved with a 2 mm sieve.

2.3. Heavy Metal Analysis

2.3.1. Reagents and Sample Digestion

All standard solutions and reagents, along with acids and chemicals are provided by Merck (Darmstadt, Germany) and MilliporeSigma (Burlington, MA, USA). All used chemical substances were of 99.99% purity level.

Soil Extraction and Determination of Fe, Mn, Cu, and Zn

Soil was weighed 10 g, and taken into a 1.25 mL dry polyethylene bottle. A DTPA (diethylenetriaminepentaacetic acid) solution of 20 mL was added with a pipette. The solution was shaken continuously for exactly 2 h on a horizontal shaker, and filtered immediately after shaking thoroughly by Whatman no. 42 filter paper into a conical flask. The contents of metals in the DTPA extract of soil were determined by AAS (Model Shimadzu AA 7000 series, Shimadzu corporation, Kyoto, Japan) using appropriate cathode lamps. Direct readings of copper and zinc were taken from AAS. For iron (Fe) and manganese (Mn), the reading was taken after the solution was diluted further. The solution was diluted after mixing 5 mL of the solution to 45 mL of distilled water [26].

Soil Extraction and Determination of Ni, Pb, Cd

Sediment sample was weighed 2 g into a 50 mL crucible, to which 10 mL concentrated nitric acid was added. The mixture was kept for 30–45 min for oxidation. After cooling, 2.5 mL of perchloric acid of 70% strength was added, and the mixture was reheated until the digest was clear. Then, the sample was filtered using Whatman no. 42 filter paper. Upon adding distilled water, the mixture was shifted to a volumetric flask, ready to be analyzed by AAS [26].

2.3.2. Analytical Technique and Quality Assurance

All of the soil matrixes were analyzed for Fe, Mn, Cu, Zn, Ni, Pb, and Cd by atomic absorption spectrophotometer (Model Shimadzu AA 7000 series). AAS conditions for analytical measurement are tabulated in Table S2 of the supplementary files. Glassware and all containers used were purified with 20% nitric acid and de-ionized water, and air-dried before usage. The quality of the data obtained from analyzed elements through AAS were thoroughly maintained. The calibration curves were maintained linear for all elements to be studied, after which the performance of the calibrated system was checked. The analytical procedure was checked using a reference soil sample provided by Soil Resource Development Institute, (SRDI, Sylhet, Bangladesh).

3. Results

3.1. Heavy Metal in Sediments

Heavy metal concentrations in river soil determined by AAS are tabulated in Table S3 in supplementary file. The mean concentrations were found as 2.68 mg/kg for Cu; 6.12 mg/kg for Zn; 291.1 mg/kg for Fe; 88.03 mg/kg for Mn, 11.73 mg/kg for Pb; 0.06 mg/kg for Cd; and 92.34 mg/kg for Ni. All metal concentration values are given Table S3 in supplementary section. The results indicate that nearly all of the studied metals failed to exceed the background values given by [27]. This suggests that the investigated area is being enriched with a low quantity of metal content in a massive volume of sediment [28]. The metal concentrations in the study area were found to be in following order: Fe > Mn > Ni > Pb > Zn > Cu > Cd. The total findings of the heavy metal from collected samples are given below in Table 1. According to this study, the river sediment has low iron concentrations, despite iron being one of the most dominant metals in the earth surface. Such a low iron concentration in the sediment can be attributed to the distinct geochemical setting of the Sylhet region. The bedrock of Sylhet region is dominated mostly with shale, nummulitic limestone, and sandstones, which have fewer Fe-oxides.

Table 1. Heavy metal concentration in the bottom sediment of Surma River with descriptive statistics.

Sample Stations	Heavy Metals (Units in mg/kg)						
	Cu	Zn	Fe	Mn	Pb	Cd	Ni
Mean	3.688	8.951	317.533	120.136	18.975	0.099	116.077
Standard Deviation	1.867	6.196	70.224	88.473	12.278	0.101	39.248
Minimum	1.590	2.350	170.000	4.200	1.080	0.015	65.560
Maximum	8.520	19.800	418.000	303.490	41.230	0.350	189.620
Surface rock average [27]	32	127	35900	750	16	0.2	49
WHO (2004)	1.5	123	NA	NA	NA	6	20
USEPA (1999)	16	110	30	30	40	0.6	16

A comparative scenario of heavy metal pollution in other major rivers around the world along with the studied river is given below in Table 2.

Table 2. Comparison of metals in sediment with other studies around the globe (units in mg/kg).

River/Date of Sampling/Country	Pb	Cd	Zn	Ni	Fe	Mn	Cu	Reference
World Average	230.75	1.4	303	102.1	57405.9	975.3	122.9	[23]
Euphrates, 1997, Iraq	19.5	0.08	30	125	-	450	-	[29]
Tigris, 1993, Iraq	17.9–30.6	0.1–1.7	8.3–47.1	105.4–125.5	-	451.3–565.6	17.4–28.9	[30]
Cauvery 2007–2009, India	4.3	1.3	93.1	27.7	11144	176.3	11.2	[31]
Bangshi River, 2014, Bangladesh	59.99	0.61	117.15	25.67	-	483.44	-	[32]
Yangtze, 2005, China	49.19	0.98	230.9	41.86	-	-	60.03	[33]
Surma River, 2019, Bangladesh	11.73	0.06	6.12	92.34	291.1	88.03	2.68	Present study

3.2. Assessment of Sediment Quality

Values from background levels (continental shale value or crustal abundance of different elements) can be used as a reference to measure the increase in concentration levels [34]. It is measured in contrast to the values from pre-industrial levels [35]. Due to the unavailability of the background values for this study area, this study utilized the world rock surface values for the assessment of pollution indices [36]. Following pollution, indices were applied to obtain a satisfactory relative ranking of samples: (i) Contamination Factor (CF), (ii) Contamination Degree (CD), (iii) Modified Degree of Contamination (MCD), (iv) Enrichment Factor (EF), (v) Pollution Load Index (PLI), (vi) Geo-Accumulation Index (IGeo).

3.2.1. Contamination Factor (CF)

Contamination Factor (CF) and Contamination Degree (CD) together are considered primary indicators of metal pollution status of the subjected soil or sediment [24]. The CF can be obtained for each of the sampling locations by dividing the metal concentrations in sediment by the background concentration values of the respective metals. The CF is the result of dividing the metal concentration in the sediment by the concentration of background value of the respective metal [37]. Ref [38] proposed the following equation to calculate CF and the proposed gradation for CF is tabulated in Table 3.

$$CF = \frac{\text{Cm (sample)}}{\text{Cm (Background)}} \tag{1}$$

where Cm_{Sample} is the metal concentration derived from river sediment, and $Cm_{\text{Background}}$ is the standard metal concentration value equal to the world surface rock average given by [27]. Contamination factors are graded into four classes. Contamination Degree is the summation of all CF values for each sample. Figure 2 contains the CF profile of the study river.

$$CD = \sum(CF) \tag{2}$$

Table 3. Description of Contamination Factor (CF) and Contamination Degree (CD) according to [38].

Contamination Factor Ranges	Description	Contamination Degree Ranges	Description
CF < 1	low contamination	CD < 8	Low degree of contamination
1 ≤ CF ≤ 3	Moderate Contamination	8 ≤ CD < 16	Moderate degree of contamination
3 ≤ CF ≤ 6	Considerable Contamination;	16 ≤ CD < 32	Considerable degree of contamination
CF ≥ 6	Very High Contamination	CD ≥ 32	Very high degree of contamination

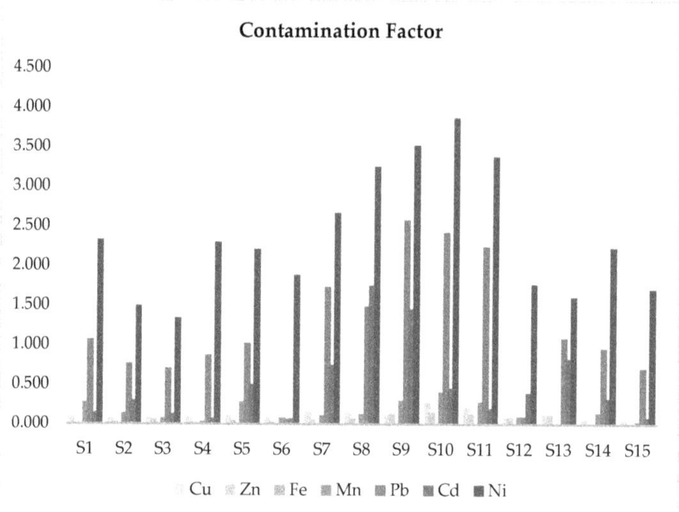

Figure 2. Contamination Factor of heavy metals at all sample stations.

3.2.2. Contamination Degree (CD)

The study area falls into CD range "CD < 8" predominantly, as most of the sampling locations have CD values below 8. However, sample stations S9 and S10 have a moderate degree of contamination, probably due to the sampling locations being situated adjacent to the industrial vicinity.

3.2.3. Modified Contamination Degree, MCD

Ref [39] gave a more simplified method of measuring Contamination Degree, previously given by Hakanson [38]. The formula is given below:

$$MCD = \frac{\sum CF}{n} \tag{3}$$

where n = number of analyzed elements, and CF = Contamination Factor.

In Figure 3, the categories of MCD are shown, which are used to describe and classify the Modified Contamination Degree.

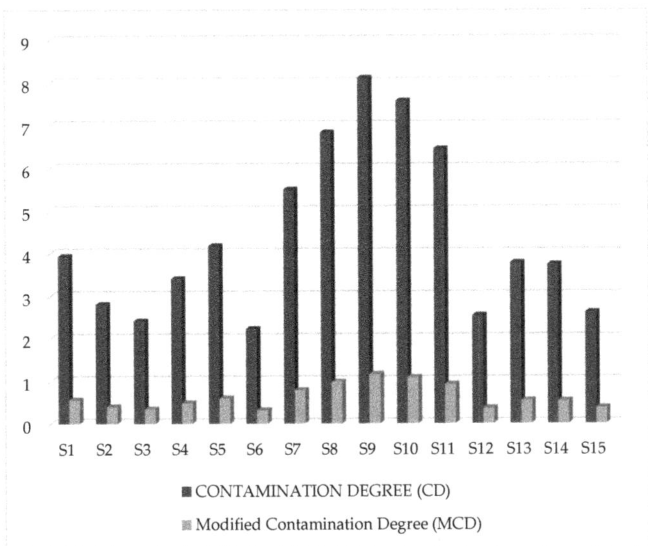

Figure 3. CD and MCD values of the heavy metals of the river sediment.

Following are the proposed MCD classes: MCD < 1.5 indicates a significantly low degree of contamination; 1.5 ≤ MCD < 2 designates a low degree of contamination in the sediment; a moderate degree of contamination occurs when MCD levels fall between 2 ≤ MCD < 4; a high degree of contamination is evident in soil when MCD levels rise as high as 4 ≤ MCD < 8; MCD values of an even higher range, 8 ≤ MCD < 16, indicate a very high degree of contamination in the sediment; 16 ≤ MCD < 32 indicates an extremely high degree of contamination in the sediment; and finally, an ultra-high degree of contamination is indicated by MCD levels in the range of ≤32 [39]. In the present study, the MCD values of all sample stations MCD are below 1.5, which indicates a nil-to-very-low degree of contamination. The MCD values are shown in Figure 3.

3.2.4. Enrichment Factor (EF)

Ref [40] designated the enrichment factor as an indicator to quantify the anthropogenic contribution to any change in the metal concentration in the sediment. The enrichment factor for the metals can be calculated by the following equation given by [41]:

$$EF = \frac{(\text{Me}/\text{Fe})\ \text{sample}}{(\text{Me}/\text{Fe})\ \text{background}} \quad (4)$$

where (Me/Fe) sample is the ratio of subjected metal and Fe of the sediment from sampling location, and, on the other hand, (Me/Fe) background denotes the environmental background value of the metal–Fe ratio. Values of metal concentrations of surface world rocks were chosen as reference, owing to the lack of background values of pre-industrial times [27]. Iron was elected as the suitable element for normalization between the two sets of values from both the metal–Fe ratio of the sample and backgrounds used previously by [36,42]. Enrichment factor is graded in five classes: EF values less than 2 indicate deficiency to minimum enrichment; moderate enrichment is expressed by EF ranging from 2 ≤ EF < 5; values ranging from 5 ≤ EF < 20 indicate significant enrichment; metal enrichment is very high when EF values fall between 20 ≤ EF < 40; and lastly, EF ≥ 40 indicates extremely high enrichment. Heavy metals in River Surma posed following metal enrichment trend shown in Figure 4.

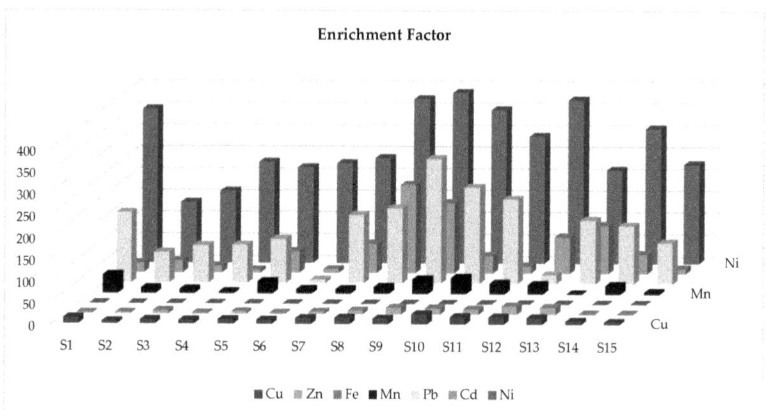

Figure 4. Metal enrichment of all heavy metals in River Surma.

3.2.5. Pollution Load Index (PLI)

Pollution Load Index is a frequently used method for estimating the quality and toxicity of sediment proposed by Tomlinson et al. [43]. The PLI of a particular site is generally estimated by calculating the nth root of the product of multiplying n-numbered CF values for all investigated elements. The following equation was used for the determination of PLI:

$$PLI = (CF1 \times CF2 \times CF3 \times \ldots \times CFn)^{\frac{1}{n}} \quad (5)$$

where CF denotes the contamination factor, and n is the considered number of metals. There are three discrete categories for pollution measurement with this index. Perfect pollution status is indicative of no pollution when the PLI values are 0 (the first category); the second category is indicative of the baseline degree of pollution when the PLI values are equal or less than 1; and the third category (when PLI is greater than 1) designates progressive decline in terms of pollution of the sites. The PLI values for respective sampling locations are shown in Figure 5.

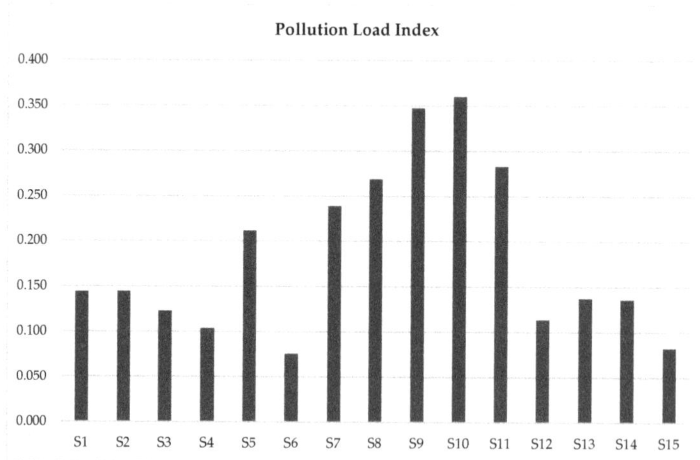

Figure 5. Pollution Load Index values of sampling sites at Surma River.

3.2.6. Geo-Accumulation Index (Igeo)

Geo-accumulation index (Igeo) is a method of estimating the enrichment of metal concentration above background values proposed by Muller [44]. The equation used to determine Igeo values is:

$$\text{Igeo} = \log_2 \frac{\text{Cm (Sample)}}{1.5 \times \text{Cm (Background)}} \tag{6}$$

where C_m Sample is the concentration of a particular element in the sample, and C_m Background is the geochemical background value of the metal. The values from world rock surface averages given by [27] are used as reference background value [36]. Geo-accumulation index produces results in seven classes of purity. These classes are portrayed in Table 4. From the achieved results, it is evident that for most of the sites and metals, the Igeo values remained below 0, depicting uncontaminated sediments, whereas nickel (Ni) and cadmium (Pb) showed some deviance from the trend and fall in class 1. Geo-accumulation Index values of the heavy metals in river sediment are shown in the following Table 4. Figure 6 shows the variability of the Igeo values for the metal concentrations in the sediment.

Table 4. Geo-accumulation Index categories [45,46].

Igeo Class	Igeo Values	Description
Class 0	Igeo < 0	uncontaminated sediments
Class I	0 < Igeo < 1	uncontaminated to moderately contaminated
Class II	1 < Igeo < 2	moderately contaminated
Class III	2 < Igeo < 3	moderately to highly contaminated
Class IV	3 < Igeo < 4	highly contaminated
Class V	4 < Igeo < 5	highly to extremely contaminated
Class VI	Igeo > 5	extremely contaminated

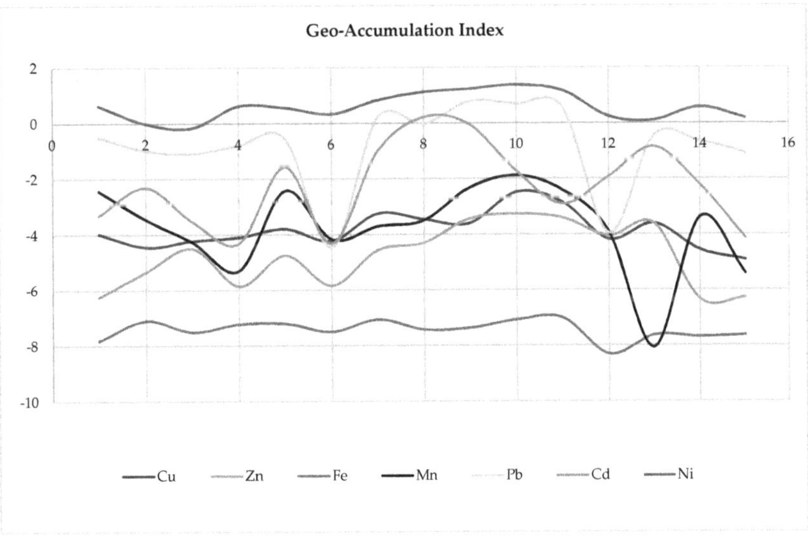

Figure 6. Igeo variation in the sediment of River Surma.

3.3. Pearson's Correlation Matrix

Pearson's Correlation (PC) was calculated for examined metal elements to investigate if there is any correspondence among the elements. Pearson's Correlation matrix corroborates inter-metal characteristics in terms of origin and behavior along their paths of transport [47]. The results are tabulated below in Table 5.

Table 5. Pearson's Correlation Matrix of heavy metals.

	Cu	Zn	Fe	Mn	Pb	Cd	Ni
Cu	1						
Zn	0.78 ***	1					
Fe	0.58 **	0.29	1				
Mn	0.66 ***	0.48 *	0.39	1			
Pb	0.78 ***	0.71 ***	0.6 **	0.69 ***	1		
Cd	0.26	0.45 *	0.06	0.15	0.49 *	1	
Ni	0.8 ***	0.61 **	0.48 *	0.74 ***	0.87 ***	0.5 *	1

* is significant at $0.05 < p \leq 0.1$ levels, ** is significant at $0.01 \leq p \leq 0.05$ levels, and *** is significant at $p < 0.01$ levels.

Existing metal concentrations in the bottom sediment of river Surma stipulate the concurrent levels of correlation with each other at significant levels of $p \leq 0.05$ and $p < 0.01$ (Table 5). In the present study, Cu, Zn, Ni, and Pb showed significant correlation coefficients, which indicates that they have common sources of origin, and could be dominated by an exclusive factor. Cu displayed soaring levels of a positive relationship with Zn, Mn, and Pb, and in a moderate degree with Fe. The correlation matrix demonstrates that Cu has a low level of relationship with Cd, which points to the possibility of a different origin of these elements. Ni is found to be corresponding and intercorrelated with Cu, Mn, and Pb significantly (at $p < 0.01$ levels), and moderately with Zn (at $0.01 \leq p \leq 0.05$ levels), which is associated with common sources of input of heavy metal to the river from municipal waste, agricultural runoff, and industrial sewage. Poor correlations between Fe and Cd could be resulting from the differential sources of origin, where Fe has a natural origin and Cd has anthropogenic origins. Copper and cadmium also deviate from the possibility of being originated from undifferentiated sources, and this difference can be ascribed to the copiousness of Cd in common anthropogenic sources, such as industrial effluents and municipal waste; on the other hand, the original sources of copper can be attributed to agricultural runoffs. Zn, Cu, and Cd also possibly have origins in natural fluvial sediment.

3.4. Potential Ecological Risk Index (PERI)

Refs. [38,48] proposed the PERI method to evaluate the environmental characteristics due to heavy metal contamination in fluvial sediments. Ref. [49] evaluated concurrent pollution levels, and the environmental response to the pollution. The equations employed to determine the ecological risk of a certain area are:

$$\mathrm{RI} = \sum(\mathrm{E}_{\mathrm{r}^{\mathrm{i}}}) \tag{7}$$

$$\mathrm{E}_{\mathrm{r}^{\mathrm{i}}} = \mathrm{T}_{\mathrm{r}^{\mathrm{i}}} \times \mathrm{CF} \tag{8}$$

Here,

RI = risk factor or summation of all individual potential ecological risk factors contributed by each meal element;
E_r^i = factor of potential ecological risk;
CF = contamination factor;
T_r^i = toxic response factor.

According to Hakanson [33], elements such as Ni, Cd, Pb, Zn, and Cu have toxic response factors (T_r^i) of 5, 30, 5, 1, and 5, respectively. As per Hakanson's suggestion [38], E_r^i and RI are two terms to be multiplied together for calculating ecological risk. According to this approach, the potential ecological risk is minimal when $E_r^i < 40$; a moderate level of risk for $40 \leq E_r^i \leq 80$; $80 \leq E_r^i \leq 160$ portrays a considerable level of risk; $160 \leq E_r^i \leq 320$ depicts a staggering level of potential ecological risk; whereas $E_r^i > 320$ is construed as a very high ecological risk. Whereas a total ecological risk (RI) value below 150 points indicates a low ecological risk; 150 < RI < 300 suggests moderate degree of ecological risk; a considerable level of ecological risk is generally designated by RI values between 300 to 600; and, ultimately, RI > 600 tends to portray a very high ecological risk of the study area. According to Table 6, the E_r^i values of Pb, Cd, Cu, Ni, Zn in all sampling sites stipulated values predominantly lower than 40, as to specify low levels of ecological risk, except for Cd. Moderate ecological risks are observed for Cd. All of the sample stations can be categorized with low ecological risk levels, as the Risk Index (RI) values are less than 150.

Table 6. Potential ecological risk and Risk index values.

Site ID	Cu	Zn	Pb	Cd	Ni	RI
S1	0.48	0.02	5.31	4.5	11.62	21.930
S2	0.34	0.04	3.81	9	7.45	20.640
S3	0.4	0.07	3.53	3.9	6.69	14.590
S4	0.44	0.03	4.34	2.25	11.5	18.560
S5	0.54	0.06	5.09	15	11.03	31.720
S6	0.4	0.03	0.34	2.25	9.42	12.440
S7	0.79	0.06	8.67	22.5	13.33	45.350
S8	0.68	0.08	7.42	52.5	16.27	76.950
S9	0.62	0.14	12.88	43.5	17.6	74.740
S10	1.33	0.16	12.12	13.5	19.35	46.460
S11	1.04	0.14	11.19	6	16.89	35.260
S12	0.41	0.09	0.48	11.85	8.83	21.660
S13	0.61	0.12	5.44	24.75	8.03	38.950
S14	0.32	0.02	4.78	9.6	11.16	25.880
S15	0.25	0.02	3.53	2.55	8.51	14.860

3.5. Monte Carlo Simulation

Generally, Monte Carlo Simulation is performed to elucidate the uncertainty issue, which is intrinsic to the calculation of potential ecological risk using absolute point values of metal concentration. In this method, a suitable dataset is developed, which agrees with a particular probability distribution [50]. The elemental concentrations of the river sediment acted as the primary dataset for finding apposite probability distribution and simulation of RI. According to the Kolmogorov–Smirnov test, the most suitable fitting was demonstrated by the log-normal probability distribution function, whereas other notable density functions with poor fitting included log-logistic, BetaPERT, Weibull, gamma, max-extreme density functions. Ten-thousand Monte Carlo iterations were carried out employing the software CrystalBall (Oracle Corporation, Santa Clara, CA, USA). Repeated calculation produced probability distribution for the Hakanson Risk Index. The output distribution for RI followed a log-normal distribution.

The results from the Monte Carlo Simulation produced probabilistic ecological risk values (E_r^i) for heavy metals. Nickel (Ni) indicates a 100% probability to fall under the E_r^i value of 40, which indicates low ecological risk shown in Figure S1. Lead (Pb) exhibited a probability of 98.25% to fall in the low-risk category, and 1.51% for moderate potential ecological risk shown in Figure S2. Cadmium (Cd) portrayed a 92.04% probability for the low-risk category, 6.22% in moderate ecological risk, and a 1.49% probability of considerable potential ecological risk shown in Figure S3. Zinc (Zn) and Copper (Cu) both depicted low-risk potential ecological risk probabilities shown in Figure S4 and Figure S5 respectively in supplementary Files. In Figure 7, 100% of the cumulative probability of Risk Index (RI) values is less than 150, which, according to Hakanson's Risk Index, is representative of low ecological risk. From the sensitivity analysis, it is evident that 67.3% of risk is contributed by Cd, followed by Pb with 2.4%, and Ni with 10.3% variance.

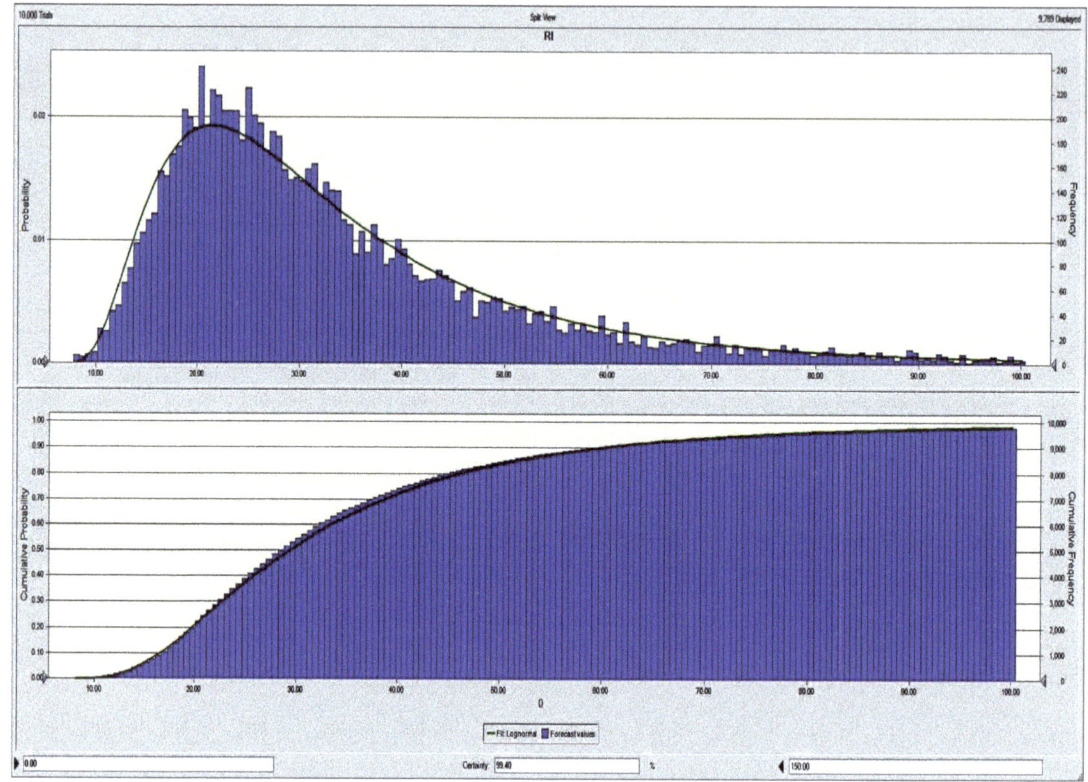

Figure 7. Probability and cumulative probability of Risk Index (RI) Value.

3.6. Principal Component Analysis

PCA was applied to determine the factor responsible for deteriorating the surface water quality. It signifies the association between components and variables. An eigenvalue greater than 1 was considered to define the components. As a result, two principal components were found whose eigenvalues were greater than 1. Figure 8 shows that the scree plot reaches a sharp decline after getting an eigenvalue of 1. PC1 has an eigenvalue of 4.40, and PC2 has 1.04. Moreover, Pb, Ni, and Cu were found to have higher PC1 values, respectively, compared to other parameters. On the other hand, Cd and Fe have lower PC1 values. Besides, PC2 dominated with a higher range of negative values. Fe, Mn, and Cu have negative PC2 values, whereas Cd has a higher positive PC2 value. However, PC1 and PC2 explain 63% and 78% cumulative variance, whereas these two components have 71% and 25% total variance, as per Figure 8. Scree plot of the metals shows Pb > Cu > Ni> Zn > Fe > Cd trend in terms of variance in Table 7.

Agglomerated hierarchical cluster analysis sorted sampling stations according to their magnitude. It clustered the sampling sites using the dendrogram approach. Four clusters were found to have identical characteristics each. Figure 9 elaborately depicts that S5, S9, S11, and S10 have different features within their cluster. S1, S8, S12, and S14 have comparatively lower values within the cluster. This indicates that the cluster of sampling stations had lower pollution. Cluster analysis implies the degree of pollution over the sampling stations.

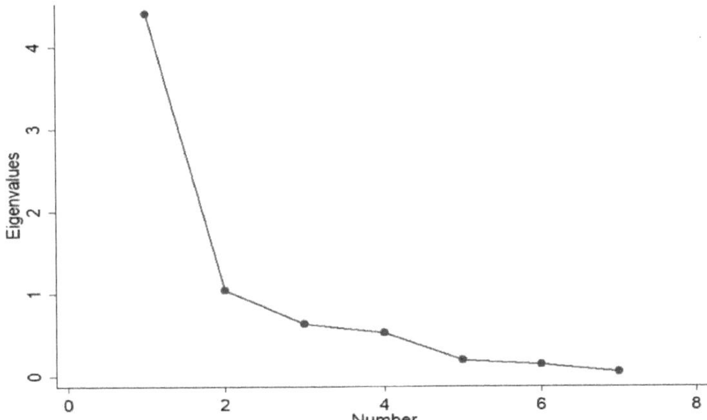

Figure 8. Scree plot of PCA.

Table 7. Principal Component Analysis.

Variables	PC1	PC2
Cu	0.43	−0.16
Zn	0.38	0.23
Fe	0.29	−0.52
Mn	0.37	−0.24
Pb	0.45	0.03
Cd	0.23	0.77
Ni	0.44	0.06
Eigenvalue	4.40	1.04
Cumulative variance (%)	63.00	78.00
Total variance (%)	71.00	25.00

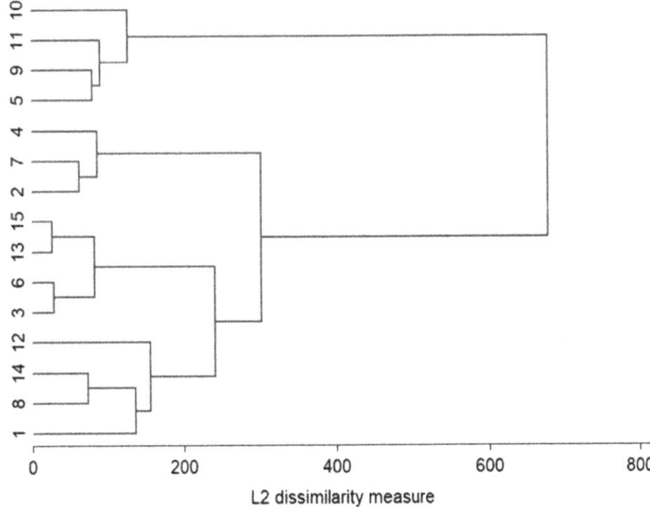

Figure 9. Cluster analysis.

4. Discussion and Conclusions

The study demonstrated the concentration of several heavy metals in the river bottom sediment, and systematically examined the ecological risk by employing PERI and Monte Carlo Simulation. However, the result of the present investigation indicates that the comprehensive ecological risk posited by heavy metals in the Surma River does not exceed the lowest limits of Hakanson's RI index for all heavy metals. This joint approach secures the lessening of the problems of underestimation and overestimation regarding the estimation of ecological risks. Analyses of heavy metal with Hakanson's RI index and Monte Carlo Simulation in the urban river sediment are very significant for the monitoring and management of river pollution in the developing world, as the urbanized and densely populated cities contribute a huge amount of domestic sewage directly discharged in the river [51]. The river sediment is reported to be marginally contaminated, and probably provides sustenance to the dependent flora and fauna without posing any ecological threat at present. However, grim reports from other similar studies [32,46] from rivers of Bangladesh provide a viewpoint from where River Surma is not far from degrading eventually. The study provides useful tools for future study combined with land use and land cover change, public health issues, and other ecological parameters, which would help decision-makers in the formulation of rules and guidelines about the sustainable management of domestic sewage disposal, and aid in minimizing negative impacts on riverine organisms and the environment. This study suggests that proper focus should be employed on monitoring the point sources of metals entering the river water from nearby cities, and also on the reduction of urban domestic sewage discharge and industrial effluent.

Supplementary Materials: The following are available online at the https://www.mdpi.com/article/10.3390/w14020180/s1. Table S1: Geographical Coordinates of Sampling Sites, Table S2: AAS conditions during analysis, Table S3: Metal Concentrations throughout the study area (units in mg/kg), Figure S1: Probability and Cumulative probability of Ecological Risk factor of Ni; Figure S2: Probability and Cumulative probability of Ecological Risk factor of Pb; Figure S3: Probability and Cumulative probability of Ecological Risk factor of Cd; Figure S4: Probability and Cumulative probability of Ecological Risk factor of Zn, Figure S5: Probability and Cumulative probability of Ecological Risk factor of Cu.

Author Contributions: Conceptualization—A.A., Z.A., J.S.-G.; methodology—A.A., Z.A., J.S.-G.; formal analysis—A.A., Z.A., J.S.-G.; writing—original draft preparation—A.A., Z.A., P.K., R.A., M.S.R., J.S.-G.; writing—review and editing—A.A., Z.A., P.K., R.A., M.S.R., J.S.-G.; funding acquisition—P.K. All authors have read and agreed to the published version of the manuscript.

Funding: The authors would be delighted to express gratitude to the authority of the Soil Resource Development Institute Laboratory (SRDI), Sylhet, Bangladesh, for provisioning expert suggestions and laboratory aptitude crucial to the fulfillment of this study. The authors are deeply ingratiated to express cordial gratitude to the staff members of SRDI Facility for the avid support and supervision throughout the sample analysis process. This publication is supported by the Asia Pacific Network for Global Change Research (APN) under Collaborative Regional Research Programme (CRRP) with project reference number CRRP2019-01MY-Kumar.

Data Availability Statement: Not applicable.

Conflicts of Interest: The authors declare no conflict of interest.

References

1. Silambarasan, K.; Senthilkumar, P.; Velmurugan, K. Studies on the distribution of heavy metal concentrations in River Adyar, Chennai, Tamil Nadu. *Eur. J. Exp. Biol.* **2012**, *2*, 2192–2198.
2. Barakat, A.; El Baghdadi, M.; Rais, J.; Nadem, S. Assessment of Heavy Metal in Surface Sediments of Day River at Beni-Mellal Region, Morocco. *Res. J. Environ. Earth Sci.* **2012**, *4*, 797–806.
3. Shazili, N.A.M.; Yunus, K.; Ahmad, A.S.; Abdullah, N.; Rashid, M.K.A. Heavy metal pollution status in the Malaysian aquatic environment. *Aquat. Ecosyst. Health Manag.* **2006**, *9*, 137–145. [CrossRef]
4. Islam, M.S.; Ahmed, M.K.; Raknuzzaman, M.; Habibullah-Al-Mamun, M.; Islam, M.K. Heavy metal pollution in surface water and sediment: A preliminary assessment of an urban river in a developing country. *Ecol. Indic.* **2015**, *48*, 282–291. [CrossRef]

5. Karaca, A.; Cetin, S.C.; Turgay, O.C.; Kizilkaya, R. Effects of Heavy Metals on Soil Enzyme Activities. In *Soil Heavy Metals*; Sherameti, I., Varma, A., Eds.; Springer: Berlin/Heidelberg, Germany, 2010; pp. 237–262.
6. Duruibe, J.; Egwurugwu, J. Heavy metal pollution and human biotoxic effects. *Int. J. Phys. Sci.* **2007**, *2*, 112–118.
7. Saleem, M.; Iqbal, J.; Shah, M.H. Geochemical speciation, anthropogenic contamination, risk assessment and source identification of selected metals in freshwater sediments—A case study from Mangla Lake, Pakistan. *Environ. Nanotechnol. Monit. Manag.* **2015**, *4*, 27–36. [CrossRef]
8. Moore, F. Assessment of heavy metal contamination in water and surface sediments of the Maharlu saline lake, SW Iran. *Iran. J. Sci. Technol. Trans. A* **2009**, *33*, A1.
9. Jaishankar, M.; Tseten, T.; Anbalagan, N.; Mathew, B.B.; Beeregowda, K.N. Toxicity, mechanism and health effects of some heavy metals. *Interdiscip. Toxicol.* **2014**, *7*, 60–72. [CrossRef] [PubMed]
10. Go'mez-Ariza, J.; Gira'ldez, I.; Sa'nchez-Rodas, D.; Morales, E. Metal readsorption and redistribution during the analytical fractionation of trace elements in toxic estuarine sediments. *Anal. Chim. Acta* **1999**, *399*, 295–307. [CrossRef]
11. Lasheen, M.; Ammar, N. Speciation of some heavy metals in River Nile sediments, Cairo, Egypt. *Environmentalist* **2009**, *29*, 8–16. [CrossRef]
12. Qu, C.; Li, B.; Wu, H.; Wang, S.; Li, F. Probabilistic ecological risk assessment of heavy metals in sediments from China's major aquatic bodies. *Stoch. Environ. Res. Risk Assess.* **2016**, *30*, 271–282. [CrossRef]
13. Duke, L.D.; Taggart, M. Uncertainty factors in screening ecological risk assessments. *Environ. Toxicol. Chem.* **2009**, *19*, 1668–1680. [CrossRef]
14. Chen, S.; Fath, B.D.; Chen, B. Information-based network environ analysis: A system perspective for ecological risk assessment. *Ecol. Indic.* **2011**, *11*, 1664–1672. [CrossRef]
15. Chen, S.; Chen, B.; Fath, B.D. Assessing the cumulative environmental impact of hydropower construction on river systems based on energy network model. *Renew. Sustain. Energy Rev.* **2015**, *42*, 78–92. [CrossRef]
16. Wang, D. Sustainable management of the future environment under uncertainties and risks. *Hum. Ecol. Risk Assess.* **2010**, *16*, 1249–1254. [CrossRef]
17. Yu, J.J.; Qin, X.S.; Larsen, O. Joint Monte Carlo and possibilistic simulation for flood damage assessment. *Stoch. Environ. Res. Risk Assess.* **2013**, *27*, 725–735. [CrossRef]
18. Bai, J.; Cui, B.; Chen, B.; Zhang, K.; Deng, W.; Gao, H.; Xiao, R. Spatial distribution and ecological risk assessment of heavy metals in surface sediments from a typical plateau lake wetland. *China Ecol. Model.* **2011**, *222*, 301–306. [CrossRef]
19. Su, L.; Liu, J.; Christensen, P. Spatial distribution and ecological risk assessment of metals in sediments of Baiyangdian wetland ecosystem. *Ecotoxicology* **2011**, *20*, 1107–1116. [CrossRef]
20. Sakib, A.N.; Rahman, A.; Iqbal, S.A.; Das, S.; Yousuf, A. Solid waste management of Sylhet city in terms of energy. In Proceedings of the International Conference on Mechanical Engineering and Renewable Energy 2011, Chittagong, Bangladesh, 22–24 December 2011.
21. Larsen, W.E.; Gilley, J.R.; Linden, D.R. Consequences of Waste Disposal on Land. *Soil Water Conserv.* **1975**, *2*, 68.
22. Ahmed, M.F. *Municipal Waste Management in Bangladesh with Emphasis on Recycling, Aspect of Solid Waste Management Bangladesh Context*; Mofizul Hoq, L., Ed.; German Cultural Institute: Dhaka, Bangladesh, 1994; p. 113.
23. Rashid, H.E. *Geography of Bangladesh*; The University Press Limited (UPL): Dhaka, Bangladesh, 1977.
24. Manoj, K.; Kumar, B.; Padhy, P.K. Characterization of Metals in Water and Sediments of Subarnarekha River along the Projects' Sites in Lower Basin. *Univ. J. Environ. Res. Technol.* **2012**, *2*, 402–410.
25. Rabee, A.M.; Al-Fatlawy, Y.F.; Abd, A.A.H.N.; Nameer, M. Using Pollution Load Index (PLI) and Geoaccumulation Index (I-Geo) for the Assessment of Heavy Metals Pollution in Tigris River Sediment in Baghdad Region. *J. Al-Nahrain Univ. Sci.* **2011**, *14*, 108–114. [CrossRef]
26. Petersen, L. *Analytical Methods: Soil, Water, Plant Material, Fertilizer*; Soil Resource Development Institute: Dhaka, Bangladesh, 2002.
27. Martin, J.M.; Meybeck, M. Elemental mass-balance of material carried by major world rivers. *Mar. Chem.* **1979**, *7*, 173–206. [CrossRef]
28. Herut, B.; Hornung, H.; Krom, M.D.; Kress, N.; Cohen, Y. Trace metals in shallow sediments from the Mediterranean coastal region of Israel. *Mar. Pollut. Bull.* **1993**, *26*, 675–682. [CrossRef]
29. Kassim, T.; Al-Saadi, H.; Al-Lami, A.; Al-Jaberi, H. Heavy Metals in Water, Suspended Particles, Sediments and Aquatic Plants of the Upper Region of Euphrates River, Iraq. *J. Environ. Sci. Health* **1997**, *32*, 2497–2506. [CrossRef]
30. Al-Juboury, A. Natural Pollution by Some Heavy Metals in the Tigris River, Northern Iraq. *Int. J. Environ. Res.* **2009**, *31*, 189–198.
31. Raju, K.V.; Somashekar, R.; Prakash, K. Heavy Metal Status of Sediment in River Cauvery, Karnataka. *Environ. Monit. Assess.* **2012**, *184*, 361–373. [CrossRef]
32. Rahman, M.S.; Saha, N.; Molla, A.H.; Al-Reza, S.M. Assessment of Anthropogenic Influence on Heavy Metals Contamination in the Aquatic Ecosystem Components: Water, Sediment, and Fish. *Soil Sediment Contam. Int. J.* **2014**, *23*, 353–373. [CrossRef]
33. Wang, Y.; Yang, Z.; Shen, Z.; Tang, Z.; Niu, J.; Gao, F. Assessment of Heavy Metals in Sediments from a Typical Catchment of the Yangtze River, China. *Environ. Monit. Assess.* **2011**, *172*, 407–417. [CrossRef]
34. Christophoridis, C.; Dedepsidis, D.; Fytianos, K. Occurrence and distribution of selected heavy metals in the surface sediments of Thermaikos Gulf, N. Greece. Assessment using pollution indicators. *J. Hazard. Mater.* **2009**, *168*, 1082–1091. [CrossRef]

35. Turekian, K.K.; Wedepohl, K.H. Distribution of the elements in some major units of the earth's crust. *Geol. Soc. Am. Bull.* **1961**, *72*, 175–192. [CrossRef]
36. Salah, E.A.M.; Zaidan, T.A.; Al-Rawi, A.S. Assessment of Heavy Metals Pollution in the Sediments of Euphrates River, Iraq. *J. Water Resour. Prot.* **2012**, *4*, 1009–1023. [CrossRef]
37. Varol, M. Assessment of heavy metal contamination in sediments of the Tigris River (Turkey) using pollution indices and multivariate statistical techniques. *J. Hazard. Mater.* **2011**, *195*, 355–364. [CrossRef]
38. Hakanson, L. An ecological risk index for aquatic pollution control. A sedimentological approach. *Water Res.* **1980**, *14*, 975–1001. [CrossRef]
39. Abrahim, G.M.S.; Parker, R.J. Assessment of heavy metal enrichment factors and the degree of contamination in marine sediments from Tamaki Estuary, Auckland, New Zealand. *Environ. Monit. Assess.* **2008**, *136*, 227–238. [CrossRef]
40. Feng, H.; Han, X.; Zhang, W.; Yu, L. A preliminary study of heavy metal contamination in Yangtze River intertidal zone due to urbanization. *Mar. Pollut. Bull.* **2004**, *49*, 910–915. [CrossRef]
41. Sinex, S.A.; Helz, G.R. Regional geochemistry of trace elements in Chesapeake Bay sediments. *Environ. Geol.* **1981**, *3*, 315–323. [CrossRef]
42. Tippie, V.K. An Environmental Characterization of Chesa-Peak Bay and a Framework for Action. In *The Estuary as a Filter*; Kennedy, V., Ed.; Academic Press: New York, NY, USA, 1984; pp. 467–487.
43. Tomlinson, D.L.; Wilson, J.G.; Harris, C.R.; Jeffrey, D.W. Problems in the assessment of heavy-metal levels in estuaries and the formation of a pollution index. *Helgol. Meeresunters.* **1980**, *33*, 566–575. [CrossRef]
44. Muller, G. The Heavy Metal Pollution of the Sediments of Neckers and Its tributary, A Stocktaking. *Chem. Zeit.* **1981**, *150*, 157–164.
45. Bhuiyan, M.A.H.; Parvez, L.; Islam, M.A.; Dampare, S.B.; Suzuki, S. Heavy metal pollution of coal mine-affected agricultural soils in the northern part of Bangladesh. *J. Hazard. Mater.* **2010**, *173*, 384–392. [CrossRef] [PubMed]
46. Bhuyan, M.S.; Bakar, M.A.; Akhtar, A.; Hossain, M.B.; Ali, M.M.; Islam, M.S. Heavy metal contamination in surface water and sediment of the Meghna River, Bangladesh. *Environ. Nanotechnol. Monit. Manag.* **2017**, *8*, 273–279. [CrossRef]
47. Suresh, G.; Ramasamy, V.; Meenakshisundaram, V.; Venkatachalapathy, R.; Ponnusamy, V. Influence of mineralogical and heavy metal composition on natural radionuclide concentrations in the river sediments. *Appl. Radiat. Isot.* **2011**, *69*, 1466–1474. [CrossRef] [PubMed]
48. Hakanson, L. Metal monitoring in coastal environments. *Met. Coast. Environ. Lat. Am.* **1988**, 239–257. [CrossRef]
49. Devanesan, E.; Gandhi, M.S.; Selvapandiyan, M.; Senthilkumar, G.; Ravisankar, R. Heavy metal and Potential Ecological Risk Assessment in sedimentscollected from Poombuhar to Karaikal Coast of Tamilnadu using Energy dispersive X-ray fluorescence (EDXRF) technique. *Beni-Suef Univ. J. Basic Appl. Sci.* **2017**, *6*, 285–292. [CrossRef]
50. Li, X.; Chi, W.; Tian, H.; Zhang, Y.; Zhu, Z. Probabilistic ecological risk assessment of heavy metals in western Laizhou Bay, Shandong Province, China. *PLoS ONE* **2019**, *14*, e0213011. [CrossRef] [PubMed]
51. Withanachchi, S.S.; Ghambashidze, G.; Kunchulia, I.; Urushadze, T.; Ploeger, A. Water Quality in Surface Water: A Preliminary Assessment of Heavy Metal Contamination of the Mashavera River, Georgia. *Int. J. Environ. Res. Public Health* **2018**, *15*, 621. [CrossRef]

 water

MDPI

Article

Microplastics in Freshwater Environment in Asia: A Systematic Scientific Review

Pankaj Kumar [1,*], Yukako Inamura [1,*], Pham Ngoc Bao [1], Amila Abeynayaka [1], Rajarshi Dasgupta [1] and Helayaye D. L. Abeynayaka [2]

[1] Institute for Global Environmental Strategies, Hayama 240-0115, Kanagawa, Japan; ngoc-bao@iges.or.jp (P.N.B.); abeynayaka@iges.or.jp (A.A.); dasgupta@iges.or.jp (R.D.)

[2] Department of Natural Science, Faculty of Education, Saitama University, 255 Shimookubo, Sakura Ward 338-8570, Saitama, Japan; hdlakmali@mail.saitama-u.ac.jp

* Correspondence: kumar@iges.or.jp (P.K.); y-inamura@iges.or.jp (Y.I.)

Citation: Kumar, P.; Inamura, Y.; Bao, P.N.; Abeynayaka, A.; Dasgupta, R.; Abeynayaka, H.D.L. Microplastics in Freshwater Environment in Asia: A Systematic Scientific Review. *Water* **2022**, *14*, 1737. https://doi.org/10.3390/w14111737

Academic Editors: Judith S. Weis and Grzegorz Nałęcz-Jawecki

Received: 15 May 2022
Accepted: 26 May 2022
Published: 28 May 2022

Abstract: Microplastics (MPs) are an emerging pollutant in the aquatic environment, and this has gradually been recognized in the Asian region. This systematic review study, using the Scopus database, provides an insightful understanding of the spatial distribution of scientific studies on MPs in freshwater conducted across the Asian region, utilized sampling methods, and a detailed assessment of the effects of MPs on different biotic components in freshwater ecosystems, with special focus on its potential risks on human health. The results of this review indicate that research on microplastics in Asia has gained attention since 2014, with a significant increase in the number of studies in 2018, and the number of scientific studies quadrupled in 2021 compared to 2018. Results indicated that despite a significant amount of research has been conducted in many Asian countries, they were not distributed evenly, as multiple studies selected specific rivers and lakes. Additionally, around two-thirds of all the papers focused their studies in China, followed by India and South Korea. It was also found that most of the studies focused primarily on reporting the occurrence levels of MPs in freshwater systems, such as water and sediments, and aquatic organisms, with a lack of studies investigating the human intake of MPs and their potential risks to human health. Notably, comparing the results is a challenge because diverse sampling, separation, and identification methods were applied to estimate MPs. This review study suggests that further research on the dynamics and transport of microplastics in biota and humans is needed, as Asia is a major consumer of seafood products and contributes significantly to the generation of plastic litter in the marine environment. Moreover, this review study revealed that only a few studies extended their discussions to policies and governance aspects of MPs. This implies the need for further research on policy and governance frameworks to address this emerging water pollutant more holistically.

Keywords: microplastics; freshwater; human health; Asia; systematic review

1. Introduction

Among various emerging pollutants, plastic is of high concern as its contamination poses a serious threat to different components of the environment as well as human well-being [1]. Although the first commercial synthetic polymer, "phenol-formaldehyde resin," also known as Bakelite, was developed by Leo Baekeland in 1907 [2], the wider use of commercial plastic in all the sectors viz. textile, packaging, personal care products, etc. around the world started in the 1950s [3]. People prefer using plastic products because of their durability, low conductivity, low corroding properties, etc. [4]. It is reported that the global production of plastic for the year 2019 was 368 million metric tons, and only 20% of it was recycled or burned properly, whereas the rest, about 80% of it, ended up either in landfills or was dumped in water bodies [5]. These large untreated pieces of plastic typically go through different decomposition pathways, and over time, small particles are formed with diameters of less than 5 mm, also called microplastic [6]. Because of their very small size, they can travel undetected through different parts of the Earth, for example, from the soil to the water and atmosphere [7]. Additionally, because of their slow degradation rates, microplastics can be present in nature for a long period (ranging between 20 to 500 years) and cause severe environmental pollution [8].

Assessing the detrimental effects of MPs on the ecosystem is quite challenging because they include a variety of physical (size, shape, colors, etc.) and chemical (polymer, adhesives, other chemicals, etc.) compounds, which regulate their fate, transport, and bioaccumulation in different ecosystems [9,10]. For instance, depending on the density, they can either float and interact with pelagic organisms on the top layer of the water surface (low-density MPs) or sink in water bodies and interact with benthic organisms (high-density MPs) [11]. Similarly, it is reported that MPs with strident ends (for instance, fibers) might have more harmful impacts than MPs with blunt edges (such as spherical ones) because they can severely injure the digestive system or other body parts upon digestion [6]. Additionally, regarding the impacts of MPs on different organisms, most studies give a snapshot of the effect of MPs on any particular species for any specific biological functions like accumulation, mortality, reproduction, etc.; however, very few or almost none of the studies discuss how MPs affect different key ecological interactions and functions at different trophic levels [12]. Similarly, very little is known about how directly or indirectly MPs affect or will affect human health [13,14].

In 2015, the United Nations and its associated members univocally recognized the different actions needed to achieve Sustainable Development Goals (SDGs). Among these, one goal is to assess emerging environmental pollutants such as plastic pollution, their environmental impacts, and different management options both in terms of adaptation and mitigation. Addressing plastic pollution will help to expedite our efforts to achieve various SDG goals, namely SDG 12 (Responsible Consumption and Production), SDG 14 (Life Below Water), SDG 15 (Life On Land), etc. Therefore, both the scientific community and policymakers are confronting this issue on an urgent basis, and many efforts are being directed to address this critical issue.

A past study found that Asian countries, particularly China, Indonesia, the Philippines, Thailand, and Vietnam, contributed to about half of the world's marine litter generation [15]. There have been a few review studies conducted in the Asian region, but they have focused on individual countries. Unfortunately, a comprehensive and comparative study focusing on several countries in this region has not been found. Considering the general pathways of plastic wastes to the ocean, the freshwater system is a critical part of the entire plastic problem because of its close connection to human life. Therefore, our review study emphasizes MPs in the freshwater system in Asia. Furthermore, few systematic review studies have investigated the human health impacts of MPs; therefore, our research also explored what studies have been carried out on MPs and their effects on human health.

With basic background information obtained through this exercise, a detailed analysis of research articles was carried out to achieve the following objectives: (a) to examine the spatial distribution of scientific studies on MPs in freshwater in Asia; (b) to identify the sampling methods of MPs for laboratory analyses (different size, color, shapes, their associated materials such as adhesives/heavy metals, etc.) in soils/sediment/water, for source identification; (c) to evaluate the effects of MPs on different organisms in the freshwater ecosystem and their ultimate impacts on human health as an end-user; (d) to understand the needs for policy improvements at institutional and governance levels in order to tackle this emerging environmental pollutant in a more holistic manner. In the methodology section, we provide information on how the literature database was constructed for this review, and the first part of the results section presents the summary of these reviewed articles. The results of the systematic review analysis are presented in the second half of the section. Here, the following items are analyzed and presented: spatio-temporal variation of research works on MPs, their target journals, various samples from the freshwater environmental system being analyzed, morphological and chemical features, aquatic organisms being analyzed, and methodological techniques being employed for the analysis of MPs, to report key findings related to MPs in the freshwater environment in Asia. Then, we discuss the gaps in the current research related to MPs in riverine systems and the way forward for future research activities to understand the consequences of MP pollution and ameliorate the situation.

2. Methodology

A systematic literature review was conducted using the Scopus database (http://www.scopus.com/ Accessed on 18 October 2021) to collect existing literature related to microplastic pollution in the freshwater environment. This means that our search was limited to references published before 18 October 2021. A few articles published in 2022 were included in our literature database because they became available online before we retrieved the data from SCOPUS. For this study, freshwater bodies included rivers, ponds, reservoirs, and wetlands. Preferred Reporting Items for Systematic Reviews and Meta-Analyses (PRISMA) guidelines were followed to document the literature review process [16]. As the search query, the Boolean string was used: TITLE-ABS-KEY (Microplastic) AND (TITLE-ABS-KEY (River) OR TITLE-ABS-KEY (Riverine) OR TITLE-ABS-KEY (Freshwater). The search did not limit the language; however, non-English articles were omitted at the review stage later. With the above search query, a total of 1335 research articles were retrieved. Next, the full texts of "article" and "review" retrieved from this search were downloaded and manually screened for peer-reviewed articles, and the screened number of articles was reduced to 1093. Again, the second round of screening was performed for the following purposes: (i) screening of review papers dealing with microplastics (MPs) in freshwater around the globe to get an overall idea about the scientific progress made and remaining gaps in this area, (ii) screening of research articles dealing with MPs in freshwater in Asian regions. As a result, the number of review papers and research articles retrieved is 83 and 166, respectively (see Tables S1 and S2 in the Supplementary File). Supplementary Table S1 is giving the information about 83 review papers being assesses in this manuscript to gather the basic information about the MPs in freshwater system around the world and the knowledge gap especially in Asia. Table S2 mainly depicts the list of 166 research papers being analyzed and their assessment result presented in this manuscript. The methodology adopted for this work is shown in the flowchart in Figure 1.

First, we analyzed above mentioned 83 accessible review articles to investigate the important research highlights or updates on microplastics and knowledge gaps around the world. The result showed that all 83 review studies were published between 2015 and 2022, and 59 were published between 2020 and 2021, showing a sharp spike in the last few years. Among them, 63 articles reviewed studies without spatial consideration, while 4 articles focused on Europe, 3 articles on Africa, and 4 articles on Latin and North America. There were nine articles that reviewed studies in Asia, out of which four focused on China, two on India, and one each on Indonesia, Iran, and Malaysia. All these review studies had different objectives and perspectives. Looking at the target water environment, 35 articles reviewed articles on MPs in all types of water such as marine, lake, reservoir, and river; 25 articles reviewed articles focusing on the freshwater system, such as a lake, reservoir, and river; and 3 articles focused on the marine system. The other 21 articles emphasized the removal of MPs from wastewater treatment plants, groundwater, and laboratory analysis, rather than looking at MP problems in specific water environments. Many of the studies analyzed MPs in water, sediment, and aquatic organisms and examined their effects on ecosystems. However, there were 6 papers that reviewed MP studies from the perspective of human health and 36 papers that reviewed MP studies from the perspective of their effects on ecosystems. Morphological analysis was one of the major objectives in the MP studies, which is well supported by 44 review articles. Overall, it was found that a review work presenting a holistic approach to depict the current status quo of MPs, particularly in the freshwater environment and its impact on different ecosystems, is still lacking. With the aforementioned gap, this review work was carried out to achieve our aforementioned objectives.

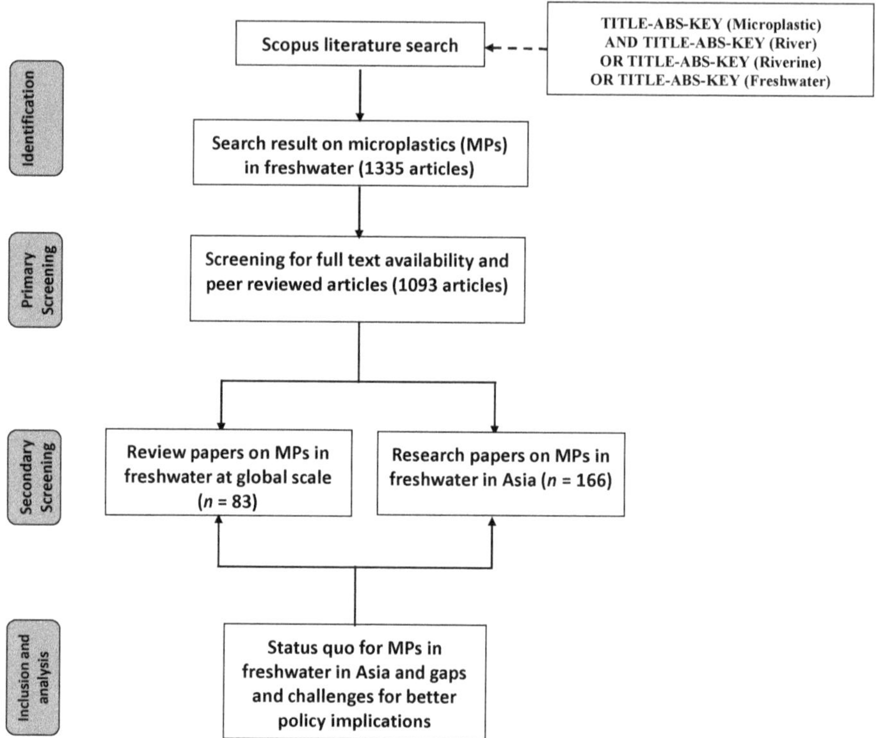

Figure 1. PRISMA flowchart of literature review work.

Moreover, these review articles bring out different lines of information based on the objectives mentioned. For example, Wang et al. [17] reviewed MP studies conducted in different parts of the world; they compared the sampling, processing, and identification methods, presented characteristics of MPs such as concentration and morphologies, and explored the sources, paths, and impacts in 53 articles. Xu et al. [18] also reviewed MP studies focusing on the source and morphologies. Similarly, Koutnik et al. [19] conducted a systematic review of 196 studies, and their review extended to finding the MP transport modeling frameworks in the literature. Gao et al. [20] reviewed 32 studies and summarized MPs found in freshwater and marine algae. Bellasi et al. [21] provided an overview of MP pollution as well as ecotoxicology. The review studies define the ongoing research and highlight crucial aspects and gaps. Many review studies reviewed laboratory analysis and the characteristics of MPs and discussed the impacts of MPs in introductions or discussions. Fewer studies extended their reviews to the impacts on ecosystems, and a few studies extended their reviews to the human health impacts.

Based on this mentioned exercise, the knowledge or information gap was identified about microplastics in the freshwater system, especially in the Asian region. Thereafter, a review of 166 research articles was conducted, and the findings are presented in the next sections, i.e., Results and Discussions.

3. Results

3.1. Spatio-Temporal Distribution of Scientific Literature

The temporal distribution of research articles is shown in Figure 2. Among all retrieved articles on MPs in the freshwater environment, the oldest one was from 2014, while the latest one is from 2022. The number of publications per year suddenly increased by many folds from 2018. This trend continues moving upwards, clearly showing that it is one of the emerging environmental pollutants. In 2021 alone, 54 articles focused their investigation on MPs in the freshwater environment.

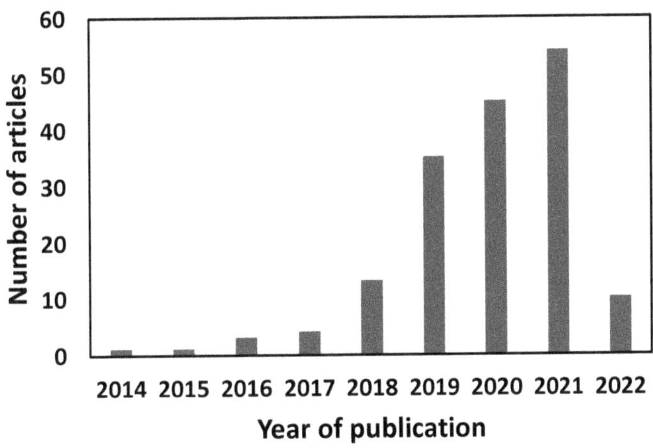

Figure 2. Yearly distribution of research articles on MPs in freshwater environments.

The spatial distribution of research work on MPs in freshwater at the country level is shown in Figures 3 and 4. The reviewed studies generally select a water system and collect samples from multiple locations in the same system. The points in Figure 3 do not indicate all sampling points, but they indicate the centers of sampling points in each study. Thus, one point represents one study. The detailed information for the GPS locations of all sampling sites is provided in Table S3 as a Supplementary File. In other words, Table S3 give the detailed information about various geographical locations from where samples being collected in total 166 research papers considered for the analysis in this manuscript. The light blue points indicate the studies that assessed MPs in multiple water systems such as rivers and lakes. Since MPs have been recognized as an emerging pollutant affecting the water environment, the MP studies were conducted in various countries, from Turkey in the west to Japan in the east of Asia. Indeed, out of 166 papers, research work was spatially distributed among 18 different countries. Also, two articles focus on two countries, for example, China and Nepal [22] and India and Bangladesh [23]. While the objective of many articles was a field-based assessment of the water environment, there were a few studies that conducted experiments in the laboratory, mainly in China but also in Bangladesh and Indonesia. It was found that 68.1% of the reviewed works are focused on China, followed by India and South Korea with 5.4% and 4.2%, respectively. There is a huge gap between China and the rest of the countries in Asia regarding the research findings on MPs in freshwater. Overlapping points in Figure 3 imply that some water bodies attract more scientific attention, such as the Pearl River, Yangtze River, etc., in China.

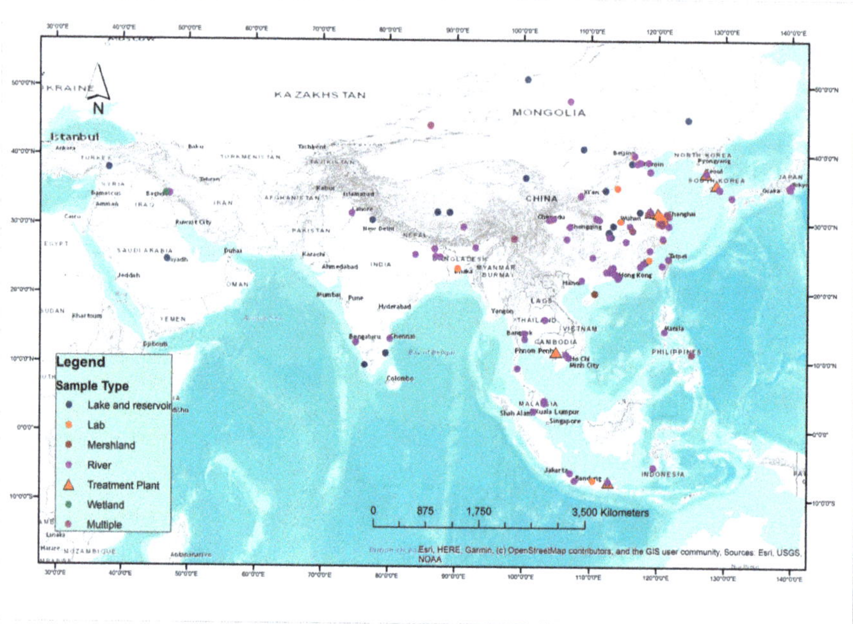

Figure 3. Map of study locations of reviewed articles.

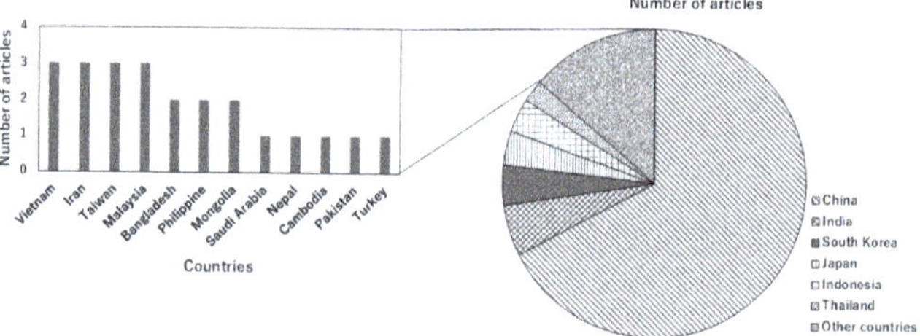

Figure 4. Spatial distribution of research articles on MPs.

Extensive scientific investigation has been performed on some key rivers/lakes, such as the Yangtze River (15 papers), Pearl River (21 papers), Haihe River (8 papers), and Taihu Lake (7 papers). For India, 9 research papers were found that focused on rivers or lakes in the high-altitude areas of Brahmaputra (2 papers), Ganga (2 papers), and Renuka Lake (1 paper). Other papers focused on rivers/lakes from coastal Southern India in areas such as the Netravathi River, Vembanada Lake, and Veeranam Lake, with one case study from each. For South Korea, seven papers were found that focused on urban rivers, such as the Han River and Nakdong River. Additionally, some papers focused on the role of wastewater treatment plants in the fate and transport of MPs in urban water bodies. For Japan, a total of six papers were found that focused on the Awano, Tsurumi, and Ayaragi Rivers, with one case each. For Indonesia, the target rivers were the Tallo River, Surabaya River, and Ciwalengke River, with one paper each. For Thailand, the main river bodies explored were the Tapi-Phumduang River system (one paper), Chi River (two papers), and Chao Phraya River (one paper).

3.2. Target Journals and Objectives of Research Papers

Academic journals publish diversified research articles taking a transdisciplinary approach, and the 166 articles on freshwater MP were published in 34 different journals. Figure 5 shows the major discipline of the journals where the 166 papers were published. Although environmental pollution could be considered part of environmental science, it is shown separately in the graph because a significant number of articles found in the journals were mainly concerned with environmental pollution. We found that two-thirds of the articles were published in journals that are strongly linked to environmental science. Furthermore, this shows that research works about microplastics are getting attention from across disciplines such as chemistry, biology, environment, sustainable resource management, disaster risk reduction, climate change, etc. Hence, there is a high demand for up-to-date information on the status of microplastics in the environment on a regional or local basis. Such need/demand makes this review work even more crucial at the present time. We further classified all the research articles based on the broad categories of the objective for the research work on MPs, and the result is shown in Figure 6. A total of seven broad categories were found. It was found that more than half, 52.4%, of the research works were carried out with the objective of reporting the concentration of MPs in freshwater (river, lake, reservoir, pond, etc.). The reason behind this is very clear. Because this is a very new research topic, no past studies and data are available, so in most cases, these are a kind of baseline study reporting the concentration of MPs for the first time. The second largest group contained 23.5% of articles that assessed the exposure of MPs in freshwater in different aquatic animals like fish, mollusks, clams, planktons, crustaceans, bacteria, amphibians, etc. In the third major category, with 10.2% articles, the objective was to assess the concentration of MPs in the sediment/soil/sludge in the freshwater environment. In addition, 5% of articles reported assessments of MPs in both water and sediment, and 4% of articles focused on the interaction of MPs with other chemicals and associates. Next were papers focusing on the removal efficiency of various treatment plants (domestic treatment plants, wastewater treatment plants) for MPs and their impact on the freshwater environment. The number of papers focused on exposure analysis to aquatic plants and human health was less than 2% of articles. This indicates that scientific information on how MPs impact human health is still in the incipient stage, especially in the Asian region. Both of these papers first reported the exposure to aquatic animals and then extended their evaluation to human beings. Hence, to prepare a robust management plan, more efforts are needed to obtain scientific evidence on MPs' status, their fate, and transport in different environments or ecosystems.

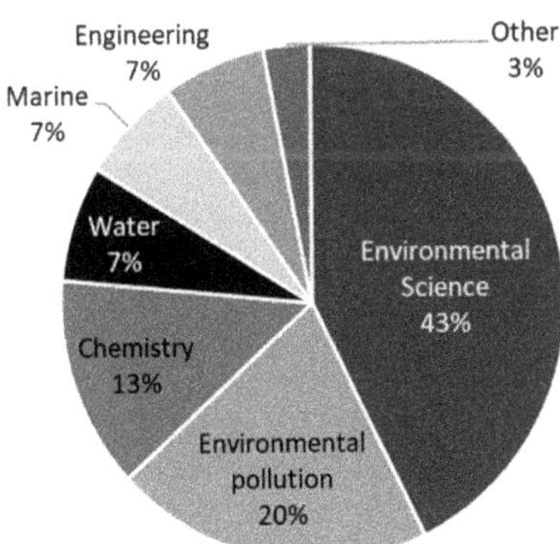

Figure 5. Main discipline of journals in which selected papers were found.

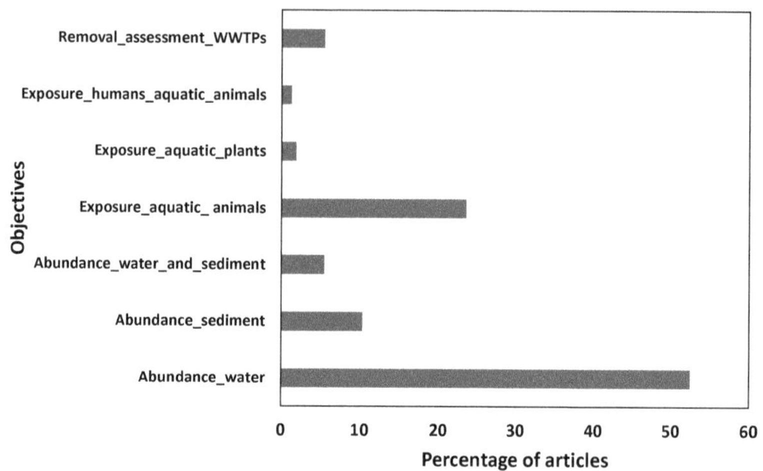

Figure 6. Classification of research articles based on the broad categories of the objective for the research work on MPs.

For this review study, we retrieved articles on MPs in the freshwater environment. Of these, 83% of selected articles assessed MPs in rivers, lakes, and reservoirs, or both water systems. The studies assessing MPs provided detailed pictures of MP pollution in the study area by estimating the concentration and analyzing the morphology. In Asia, MPs in the freshwater system are intensively studied in China, as supported by Figures 3 and 4, and MPs in lakes and reservoirs gained more attention there than in other countries. Figure 7 shows that 72% of studies analyzing lakes and reservoirs came from China. While the major focus was on the assessment of MPs in the natural environment, 13% of the reviewed articles performed exposure analysis, carried out experiments in the laboratory to understand the mechanisms of MP absorption by organisms, or explored MP removal mechanisms by aquatic organisms [24,25].

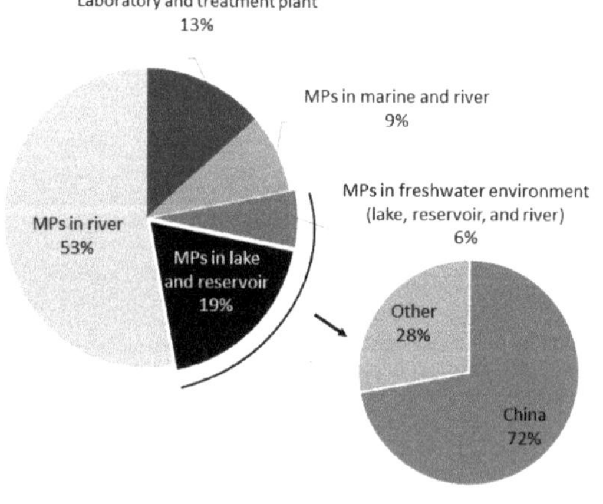

Figure 7. Water systems focused on by the selected studies.

The reviewed articles had diversified objectives; hence, different types of analyses were conducted. The studies assessing MPs in the freshwater system generally sampled water, sediment, and sometimes organisms and analyzed the presence of MPs in the samples.

Among 166 articles, half aimed to assess the abundance of MPs, as Figure 6 shows, and 70% of them analyzed MPs in either water, sediment, or both, as seen in Figure 8. In addition to sampling water for the assessment, some studies collected water to use in a laboratory experiment [26–28]. Studies focusing on MPs in sediment investigated if the received water played a role as a pathway to other waterbodies or a final destination, such as the sink [29], as well as if the protected area would make any differences in MP occurrence [30,31]. Additionally, 19% of the studies extended the assessment to aquatic organisms by sampling plants, fishes, and other species. A few reviewed articles did not take any samples from the natural environment because they conducted experiments by purchasing what was needed for the experiment and artificially creating the environment in the laboratory [32–34].

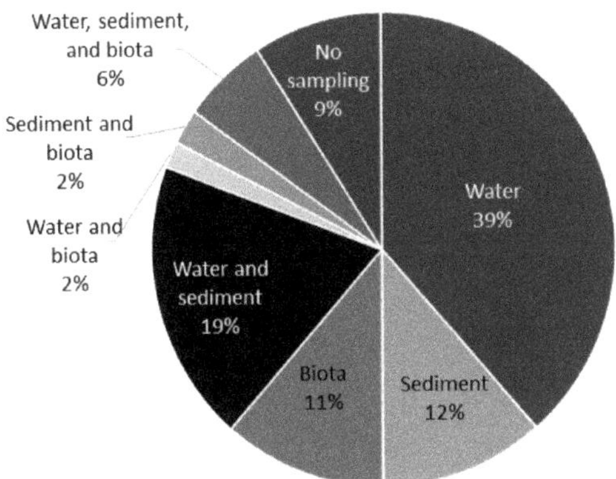

Figure 8. Sample categories investigated in the studies.

3.3. *Different Key Features of the Reviewed Papers*

3.3.1. MP Shapes, Polymer Types, and Color

The articles that assessed MPs conducted morphology analysis and identified shape, color, and polymer types. Figure 9 shows the shapes of MPs found in the studies. Fiber was found the most in the studies, as 95% of assessment studies found it, followed by fragments (86%) and film (74%). While the dominant shape was fiber in some studies [35–38], other shapes were dominant in a few studies [39–41]. The major sources of fiber were found to be the textile industry, households, and wastewater treatment plants [ibid.]. Polypropylene, polyethylene, and polystyrene were identified in many studies and were dominant in some studies [42–45].

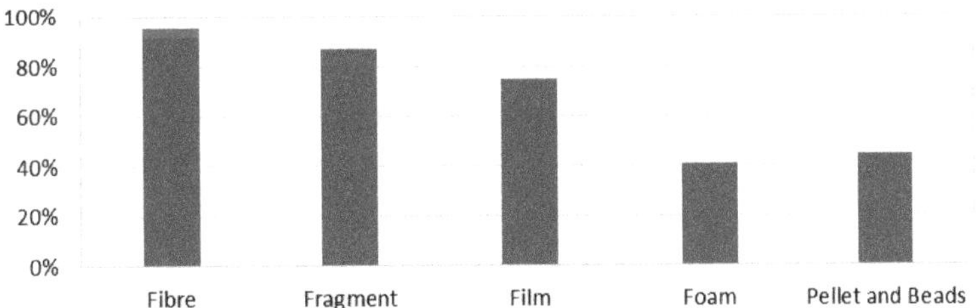

Figure 9. Shape of MP identified in studies.

The other important attribute when assessing the occurrence of MPs in any environment is their variety of colors. Color is often identified in the morphology analysis. More than half (84 of 166 articles) of the studies analyzed the color of the MPs found in their analyzed samples, and 48% of them differentiated 6 colors or more, as shown in Figure 10.

Typical colors identified were transparent, white, red, black, blue, green, and yellow. Some studies identified brown, grey, pink, purple, or violet [46–50]. Dominant colors differed from study to study, such as white and transparent [24,51] or black [36], depending on the location of the study conducted.

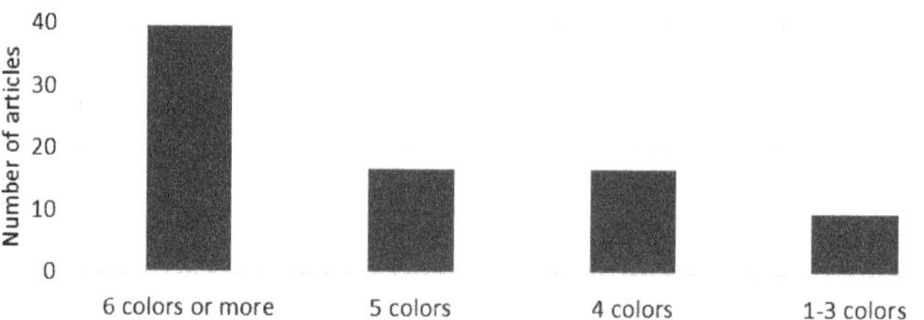

Figure 10. MP colors identified in the analysis.

3.3.2. Target Organisms

Further, target organisms in the reviewed papers were explored, and the summary is presented in Figure 11. It was found that some studies assessed the impacts of MPs on aquatic organisms by capturing them in the field for the MP assessment, and others assessed the impacts by exposing them to MPs in the laboratory environment. China conducted a significant number of studies on the impacts of MP on biota; 66% of 47 studies that analyzed the impacts on biota were found, followed by South Korea at 6%, Bangladesh, Indonesia, Iran, Taiwan, and Thailand at 4%. Of these, 88% of the studies analyzing biota investigated the impacts on fauna, and 6% flora and bacteria. Freshwater fishes were the most investigated organisms, followed by amphibians. The impacts on flora were all investigated in China [52–57], and the impacts on bacteria were investigated in China and Indonesia [33,45,58].

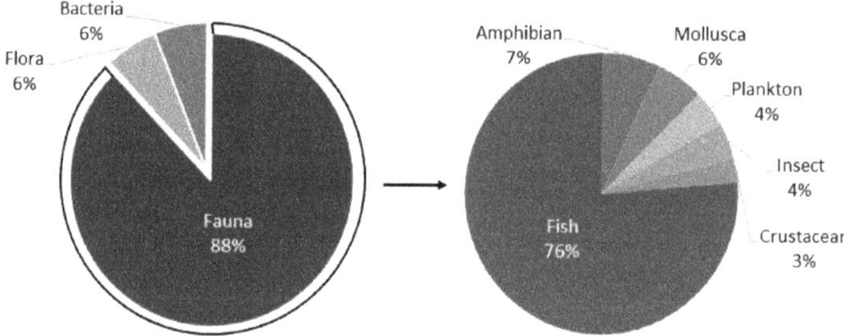

Figure 11. Aquatic organisms analyzed in the studies.

3.3.3. Methodological Techniques

To understand how the literature investigates MPs differently, we compared the following four steps and items: (a) sample collection, (b) density separation, (c) organic matter digestion, and (d) identification. The findings are shown in Figure 12a–d, respectively. Here, both organic matter digestion and density separation can also be understood as extraction steps. The first step for the MPs' investigation/study of the environmental components is sample collection, and the result is shown in Figure 12a. All the processes for sample collection are divided into four categories, and the order for them is grab > net > pump > hand. Out of 108 papers that reported sample collection techniques, 55 opted for the grab sampling methodology. This is followed by net, pump, and hand, represented in 32, 20, and 1 paper, respectively. Figure 12b shows the result for the summary of chemical treatments authors have used for density separation to obtain MPs of different shapes/ sizes. It was found that low-density treatment using salts such as NaCl is a very common practice, as shown in 49 out of a total 94 articles which mentioned it. This is followed by high-density treatment using salts such as $ZnCl_2$ to separate different MPs species, as found in 29 out of 94 articles. Figure 12c shows the methodology to remove organic matter from the sample being collected. The most common method is to treat the sample using H_2O_2, as found in 93 out of 111 papers that mentioned it. This was followed by other methods such as using H_2O_2 plus Iron (Fe) salts and using other chemicals.

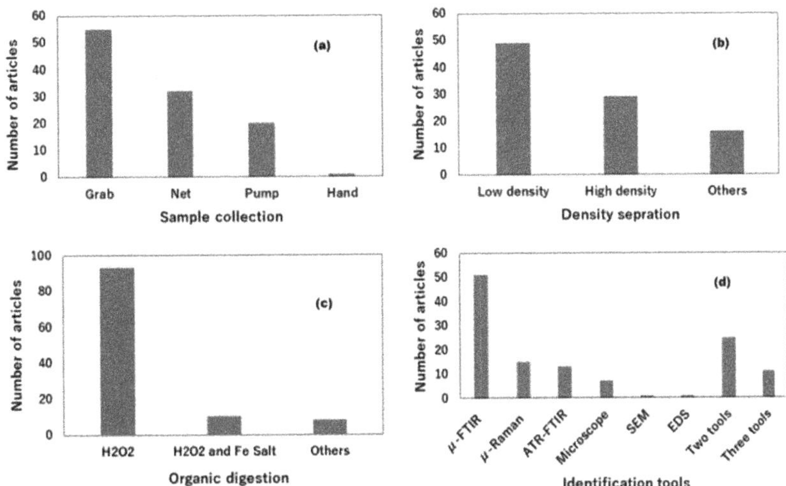

Figure 12. Summary for different steps of methodology (**a**) Sample collection, (**b**) Density separation, (**c**) Digestion of organic matter, and (**d**) Identification tools used to study MPs in the reviewed research articles.

Figure 12d describes different technologies or tools used to identify MPs in the reviewed research work, as shown in Figure 10. A total of 124 articles out of 166 were identified where authors mentioned any type of tools used to identify MPs in their studies. The results show that six different standalone technologies, namely Micro Fourier transform infrared (μ-FTIR) spectroscopy, Attenuated Total Reflection-Fourier transform infrared (ATR-FTIR) spectroscopy, μ-Raman imaging microscope, Microscope, Energy-dispersive X-ray spectroscopy (EDS), and Scanning electron microscope (SEM), were used to identify the MPs in the reviewed research works. In addition, many articles used multiple (two or three) technologies to identify MPs. The most common technology used was μ-FTIR spectrophotometer, with 51 articles. The order of frequency for different technologies employed in the reviewed research works were in the order of μ-FTIR > Two tools > μ-Raman > ATR-FTIR > Three tools > Microscope > SEM > EDS. Among two tools, different combinations were used, such as μ-FTIR and SEM, μ-Raman and stereomicroscope, chromatography and stereomicroscope, fluorescence microscopy and SEM, etc. On the other hand, among the three tools, the most common combinations used were stereo microscope, μ-FTIR, and EDS; stereo microscope, ATR-FTIR spectrophotometer, and SEM; and stereo microscope, μ-Raman, and SEM, etc.

4. Discussions

For measuring the concentration of MPs, several units were adopted by the scientific communities. First, this study analyzed the trend in the number of studies per year and found that since the year 2018, the numbers increased by almost four times by 2021. Regarding spatial distribution, it was found that about two-thirds of the total publications are focused on China, followed by India and South Korea. Furthermore, the gap between percentage shares of papers between China and India was around 61% of total papers, meaning China is leading the scientific investigation on MPs in freshwater systems in

Asia. However, on the other hand, the number of scientific publications may not provide a true picture of the available data within a country. For example, 2 freshwater MPs-related publications cover more than 150 sampling locations in Japan, with wide spatial distribution over the whole country [59,60]. Based on the objectives of the papers analyzed, it was found that estimating the concentration of MPs in freshwater environments, whether in water or sediment, is the primary focus, followed by exposure analysis on different organisms. This shows that baseline studies on MPs in freshwater systems are still lacking in most countries. Regarding units of measurement, the most common units for water, sediment, and biota found were items/L [61], items/Kg dw [62], and items/individuals [63], respectively. Looking at the objectives of these reviewed articles, it was found that most of the articles were reporting for the first time about concentration and accumulation in the freshwater environment (water, soil/sediment) and biota, respectively. Furthermore, looking at the exposure analysis, a few articles presented the effect of MPs pollution and its impact on biota [52,64,65]. According to the Food and Agriculture Organization of the United Nations, per capita consumption of fish in Asia is the second highest, 24.1 kg/year, following the Oceania region, 24.2 kg/year [66]. Many articles indeed studied the impacts of MPs on freshwater fishes but did not represent the real picture for the whole aquatic ecosystem. Carbery et al. [67] reviewed articles to investigate trophic transfers in the marine food web and found eight articles that presented MP ingestion by the transfer. Furthermore, Hasegawa and Nakaoka [68] investigated how MPs were ingested by trophic transfer, and they found that aquatic organisms ingest more MPs through it than from the water. Considering the high consumption of fish and the trophic transfer, exposure analysis of the first trophic level as well as multiple levels is important in this region, although more attention is currently paid to aquatic organisms at the single and higher trophic level in the reviewed articles. Assessing MPs in multiple trophic levels and understanding trophic transfer in organisms is important in considering MP transfer to humans and its impacts on health in this region. When referring to the exposure to human health, only two papers focused on the effect of MP pollution on human health [69,70]. Li et al. [69] investigated the joint cytotoxicity of two different MPs co-exposed with diverse ionic pollutants in two cell lines from the human digestive system: human gastric epithelium (GES-1) and colorectal mucosa (FHC) cell lines. Also, their finding indicated that the cytotoxicity of cationic pollutants was alleviated by MPs more significantly than that of anionic pollutants in both culture medium and river water. Additionally, the electrostatic attraction between negatively charged MPs and cations was a key factor in determining the ultimate joint toxicity. On the other hand, Ajay et al. [70] investigated MPs and phthalic acid esters in the aquatic system. However, they mentioned that PAEs end up accumulating in the human body and cause various health effects, including a hormonal imbalance in adults and changing levels of urinary thyroid hormones in children. This shows that the scientific works in this domain or direction are still in the early days, and it will take some time to further develop a clear understanding of the fate and transmission mechanism through which MPs in the ambient environment could impact human health. Color was analyzed in about half of the studies as it plays an important role in identifying the source and the original plastic before decomposition into microplastic. There were 40 studies that identified more than 6 colors. However, Xu et al. [50] mentioned the possibility of discoloration at the digestion of organic matter by hydrogen peroxide solution and found fading fibers from blue to transparent and fragments from green to light blue. UV light decomposes plastic into microplastic as well as changes the color. Therefore, it may be difficult to rule out that the colors identified in the analysis are the original color. Further studies and discussions are needed to understand the discoloration of MP and usage of the information.

It was found that three main types of sampling methodology were opted for in most of the articles, i.e., grab, net, and pump sampling. Here, both pump and net sampling can be kept together in a category called on-site filtration [29]. Grab sampling is mainly used to collect sediment samples or large volumes of water [50,70], whereas on-site filtration is used for water bodies where a net or pump is used to pass the water through the net and collect the filtrate for MPs analysis [42,71]. After collection, samples go through an extraction process to remove undesirable materials. Here, the extraction step again comprised two steps, i.e., density separation and organic matter degradation (especially for sediment and sludge samples). For density separation, the majority of research papers opted for low-density separation using saturated NaCl solutions [65]. As the name suggests, this method is especially effective for removing debris with low-density polymers. However, to separate high-density polymers like Polyethylene Terephthalate (PET), Poly Vinyl Chlorate (PVC), etc., $ZnCl_2$ or KI solutions are used [35]. The reason for fewer occurrences of the high-density separation method is because both $ZnCl_2$ and KI solutions are comparatively more expensive than NaCl, and they need special disposal measures, which constrain their common use [38]. Steps for organic materials decomposition are used when samples have too many biofilms, such as in the case of sediment and sludge. The most commonly used chemical for this purpose is H_2O_2 [72,73], as shown in about 90% of the reviewed articles; however, Iron (Fe) salts are added to H_2O_2 if there is a high amount of clay present in the sample [74]. For the identification phase, the main objective is to classify different MPs on the basis of their physico-chemical properties. Here physical properties refer to their shape, color and mass, whereas chemical properties refer to polymer types. Obtained results show that the most common tools used in the reviewed articles were μ-FTIR followed by the use of two tools > μ-Raman > ATR-FTIR > Three tools > Microscope > SEM > EDS. However, when we looked closely at the single instrument used for the identification, it was found that μ-FTIR was followed by μ-Raman, because of their ability to identify various polymers from MPs [75,76]. On the other hand, microscopes were also used in a significant number of studies, and their usage is often associated with the application of fluorescence dye to identify different MP polymers as well as their relatively low price [77]. The use of SEM was quite low, maybe because of the high price and low affordability [78].

There were no publications concerning the policy aspects of MP pollution in the Asian region. However, there were few studies that extended the discussion on MP pollution, but their geographical focus was either on the EU region or global in general [79–81]. Considering this policy gap, one of the important aspects for future studies is to discuss possible options to mitigate MP pollution.

While the issue of the marine plastic problem has been well studied and management policies are well established, there is a huge knowledge gap in terms of MPs in the freshwater environment. The number of research works on MPs in the freshwater environment is also on the rise, but addressing this problem is still in the incipient stage. This systematic review revealed spatial unevenness and differences in field sampling methods as well as laboratory analyses in the scientific studies; they are some of the key challenges for designing robust management strategies for MPs in freshwater environments [82,83]. Hence, to address the knowledge and information gap mentioned above, further engagement from all the relevant stakeholders, such as scientific communities, policymakers, local communities, industries, etc., need to work together to co-design and co-deliver holistic management options for MPs from cradle to grave. Although this is an exclusive study showing up-to-date scientific information on MPs in freshwater environments presented through research papers from Asia, there might be a possibility that it missed some vital information covered in the grey literature, which can be considered as the limitation of this study. This also calls for a more integrated and holistic approach to designing future scientific studies to investigate this pressing issue.

5. Conclusions

The increase in the number of freshwater MP studies indicates a high level of interest in this emerging issue, which is increasingly recognized in Asia, although there is a regional bias in that many studies have been conducted in China. The finding of this study indicates that most of the papers are primarily focused on reporting the level of occurrence of MPs in the freshwater system, whether in water or sediment (majority of the reviewed papers) and aquatic organisms (relatively fewer papers). While the assessment of MPs reveals pollution in the water environment, it does not reach to understand the amount of ingestion and the impact on human health in the region. Considering the various methods used to assess MPs in freshwater, diversified samples, and different approaches and presentations in the morphologic analysis in the reviewed articles, establishing a standard method for the examination of MPs would help to manage the MP pollution in the region. The establishment of sample collection and separation methods are particularly important because different methods lead to different results. MPs in the freshwater environment are greatly influenced by solid waste management on the ground; hence, the pollution level is expected to be high in countries where waste management is not properly performed. The ability to identify various polymers as well as smaller sizes depends on lab technologies, but countries that do not have proper waste management are often developing countries and have difficulties procuring expensive equipment and materials. As MP pollution has become ubiquitous, further studies are needed in various locations, but at the same time, the issues found in this review study also need to be discussed. Results from this study also revealed that only a few studies extended their discussions to policies and governance aspects of MPs. Based on the remaining gaps in scientific understanding, the following points should be considered for future studies:

- To have a better understanding of the relationships between MP pollution and its potential risks to human health, it is vital to build a robust inventory (big data both on the temporal and spatial scale) on MP pollution and transport in the agroecosystem.
- Because MPs have huge, diverse morphological characteristics, understanding their physio-chemical dynamics and evolution in different ecosystems is of utmost importance. Most of the existing studies are focused on the ecotoxicity of a particular target organism; hence it is very important to understand the comprehensive effects of mixtures of MPs (mimicking the natural condition) on different trophic levels.
- Because the life cycle of MPs is very long, it is imperative to conduct field-based experiments to understand the interaction of different MPs with various environmental components.
- Further research to address policy and governance aspects of MPs for effective management and control of this emerging pollutant is also needed.

Supplementary Materials: The following supporting information can be downloaded at: https://www.mdpi.com/article/10.3390/w14111737/s1, Table S1: List of review papers on MPs in freshwater; Table S2: List of research articles on MPs in Asia; Table S3: Detailed information for the GPS locations of all sampling sites collected from reviewed papers.

Author Contributions: Conceptualization, P.K. and Y.I.; methodology, P.K. and Y.I.; formal analysis, P.K. and Y.I.; investigation, P.K. and Y.I.; data curation, P.K., Y.I., P.N.B., A.A., R.D. and H.D.L.A.; writing—original draft preparation, P.K. and Y.I.; writing—review and editing, P.K., Y.I., P.N.B., A.A., R.D. and H.D.L.A.; funding acquisition, P.N.B. All authors have read and agreed to the published version of the manuscript.

Funding: This research was funded by the Institute for Global Environmental Strategies (IGES) under the Strategic Research Fund FY2021 for the project entitled: 'Introducing decentralized wastewater treatment and management approach for multi-benefits creation in addressing emerging pollutants in Asian cities' (FY2021) led by Pham Ngoc Bao, Deputy Director of Adaptation and Water Area, IGES.

Institutional Review Board Statement: Not applicable.

Informed Consent Statement: Not applicable.

Data Availability Statement: Data available upon request to authors.

Conflicts of Interest: The authors declare no conflict of interest.

References

1. Stanton, T.; Johnson, M.; Nathanail, P.; MacNaughtan, W.; Gomes, R.L. Freshwater microplastic concentrations vary through both space and time. *Environ. Pollut.* **2020**, *263*, 114481. [CrossRef]
2. Baekeland, L.H. Address of acceptance: The chemical constitution of resinous phenolic condensation products. *Ind. Eng. Chem.* **1913**, *5*, 506–511. [CrossRef]
3. Alfonso, M.B.; Arias, A.H.; Ronda, A.C.; Piccolo, M.C. Continental microplastics: Presence, features, and environmental transport pathways. *Sci. Total Environ.* **2021**, *799*, 149447. [CrossRef] [PubMed]
4. Yang, S.; Zhou, M.; Chen, X.; Hu, L.; Xu, Y.; Fu, W.; Li, C. A comparative review of microplastic in lake systems from different countries and regions. *Chemosphere* **2022**, *286*, 131806. [CrossRef] [PubMed]
5. PlasticsEurope, E. Plastics—The Facts 2019. An Analysis of European Plastics Production. Available online: https://www.plasticseurope.org/en/resources/market-data (accessed on 15 November 2021).
6. Sun, J.; Zhu, Z.R.; Li, W.H.; Yan, X.; Wang, L.K.; Zhang, L.; Jin, J.; Dai, X.; Ni, B.J. Revisiting microplastics in landfill leachate: Unnoticed tiny microplastics and their fate in treatment works. *Water Res.* **2021**, *190*, 116784. [CrossRef]
7. Peeken, I.; Primpke, S.; Beyer, B.; Gutermann, J.; Katlein, C.; Krumpen, T.; Bergmann, M.; Hehemann, L.; Gerdts, G. Arctic sea ice is an important temporal sink and means of transport for microplastic. *Nat. Commun.* **2018**, *9*, 1–12. [CrossRef]
8. Wright, S.L.; Kelly, F.J. Plastic and human health: A micro issue? *Environ. Sci. Technol.* **2017**, *51*, 6634–6647. [CrossRef]
9. Rochman, C.M.; Brookson, C.; Bikker, J.; Djuric, N.; Earn, A.; Bucci, K.; Athey, S.; Huntington, A.; McIlwraith, H.; Munno, K.; et al. Rethinking microplastics as a diverse contaminant suite. *Environ. Toxicol. Chem.* **2019**, *38*, 703–711. [CrossRef]
10. Wright, S.L.; Thompson, R.C.; Galloway, T.S. The physical impacts of microplastics on marine organisms: A review. *Environ. Pollut.* **2013**, *178*, 483–492. [CrossRef]
11. Ockenden, A.; Tremblay, L.A.; Dikareva, N.; Simon, K.S. Towards more ecologically relevant investigations of the impacts of microplastic pollution in freshwater ecosystems. *Sci. Total Environ.* **2021**, *792*, 148507. [CrossRef]
12. Ma, H.; Pu, S.; Liu, S.; Bai, Y.; Mandal, S.; Xing, B. Microplastics in aquatic environments: Toxicity to trigger ecological consequences. *Environ. Pollut.* **2020**, *261*, 114089. [CrossRef] [PubMed]
13. Bradney, L.; Wijesekara, H.; Palansooriya, K.N.; Obadamudalige, N.; Bolan, N.S.; Ok, Y.S.; Rinklebe, J.; Kim, K.H.; Kirkham, M.B. Particulate plastics as a vector for toxic trace-element uptake by aquatic and terrestrial organisms and human health risk. *Environ. Int.* **2019**, *131*, 104937. [CrossRef] [PubMed]
14. Eder, M.L.; Teles, L.O.; Pinto, R.; Carvalho, A.P.; Almeida, C.M.R.; Gaustere, R.H.; Guimares, L. Microplastics as a vehicle of exposure to chemical contamination in freshwater systems: Current research status and way forward. *J. Hazard. Mater.* **2021**, *417*, 125980. [CrossRef]
15. Geyer, R.; Jambeck, J.R.; Law, K.L. Production, use, and fate of all plastics ever made. *Sci. Adv.* **2017**, *3*, 3–8. [CrossRef] [PubMed]
16. Moher, D.; Liberati, A.; Tetzlaff, J.; Altman, D.G. Preferred reporting items for systematic reviews and meta-analyses: The PRISMA statement. *Bio. Med. J.* **2009**, *339*, b2535. [CrossRef] [PubMed]
17. Wang, T.; Wang, J.; Lei, Q.; Zhao, Y.; Wang, L.; Wang, X.; Zhang, W. Microplastic pollution in sophisticated urban river systems: Combined influence of land-use types and physicochemical characteristics. *Environ. Pollut.* **2021**, *287*, 117604. [CrossRef]
18. Xu, C.; Zhang, B.; Gu, C.; Shen, C.; Yin, S.; Aamir, M.; Li, F. Are we underestimating the sources of microplastic pollution in terrestrial environment? *J. Hazard. Mater.* **2020**, *400*, 123228. [CrossRef]
19. Koutnik, V.S.; Leonard, J.; Alkidim, S.; DePrima, F.J.; Ravi, S.; Hoek, E.M.; Mohanty, S.K. Distribution of microplastics in soil and freshwater environments: Global analysis and framework for transport modeling. *Environ. Pollut.* **2021**, *274*, 116552. [CrossRef]
20. Gao, G.; Zhao, X.; Jin, P.; Gao, K.; Beardall, J. Current understanding and challenges for aquatic primary producers in a world with rising micro-and nano-plastic levels. *J. Hazard. Mater.* **2021**, *406*, 124685. [CrossRef]
21. Bellasi, A.; Binda, G.; Pozzi, A.; Galafassi, S.; Volta, P.; Bettinetti, R. Microplastic contamination in freshwater environments: A review, focusing on interactions with sediments and benthic organisms. *Environments* **2020**, *7*, 30. [CrossRef]
22. Yang, L.; Luo, W.; Zhao, Y.; Kang, S.; Giesy, J.P.; Zhang, F. Microplastics in the Koshi River, a remote alpine river crossing the Himalayas from China to Nepal. *Environ. Pollut.* **2021**, *290*, 118121. [CrossRef]
23. Napper, I.E.; Baroth, A.; Barrett, A.C.; Bhola, S.; Chowdhury, G.W.; Davies, B.F.R.; Duncan, E.M.; Kumar, S.; Nelms, S.E.; Niloy, M.N.H.; et al. The abundance and characteristics of microplastics in surface water in the transboundary Ganges River. *Environ. Pollut.* **2021**, *274*, 116348. [CrossRef] [PubMed]
24. Hu, J.; Zuo, J.; Li, J.; Zhang, Y.; Ai, X.; Zhang, J.; Gong, D.; Sun, D. Effects of secondary polyethylene microplastic exposure on crucian (Carassius carassius) growth, liver damage, and gut microbiome composition. *Sci. Total Environ.* **2022**, *802*, 149736. [CrossRef] [PubMed]
25. Wu, X.; Wu, H.; Zhang, A.; Sekou, K.; Li, Z.; Ye, J. Influence of polystyrene microplastics on levofloxacin removal by microalgae from freshwater aquaculture wastewater. *J. Environ. Manag.* **2022**, *301*, 113865. [CrossRef] [PubMed]
26. Li, C.; Liu, J.; Wang, D.; Kong, L.; Wu, Y.; Zhou, X.; Jia, J.; Zhou, H.; Yan, B. Electrostatic attraction of cationic pollutants by microplastics reduces their joint cytotoxicity. *Chemosphere* **2021**, *282*, 131121. [CrossRef]
27. Liu, P.; Wu, X.; Huang, H.; Wang, H.; Shi, Y.; Gao, S. Simulation of natural aging property of microplastics in Yangtze River water samples via a rooftop exposure protocol. *Sci. Total Environ.* **2021**, *785*, 147265. [CrossRef]
28. Wu, X.; Pan, J.; Li, M.; Li, Y.; Bartlam, M.; Wang, Y. Selective enrichment of bacterial pathogens by microplastic biofilm. *Water Res.* **2019**, *165*, 114979. [CrossRef]

29. Zhang, Q.; Liu, T.; Liu, L.; Fan, Y.; Rao, W.; Zheng, J.; Qian, X. Distribution and sedimentation of microplastics in Taihu Lake. *Sci. Total Environ.* **2021**, *795*, 148745. [CrossRef]
30. Duan, Z.; Zhao, S.; Zhao, L.; Duan, X.; Xie, S.; Zhang, H.; Liu, Y.; Peng, Y.; Liu, C.; Wang, L. Microplastics in Yellow River Delta wetland: Occurrence, characteristics, human influences, and marker. *Environ. Pollut.* **2020**, *258*, 113232. [CrossRef]
31. Yin, L.; Wen, X.; Du, C.; Jiang, J.; Wu, L.; Zhang, Y.; Hu, Z.; Hu, S.; Feng, Z.; Zhou, Z.; et al. Comparison of the abundance of microplastics between rural and urban areas: A case study from East Dongting Lake. *Chemosphere* **2020**, *244*, 125486. [CrossRef]
32. Shi, X.; Zhang, X.; Gao, W.; Zhang, Y.; He, D. Removal of microplastics from water by magnetic nano-Fe3O4. *Sci. Total Environ.* **2022**, *802*, 149838. [CrossRef] [PubMed]
33. Khoironi, A.; Anggoro, S. Evaluation of the interaction among microalgae Spirulina sp, plastics polyethylene terephthalate and polypropylene in freshwater environment. *J. Ecol. Eng.* **2019**, *20*, 161–173. [CrossRef]
34. Kim, S.W.; Chae, Y.; Kim, D.; An, Y.J. Zebrafish can recognize microplastics as inedible materials: Quantitative evidence of ingestion behavior. *Sci. Total Environ.* **2019**, *649*, 156–162. [CrossRef]
35. Abbasi, S. Prevalence and physicochemical characteristics of microplastics in the sediment and water of Hashilan Wetland, a national heritage in NW Iran. *Environ. Technol. Innov.* **2021**, *23*, 101782. [CrossRef]
36. Liu, Y.; You, J.; Li, Y.; Zhang, J.; He, Y.; Breider, F.; Tao, S.; Liu, W. Insights into the horizontal and vertical profiles of microplastics in a river emptying into the sea affected by intensive anthropogenic activities in Northern China. *Sci. Total Environ.* **2021**, *779*, 146589. [CrossRef]
37. Strady, E.; Kieu-Le, T.C.; Gasperi, J.; Tassin, B. Temporal dynamic of anthropogenic fibers in a tropical river-estuarine system. *Environ. Pollut.* **2020**, *259*, 13897. [CrossRef]
38. Kabir, A.; Sekine, M.; Imai, T.; Yamamoto, K.; Kanno, A.; Higuchi, T. Assessing small-scale freshwater microplastics pollution, land-use, source-to-sink conduits, and pollution risks: Perspectives from Japanese rivers polluted with microplastics. *Sci. Total Environ.* **2021**, *768*, 144655. [CrossRef]
39. Park, T.; Lee, S.; Lee, M.; Lee, J.; Park, J.; Zoh, K. Distributions of microplastics in surface water, fish, and sediment in the vicinity of a sewage treatment plant. *Water* **2020**, *12*, 3333. [CrossRef]
40. Liu, Y.; Zhang, J.; Tang, Y.; He, Y.; Li, Y.; You, J.; Breider, F.; Tao, S.; Liu, W. Effects of anthropogenic discharge and hydraulic deposition on the distribution and accumulation of microplastics in surface sediments of a typical seagoing river: The Haihe River. *J. Hazard. Mater.* **2021**, *404*, 124180. [CrossRef]
41. Hu, H.; Jin, D.; Yang, Y.; Zhang, J.; Ma, C.; Qiu, Z. Distinct profile of bacterial community and antibiotic resistance genes on microplastics in Ganjiang River at the watershed level. *Environ. Res.* **2021**, *200*, 111363. [CrossRef]
42. Cheng, Y.; Mai, L.; Lu, X.; Li, Z.; Guo, Y.; Chen, D.; Wang, F. Occurrence and abundance of poly-and perfluoroalkyl substances (PFASs) on microplastics (MPs) in Pearl River Estuary (PRE) region: Spatial and temporal variations. *Environ. Pollut.* **2021**, *281*, 117025. [CrossRef] [PubMed]
43. Wicaksono, E.; Werorilangi, S.; Galloway, T.; Tahir, A. Distribution and Seasonal Variation of Microplastics in Tallo River, Makassar, Eastern Indonesia. *Toxics* **2021**, *9*, 129. [CrossRef] [PubMed]
44. Cabansag, J.; Olimberio, R.; Villanobos, Z. Microplastics in some fish species and their environs in Eastern Visayas, Philippines. *Mar. Pollut. Bull.* **2021**, *167*, 112312. [CrossRef] [PubMed]
45. Niu, L.; Li, Y.; Li, Y.; Hu, Q.; Wang, C.; Hu, J.; Zhang, W.; Wang, L.; Zhang, C.; Zhang, H. New insights into the vertical distribution and microbial degradation of microplastics in urban river sediments. *Water Res.* **2021**, *188*, 116449. [CrossRef]
46. Turhan, D.Ö. Evaluation of Microplastics in the Surface Water, Sediment and Fish of Sürgü Dam Reservoir (Malatya) in Turkey. *Turk. J. Fish. Aquat. Sci.* **2022**, *22*, 7. [CrossRef]
47. Lechthaler, S.; Waldschläger, K.; Sandhani, C.G.; Sannasiraj, S.A.; Sundar, V.; Schwarzbauer, J.; Schüttrumpf, H. Baseline study on microplastics in Indian rivers under different anthropogenic influences. *Water* **2021**, *13*, 1648. [CrossRef]
48. Parvin, F.; Jannat, S.; Tareq, S.M. Abundance, characteristics and variation of microplastics in different freshwater fish species from Bangladesh. *Sci. Total Environ.* **2021**, *784*, 147137. [CrossRef]
49. Wang, Z.; Zhang, Y.; Kang, S.; Yang, L.; Shi, H.; Tripathee, L.; Gao, T. Research progresses of microplastic pollution in freshwater systems. *Sci. Total Environ.* **2021**, *795*, 148888. [CrossRef]
50. Xu, Y.; Chan, F.K.S.; Johnson, M.; Stanton, T.; He, J.; Jia, T.; Wang, J.; Wang, Z.; Yao, Y.; Yang, J.; et al. Microplastic pollution in Chinese urban rivers: The influence of urban factors. *Resour. Conserv. Recycl.* **2021**, *173*, 105686. [CrossRef]
51. Picó, Y.; Soursou, V.; Alfarhan, A.H.; El-Sheikh, M.A.; Barceló, D. First evidence of microplastics occurrence in mixed surface and treated wastewater from two major Saudi Arabian cities and assessment of their ecological risk. *J. Hazard. Mater.* **2021**, *416*, 125747. [CrossRef]
52. Yu, H.; Qi, W.; Cao, X.; Wang, Y.; Li, Y.; Xu, Y.; Zhang, X.; Peng, J.; Qu, J. Impact of microplastics on the foraging, photosynthesis and digestive systems of submerged carnivorous macrophytes under low and high nutrient concentrations. *Environ. Pollut.* **2022**, *292*, 118220. [CrossRef] [PubMed]
53. Sun, T.; Zhan, J.; Li, F.; Ji, C.; Wu, H. Effects of microplastics on aquatic biota: A hermetic perspective. *Environ. Pollut.* **2021**, *285*, 117206. [CrossRef] [PubMed]
54. Yin, K.; Wang, D.; Zhao, H.; Wang, Y.; Guo, M.; Liu, Y.; Li, B.; Xing, M. Microplastics pollution and risk assessment in water bodies of two nature reserves in Jilin Province: Correlation analysis with the degree of human activity. *Sci. Total Environ.* **2021**, *799*, 149390. [CrossRef] [PubMed]

55. Zuo, L.; Sun, Y.; Li, H.; Hu, Y.; Lin, L.; Peng, J.; Xu, X. Microplastics in mangrove sediments of the Pearl River Estuary, South China: Correlation with halogenated flame retardants' levels. *Sci. Total Environ.* **2020**, *725*, 138344. [CrossRef] [PubMed]
56. Li, S.; Wang, P.; Zhang, C.; Zhou, X.; Yin, Z.; Hu, T.; Hu, D.; Liu, C.; Zhu, L. Influence of polystyrene microplastics on the growth, photosynthetic efficiency and aggregation of freshwater microalgae Chlamydomonas reinhardtii. *Sci. Total Environ.* **2020**, *714*, 136767. [CrossRef]
57. Wu, Y.; Guo, P.; Zhang, X.; Zhang, Y.; Xie, S.; Deng, J. Effect of microplastics exposure on the photosynthesis system of freshwater algae. *J. Hazard. Mater.* **2019**, *374*, 219–227. [CrossRef]
58. Wu, Y.; Peiyong, G.; Xiaoyan, Z.; Yuxuan, Z.; Shuting, X.; Jun, D. Effect of microplastics exposure on the photosynthesis sys-tem of freshwater algae. *J. Hazard. Mater.* **2019**, *374*, 219–227. [CrossRef]
59. Abeynayaka, A.; Kojima, F.; Miwa, Y.; Ito, N.; Nihei, Y.; Fukunaga, Y.; Yashima, Y.; Itsubo, N. Rapid sampling of suspended and floating microplastics in challenging riverine and coastal water environments in Japan. *Water* **2020**, *12*, 1903. [CrossRef]
60. Nihei, Y.; Yoshida, T.; Kataoka, T.; Ogata, R. High-resolution mapping of Japanese microplastic and macroplastic emissions from the land into the sea. *Water* **2020**, *12*, 951. [CrossRef]
61. Chen, H.L.; Gibbins, C.N.; Selvam, S.B.; Ting, K.N. Spatio-temporal variation of microplastic along a rural to urban transition in a tropical river. *Environ. Pollut.* **2021**, *289*, 117895. [CrossRef]
62. Tsering, T.; Sillanpää, M.; Sillanpää, M.; Viitala, M.; Reinikainen, S.P. Microplastics pollution in the Brahmaputra River and the Indus River of the Indian Himalaya. *Sci. Total Environ.* **2021**, *789*, 147968. [CrossRef]
63. Xu, Q.; Deng, T.; LeBlanc, G.A.; An, L. An effective method for evaluation of microplastic contaminant in gastropod from Taihu Lake, China. *Environ. Sci. Pollut. Res.* **2020**, *18*, 22878–22887. [CrossRef] [PubMed]
64. Yin, L.; Wen, X.; Huang, D.; Zeng, G.; Deng, R.; Liu, R.; Zhou, Z.; Tao, J.; Xiao, R.; Pan, H. Microplastics retention by reeds in freshwater environment. *Sci. Total Environ.* **2021**, *790*, 148200. [CrossRef] [PubMed]
65. Zhang, K.; Chen, X.; Xiong, X.; Ruan, Y.; Zhou, H.; Wu, C.; Lam, P.K. The hydro-fluctuation belt of the Three Gorges Reservoir: Source or sink of microplastics in the water? *Environ. Pollut.* **2019**, *248*, 279–285. [CrossRef] [PubMed]
66. FAO. *The State of World Fisheries and Aquaculture 2020: Sustainability in Action*; FAO: Rome, Italy, 2020. [CrossRef]
67. Carbery, M.; O'Connor, W.; Palanisami, T. Trophic transfer of microplastics and mixed contaminants in the marine food web and implications for human health. *Environ. Int.* **2018**, *115*, 400–409. [CrossRef]
68. Hasegawa, T.; Nakaoka, M. Trophic transfer of microplastics from mysids to fish greatly exceeds direct ingestion from the water column. *Environ. Pollut.* **2021**, *273*, 116468. [CrossRef] [PubMed]
69. Li, Y.; Liu, S.; Liu, M.; Huang, W.; Chen, K.; Ding, Y.; Wu, F.; Ke, H.; Lou, L.; Lin, Y.; et al. Mid-Level Riverine Outflow Matters: A Case of Microplastic Transport in the Jiulong River, China. *Front. Mar. Sci.* **2021**, *973*, 1–11. [CrossRef]
70. Ajay, K.; Behera, D.; Bhattacharya, S.; Mishra, P.K.; Ankit, Y.; Anoop, A. Distribution and characteristics of microplastics and phthalate esters from a freshwater lake system in Lesser Himalayas. *Chemosphere* **2021**, *283*, 131132. [CrossRef]
71. Mao, Y.; Li, H.; Gu, W.; Yang, G.; Liu, Y.; He, Q. Distribution and characteristics of microplastics in the Yulin River, China: Role of environmental and spatial factors. *Environ. Pollut.* **2020**, *265*, 115033. [CrossRef]
72. Pariatamby, A.; Hamid, F.S.; Bhatti, M.S.; Anuar, N.; Anuar, N. Status of microplastic pollution in aquatic ecosystem with a case study on cherating river, Malaysia. *J. Eng. Technol. Sci.* **2020**, *52*, 222–241. [CrossRef]
73. Sarkar, D.J.; Das, S.S.; Das, B.K.; Manna, R.K.; Behera, B.K.; Samanta, S. Spatial distribution of meso and microplastics in the sediments of river Ganga at eastern India. *Sci. Total Environ.* **2019**, *694*, 133712. [CrossRef] [PubMed]
74. Wang, Z.; Su, B.; Xu, X.; Di, D.; Huang, H.; Mei, K.; Dahlgren, R.A.; Zhang, M.; Shang, X. Preferential accumulation of small (<300 Mm) microplastics in the sediments of a coastal plain river network in eastern China. *Water Res.* **2018**, *144*, 393–401. [CrossRef] [PubMed]
75. Liang, T.; Lei, Z.; Fuad, M.T.I.; Wang, Q.; Sun, S.; Fang, J.K.H.; Liu, X. Distribution and potential sources of microplastics in sediments in remote lakes of Tibet, China. *Sci. Total Environ.* **2022**, *806*, 150526. [CrossRef] [PubMed]
76. Xiong, X.; Zhang, K.; Chen, X.; Shi, H.; Luo, Z.; Wu, C. Sources and distribution of microplastics in China's largest inland lake—Qinghai Lake. *Environ. Pollut.* **2018**, *235*, 899–906. [CrossRef] [PubMed]
77. Wen, X.; Du, C.; Xu, P.; Zeng, G.; Huang, D.; Yin, L.; Yin, Q.; Hu, L.; Wan, J.; Zhang, J.; et al. Microplastic pollution in surface sediments of urban water areas in Changsha, China: Abundance, composition, surface textures. *Mar. Pollut. Bull.* **2018**, *136*, 414–423. [CrossRef]
78. Zhou, Y.; He, G.; Jiang, X.; Yao, L.; Ouyang, L.; Liu, X.; Liu, W.; Liu, Y. Microplastic contamination is ubiquitous in riparian soils and strongly related to elevation, precipitation and population density. *J. Hazard. Mater.* **2021**, *411*, 125178. [CrossRef]
79. Igalavithana, A.D.; Mahagamage, M.G.Y.; Gajanayake, P.; Abeynayaka, A.; Gamaralalage, P.J.D.; Ohgaki, M.; Takenaka, M.; Fukai, T.; Itsubo, N. Microplastics and Potentially Toxic Elements: Potential Human Exposure Pathways through Agricultural Lands and Policy Based Countermeasures. *Microplastics* **2022**, *1*, 102–120. [CrossRef]
80. Deme, G.G.; Ewusi-Mensah, D.; Olagbaju, O.A.; Okeke, E.S.; Okoye, C.O.; Odii, E.C.; Ejeromedoghene, O.; Igun, E.; Onyekwere, J.O.; Oderinde, O.K.; et al. Macro Problems from Microplastics: Toward a Sustainable Policy Framework for Managing Microplastic Waste in Africa. *Sci. Total Environ.* **2022**, *804*, 150170. [CrossRef]
81. Milojevic, N.; Cydzik-Kwiatkowska, A. Agricultural Use of Sewage Sludge as a Threat of Microplastic (MP) Spread in the Environment and the Role of Governance. *Energies* **2021**, *14*, 6293. [CrossRef]

82. Manzoor, S.; Kaur, H.; Singh, R. Existence of Microplastic as Pollutant in Harike Wetland: An Analysis of Plastic Compositionand First Report on Ramsar Wetland of India. *Curr. World Env.* **2021**, *16*, 591–598. [CrossRef]
83. Naqash, N.; Prakash, S.; Kapoor, D.; Singh, R. Interaction offreshwater microplastics with biota and heavy metals: A review. *Environ. Chem. Lett.* **2020**, *18*, 1813–1824. [CrossRef]

 water

Article

A Model-Based Approach for Improving Surface Water Quality Management in Aquaculture Using MIKE 11: A Case of the Long Xuyen Quadangle, Mekong Delta, Vietnam

Huynh Vuong Thu Minh [1], Van Pham Dang Tri [2], Vu Ngoc Ut [3], Ram Avtar [4], Pankaj Kumar [5,*], Trinh Trung Tri Dang [6], Au Van Hoa [3], Tran Van Ty [7] and Nigel K. Downes [1]

1 Department of Water Resources, College of Environment and Natural Resources, Can Tho University, Can Tho 900000, Vietnam; hvtminh@ctu.edu.vn (H.V.T.M.); nkdownes@ctu.edu.vn (N.K.D.)
2 Research Institute for Climate Change, Can Tho University, Can Tho 900000, Vietnam; vpdtri@ctu.edu.vn
3 College of Aquaculture and Fisheries, Can Tho University, Can Tho 900000, Vietnam; vnut@ctu.edu.vn (V.N.U.); avhoa@ctu.edu.vn (A.V.H.)
4 Faculty of Environmental Earth Science, Hokkaido University, Sapporo 060-0810, Japan; ram@ees.hokudai.ac.jp
5 Institute for Global Environmental Strategies, Hayama 240-0115, Japan
6 Institute of Environmental Technology Sciences, Tra Vinh University, Tra Vinh 87000, Vietnam; tttdang247@gmail.com
7 College of Technology, Can Tho University, Can Tho 900000, Vietnam; tvty@ctu.edu.vn
* Correspondence: kumar@iges.or.jp; Tel.: +81-7014124622

Citation: Thu Minh, H.V.; Tri, V.P.D.; Ut, V.N.; Avtar, R.; Kumar, P.; Dang, T.T.T.; Hoa, A.V.; Ty, T.V.; Downes, N.K. A Model-Based Approach for Improving Surface Water Quality Management in Aquaculture Using MIKE 11: A Case of the Long Xuyen Quadangle, Mekong Delta, Vietnam. *Water* **2022**, *14*, 412. https://doi.org/10.3390/w14030412

Academic Editor: Domenico Cicchella

Received: 19 November 2021
Accepted: 26 January 2022
Published: 29 January 2022

Abstract: This study utilized MIKE 11 to quantify the spatio-temporal dynamics of water quality parameters (Biochemical Oxygen Demand (BOD_5), Dissolved Oxygen (DO) and temperature) in the Long Xuyen Quadrangle area of the Vietnamese Mekong Delta. Calibrated for the year of 2019 and validated for the year of 2020, the developed model showed a significant agreement between the observed and simulated values of water quality parameters. Locations near to cage culture areas exhibited higher BOD_5 values than sites close to pond/lagoon culture areas due to the effects of numerous point sources of pollution, including upstream wastewater and out-fluxes from residential and tourism activities in the surrounding areas, all of which had a direct impact on the quality of the surface water used for aquaculture. Moreover, as aquacultural effluents have intensified and dispersed over time, water quality in the surrounding water bodies has degraded. The findings suggest that the effective planning, assessment and management of rapidly expanding aquaculture sites should be improved, including more rigorous water quality monitoring, to ensure the long-term sustainable expansion and development of the aquacultural sector in the Long Xuyen Quadrangle in particular, and the Vietnamese Mekong Delta as a whole.

Keywords: EcoLab module; hydrodynamics modeling; surface water quality; one dimension; cage culture; pond/lagoon culture

1. Introduction

Globally, demand for freshwater resources continues to increase. Freshwater sources are increasingly required to meet growing domestic, agriculture, aquaculture and industrial uses, while at the same time they suffer from increased pollution and natural and anthropogenic interventions and changes to the environment driven by strong population and economic growth [1–4]. In Southeast Asia, poorly managed aquacultural and agricultural activities are one of the most prominent sources of water pollution [2,5–9]. Here, countries find it challenging to manage surface water quality due to both point and non-point sources of pollutants, as well as the increased widespread use of chemicals and drugs in the context of limited land and water resources and more intensive culture methods [10–12]. Due of the cumulative and synergistic impacts on water resources, water quality management requires a fundamental understanding of the spatial and temporal variations in water characteristics, including the hydro-morphological, chemical and biological parameters [13]. Previously, several methods have been developed to predict, monitor and assess water quality. These include using hydrogeochemical analysis [14–17], the use of various quality indexes [2,18,19], using numerical modeling for scenario development [20,21] and using socio-hydrological approaches to assess the nexus between water and human well-being [22]. All these approaches seek to provide a better understanding of the drivers and interlinkages at work, and to help decision makers take evidence-based actions with regard to improved water resource management.

However, using traditional hydro-chemical analysis and statistical approaches for analyzing the dynamics of water quality have inherent limitations when considering the different environmental components in a holistic manner. Moreover, they tend to be resource (money and human power) intensive. As a result, numerical simulation models or tools, including environmental modeling, that are capable of detecting regional and temporal changes in current and future water quality or quantity parameters are currently gaining increased popularity among scientists and water management practitioners [23,24]. In addition, the application of numerical simulations can also save labor, time and money [24,25].

Various hydraulics models, such as the Hydrological Engineering Centre—River Analysis System (HEC-RAS), MIKE 11 and Vietnamese River System and Plain (VRSAP) have been previously applied in the VMD to assess the changes in the quantity and quality of water in rivers, as well as quantify the impact of land management practices on water quality [26–28]. These models are holistic in nature and attempt to take into account the all-important environmental processes [16,20,21]. Because of its robust nature, the MIKE 11 model has previously been extensively used to investigate water security issues, especially in Asia [29].

As a rapidly developing and lower riparian country, Vietnam is particularly susceptible to water resource changes [30,31]. Southeast Asia's longest river, the Mekong River, appears to have its source on the Tibetan plateau, and runs through China, Myanmar, Laos, Thailand and Cambodia before reaching Vietnam. At its end, the Vietnamese Mekong Delta (VMD) is one of the world's largest river deltas, with a dense network of rivers, canals and ditches, and covering over 4 million hectares, which is approximately 12 percent of Vietnam's natural land area [32]. The VMD is also the largest agricultural and aquacultural hub in Vietnam, accounting for 50% of rice, 65% of aquaculture and 70% of fruit production, as well as including 95% of exported rice and 60% of exported fish [33].

The Long Xuyen Quadrangle (LXQ), as shown in Figure 1, is the first area in Vietnam to collect and use water from the Mekong River via two main branches, the Bassac River and the Mekong River. However, in the last few decades, water pollution has been an increasing problem in the delta [1,2,8,34]. The principal sources of pollution are non-point sources, such as wastewater discharged without treatment from industrial zones and commercial activities along the banks of rivers or canals. Furthermore, within the LXQ, intensive freshwater farming activities in the provinces of An Giang, Kien Giang and Can Tho city affect directly the water quality. In spite of its high regional importance and socio-economic significance, very few studies have investigated the water resources of LXQ in an in-depth, comprehensive and holistic manner.

Against this background, our study aims to apply a hydrological simulation to investigate the spatio-temporal dynamics of key water quality parameters using MIKE 11. The hydrodynamic module, the structural operation module, the advection/dispersion module and the EcoLab module are all inclusive of the MIKE 11 model system. The outcomes of this study will firstly help decision makers to understand the current situation, and secondly help the design of future management options for improved water resource management.

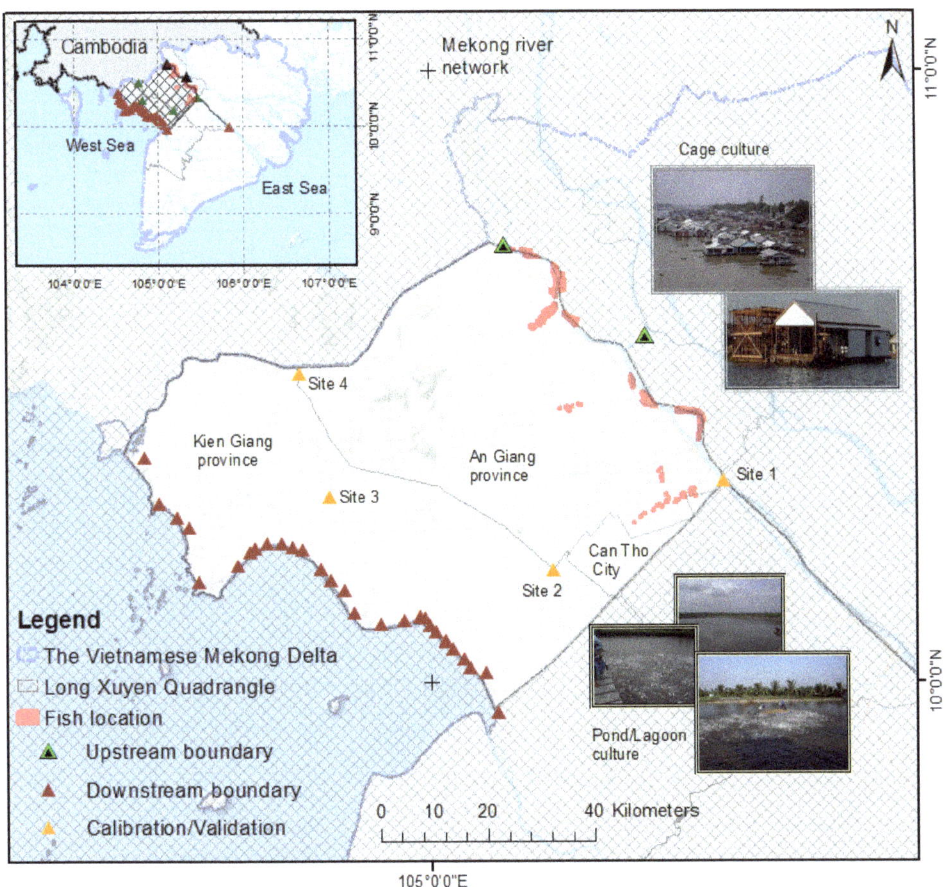

Figure 1. Map of the Vietnamese Mekong Delta and the Long Xuyen Quadrangle (LXQ) with its upstream and downstream boundaries. Calibration and validation samples were collected in the provinces of An Giang, Kien Giang and Can Tho. Long Xuyen Canal = 1, Vinh Tre Canal = 2, Rach Gia-Long Xuyen Canal = 3 and Tam Ngan Canal = 4.

2. Materials and Methods

The LXQ covers a large portion of the An Giang and Kien Giang provinces, and a small portion of Can Tho city, with a total area of approximately 0.5 million hectares. It is bordered to the north by the Bassac River and the Vietnamese–Cambodian border, to the south by the Cai San Canal, and to the west by the West Sea of Vietnam (Gulf of Thailand) [23]. Like the VMD as a whole, the topography of the LXQ, is relatively low and flat, with ground elevations of 0–1.0 m above mean sea level accounting for over 80% of the area [35,36]. The tropical monsoon climate of LXQ, has two primary seasons, dry and wet, and it is hot and humid all year round. As a result, LXQ's average annual temperature, rainfall and humidity are approximately 27°C, 1200 mm and 80%, respectively [36].

The LXQ is significantly affected by its geographical location, its monsoon climate and the upstream river network and the tidal regime [35,37]. Within the LXQ, the yearly average flow of the river systems is around 14,000 $m^3 \cdot s^{-1}$. However, this can reach up to 24,000 $m^3 \cdot s^{-1}$ during the rainy season, and recede to only 5000 $m^3 \cdot s^{-1}$ during the dry season [36]. During the rainy season, annual floods inundate roughly 70% of the total LXQ area with water levels of 1.0–2.5 m for 3–5 months a year. This episodic flooding has both positive and negative effects to the socio-economic signature of this region. On the positive side, floodwaters bring large volumes of water for agriculture, aquaculture, domestic uses and industrial operations, and provide the region with nutrient-rich sediments, as well as helping to wash out pollutants and salinity from the soil. Flood disasters, on the other hand, severely damage infrastructure, interrupt both community and livelihood activities and jeopardize agricultural and fishery production [38,39]. As a result, both the provincial and national governments have made significant investments in local infrastructure to regulate water levels and protect the region through a series of full-dyke and semi-dyke systems [36]. However, this has resulted in changes to the hydrometeorological regime and fluxes, and lessened the potential for water pollutant dispersion.

To simulate water quality in the complex and dense river network of LXQ, this study utilized the hydrodynamics and EcoLab modules, which are the foundation of the one-dimensional (1-D) MIKE 11 model.

Various hydro-meteorological data used as input for the modeling were kindly provided from a variety of sources, including the Southern Institute of Water Resources Research (SIWRR), the Southern Region Hydro-Meteorological Centre (SRHMC), the Department of Natural Resources and Environment (DoNRE), and the Department of Agriculture and Rural Development (DARD) (Table 1). Additional data regarding the population, wastewater discharge, pollution load of BOD_5 and the current land use map were collected from residential, industrial, aquacultural and agricultural areas to estimate the pollution load discharges.

Table 1. Summary of hydrology and water quality data collection and sources.

Data	Sources	Period	Remarks
Water level	The Southern Region Hydro-Meteorological Centre (SRHMC)	Jan.–May, 2019 Jan.–May, 2020	Time-step: Hourly data
Discharge	The Southern Region Hydro-Meteorological Centre (SRHMC)	Jan.–May, 2019 Jan.–May, 2020	Time-step: Hourly data
Cross-section	GIZ	-	-
DO, Temperature, BOD_5	DoNRE, DARD	Jan.–May, 2019 Jan.–May, 2020	Time-step: Monthly data

Within the study area, hourly observations of the discharge were conducted at two stations at Chau Doc and Vam Nao, and these were used as the upstream boundary conditions, and hourly observations of the water level were conducted at two stations at Rach Gia and Can Tho, and these were used as the downstream boundary conditions. The hourly discharge and water level at Long Xuyen were obtained for calibration and verification for the years 2019 and 2020 for the period from 00:00 a.m. on 1st January to 23:59 p.m. on 31st May. Average monthly water quality data, such as BOD_5, DO and temperature, were also obtained at eight stations from January to May in 2019 and 2020 for calibration and validation, respectively.

The total estimated fishery production for the year 2019 was 532.6 thousand tons, with pangasius production accounting for 412 tons in the study area. In 2020, due to the impact of the COVID-19 pandemic, the export of pangasius encountered many difficulties with the selling price becoming low, so farmers and businesses cut back on feed to prolong farming time, waiting for the price to increase, leading to a decrease in the harvest, with 211 thousand tons produced from the month of January to May 2020 [40]. Fish are mostly cultured in ponds and lagoons on both banks of the Bassac River, as well as along the tributaries Bay Tre, Xa Doi Canal, Cai Sao Canal, Don Dong Canal and Moi Canal (Figure 1 and Table A1).

2.1. Estimation of Pollution Load

Previous studies have estimated pollution load in various Vietnamese river networks, and found the ratio of BOD_5/Chemical Oxygen Demand (COD) to be around 0.65 [41,42]. This suggests that the majority of organic pollutants are soluble and easily decomposable. However, because of a lack of observed data in the study area, the pollutant load was calculated using the BOD_5 concentration as an indicator. Pollution load was computed using the discharge estimate, using Equation (1) as shown below:

$$Q_M = Q \times C_i \tag{1}$$

where Q_M denotes the pollution load from the Mekong River (tons·year^{-1}), Q denotes the discharge (m^3·s^{-1}) and C_i is the concentration of parameter i (mg·L^{-1}).

For domestic sources, pollution load was calculated based on the population statistics in the study area. The pollution emission coefficient per capita was calculated using Equation (2), as shown below:

$$Q_D = P \times Q_i \tag{2}$$

where Q_D is pollution load from the population (tons·year^{-1}), P is the population of the area (persons) and Q_i is the domestic waste load of parameter i (kg·person^{-1}·year^{-1}) (Table 2).

Table 2. Pollution load estimation.

Pollution Load	BOD_5 (mg·L^{-1})
Domestic waste load (kg/person/year)	10–25
Poultry (kg/unit/year)	2.73
Cow, buffalo (kg/unit/year)	233.6
Pig (kg/unit/year)	73
Sutchi catfish farming (kg/unit/year)	8.1

For industrial areas, the pollution load was estimated by multiplying the industrial discharge (Q) by the pollutant emission coefficient of the industrial type, using Equation (3) as shown below:

$$Q_I = \sum_{j=1}^{n} V_j \times C_{i,j} \tag{3}$$

where Q_I is the pollution load from industries, V_j is the volume of annual wastewater discharged from industry j (m^3·year^{-1}), $C_{i,j}$ is the concentration of substance i in the wastewater of industry j (mg·L^{-1}) and n is the number of industries in the region.

The pollution load from livestock production activities was calculated using the total annual livestock herd and the unit of discharge load for livestock and poultry, using Equation (4) as shown below:

$$Q_L = n \times Q_i \tag{4}$$

where Q_L is the pollution load from livestock (tons·year^{-1}), n is the number of livestock and poultry (unit) and Q_i is the load of parameter i (kg·unit^{-1}·year^{-1}).

The pollution load arising from aquaculture sources was calculated based on the aquacultural area and the coefficient of each waste generation for each different form of aquaculture, using Equation (5) as shown below:

$$Q_A = Q_i \times S \times t \tag{5}$$

where Q_A is the pollution load from aquacultural activities (tons·year^{-1}), Q_i is the load of the pollution source (kg·ha^{-1}·day^{-1}), S is the area of land used for farming (ha), and t is the time of farming in the year (day).

Table 2 shows the calculation of the waste load generated on the basis of system emissions according to UNEP (1984) [43], San Diego-McGlone (2000) [41]. This estimation method has been successfully applied to many studies [44,45].

2.2. Model Setup

MIKE is a suite of software applications developed by the Danish Hydraulic Institute (DHI), consisting of different models (MIKE 11, MIKE 21, MIKE 3, MIKE SHE, Mouse and MIKE Basin), to accurately analyze, model and simulate rivers, lakes, estuaries and coastal environments. MIKE 11 includes the hydrodynamic module (HD) and EcoLab. The HD module permits the simulation of water levels, discharge and the discharge of wastewaters, while the EcoLab module (WQ module) describes how pollutants travel and disperse along rivers or channels over time. The HD module, applied on open-channel flows, solves finite differences from Saint-Venant equations consisting of the mass conservation and fluid momentum conservation, based on the following assumptions: (i) the flow is a dimension, with depth and velocity varying in the longitudinal direction of the channel; (ii) the bottom slope is small, and scour and deposition are negligible and the channel bed is fixed; (iii) flow everywhere is parallel to the bottom (i.e., wavelengths are large compared with water depths); (iv) the flow is sub-critical; (v) the water is incompressible and homogeneous, i.e., without significant variation in density; and (vi) the lateral inflow does not affect velocity in the channel.

The EcoLab module is based on the conservation of mass, which is a basic principle of the water quality model. It involves performing a mass balance for a defined control volume over a specified period of time. The EcoLab module is coupled to the AD module; while the EcoLab module deals with the transforming processes of compounds in the river, the AD module is used to simulate the simultaneous transport process. The AD equation is based on the following main assumptions: (i) the considered substance is completely mixed over the cross-section, (ii) the substance is conservative or subject to a first-order reaction (linear decay); and (iii) Fick's diffusion law is applied. Fick's law assumes that the mass flux is proportional to the gradient of the mean concentration and that the flux is in the direction of decreasing concentration. Dependent on the nature of the water quality problem under consideration, the model can be adjusted to different levels of detail. The complexity of the model ranges from the most simple version, which includes only BOD_5 and DO, through the introduction of sediment/water interactions and the inclusion of inorganic nitrogen (ammonia and nitrate), to the most complex level, where the BOD_5 is divided into three forms: dissolved, suspended and deposited.

The MIKE 11 model was selected due to its robustness to represent a complex system, its flexibility to include possible future changes and its ability to build different plausible scenarios at different spatial scales. This study used a modified version of a MIKE 11 model to simulate hydrological dynamics over the entire LXQ area in consideration of different components, such as dykes, drainage, sluice gate operations, tidal influences and flood waters [23,46].

The following diagram depicts the modeling process used to replicate the area hydrodynamics and water quality (Figure 2). In the MIKE 11 model, the hydrodynamic (HD), advection–dispersion (AD) and the ecological (EcoLab) modules are the three basic components.

The HD module is built upon the Saint-Venant Abbott's continuity (Equation (6)) and momentum (Equation (7)) equations. The LXQ stations in both the upstream and downstream sections are strongly affected by a diurnal tide in the West Sea and a semi-lunar diurnal tide regime in the East Sea. In the East Sea, the maximum tidal range is quite high (3.0–3.5 m) and the average tide range is approximately 2.5 m, while the tidal range in the West Sea is approximately 1 m [47]. The high average tidal amplitude in the East Sea varies from 2.2 m to 3.8 m, while the low average tidal amplitude in the West Sea is around 0.5 m [48]. As a result, the high tide period was maintained for a short duration, but the low tide period was maintained for a longer duration. The model includes 1581 river or canal segments, 1193 water storage sluice gate structures, over 7500 simulated water level nodes and 4700 simulated discharge nodes (flow). The majority of the river and canal network is seen in Figure 3.

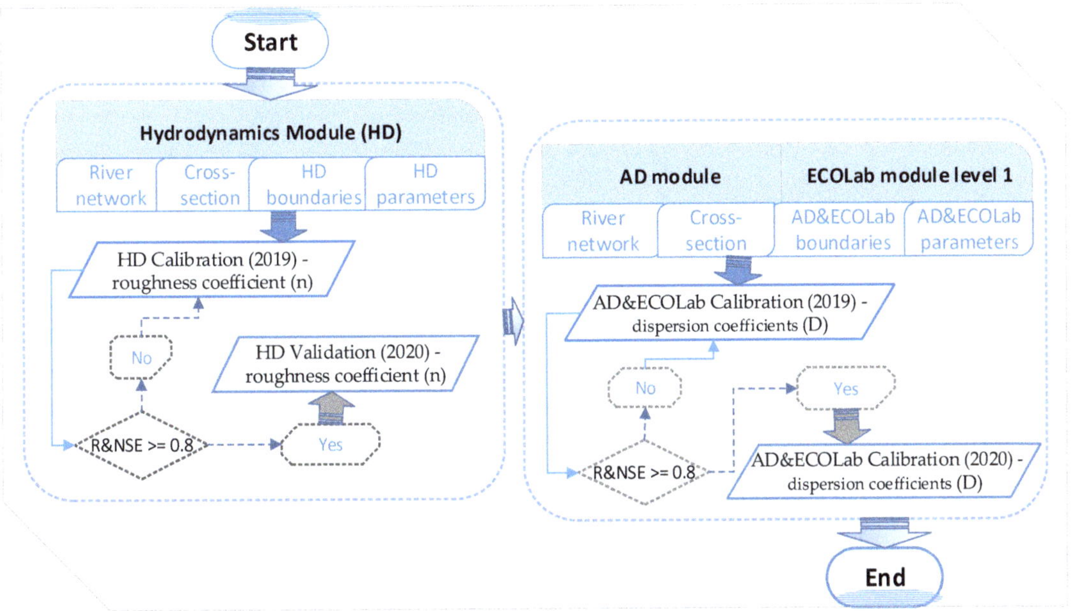

Figure 2. The hydrodynamics and water quality modeling procedure phases utilizing one-dimensional MIKE 11.

$$\frac{\partial Q}{\partial x} + \frac{\partial A}{\partial t} = q \tag{6}$$

$$\frac{\partial Q}{\partial t} + \frac{\partial\left(\alpha \frac{Q^2}{A}\right)}{\partial x} + gA\frac{\partial h}{\partial x} + \frac{n^2 gQ|Q|}{AR^{4/3}} = 0 \tag{7}$$

where Q is the discharge ($m^3 \cdot s^{-1}$), t is time (sec), A is the flow cross-sectional area (m^2), q represents lateral inflow per unit length ($m^3 \cdot s^{-1} \cdot m^{-1}$), g is the gravitational acceleration ($m \cdot s^{-2}$), h represents the height of the water level above sea level, n is the resistance coefficient ($s \cdot m^{-1/3}$), x is the direction, R is the hydraulic or resistance radius (m) and α is the momentum distribution coefficient (e).

The AD module of the MIKE 11 model simulates transportation based on the one-dimensional equation of mass conservation for dissolved or suspended material (Equation (8)). As a result, this module requires the HD module's outputs, such as discharge and water level, cross-section area and hydraulic radius.

$$\frac{\partial AC}{\partial t} + \frac{\partial QC}{\partial x} - \frac{\partial}{\partial x}\left(AD\frac{\partial A}{\partial x}\right) = -AKC + C_2 q \tag{8}$$

where C is concentration ($mg \cdot L^{-1}$), D is the dispersion coefficient ($m^2 \cdot s^{-1}$), A is the cross-sectional area (m^2), K is the linear decay coefficient, C_2 is the source or sink concentration, q is the lateral discharge ($m^3 \cdot s^{-1} \cdot m^{-1}$), x is the space coordinate (m) and t is the time coordinate (sec).

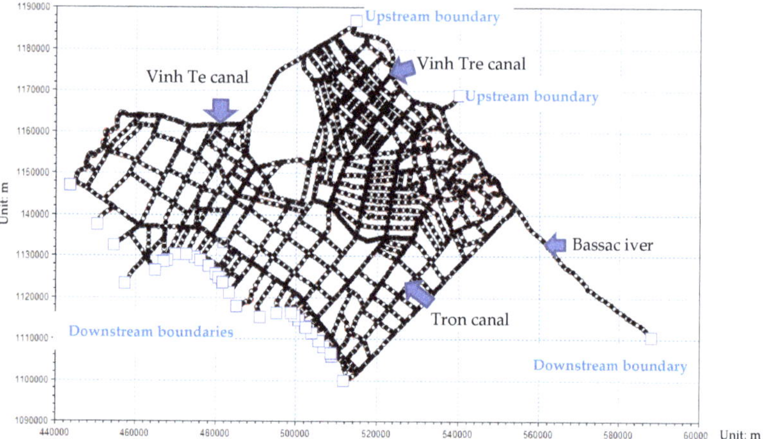

Figure 3. The river and canal network in the LXQ. Blue squares denote the HD (discharge in the upstream and water levels in the downstream) and EcoLab (BOD_5, DO and temperature) boundaries.

The EcoLab module, which is based on a traditional water quality model, can simulate six levels of natural processes, ranging from the simplest BOD_5–COD relationship to complex water quality processes, such as nitrification, denitrification, sediment precipitation and resuspension and sediment oxidation reduction, etc. Based on data availability, this study considered DO, BOD_5 and temperature as water quality indicators. The release of organic waste into rivers causes a drop in DO. As a result, the model explains the relationship between BOD_5 and DO using Equation (9):

$$\frac{dBOD_d}{d_t} = K_d \cdot BOD_d \cdot \theta^{(T-20)} \quad (9)$$

where K_d is the liner decay coefficient, BOD_d is the BOD decay, θ is the temperature coefficient for BOD decay and T is the temperature.

The above equation describes the degradation of dissolved organic materials. The biological activities of aquatic habitats are determined by the amount of DO in the water. The DO regime indicators reveal the degree of organic load and the intensity of the breakdown and mineralization events.

2.3. Calibration and Validation

Changes in the model's parameters, such as Manning's hydraulic roughness coefficient (n) for the HD module, the diffusion coefficient for the AD module, and other parameters in the EcoLab module, such as the degradation of the BOD coefficient (level 1), were used to calibrate and validate the model. Furthermore, both were carried out to assure dependable performance by trial and error until the computed data matched the observed data. Hourly water level data from the Long Xuyen, Rach Gia and Chau Doc stations, as well as hourly discharge data from the Vam Nao station, were used for calibration. The study used water level data from 2020 to validate the model. Because of frequent use in the scientific works, the Nash–Sutcliffe efficiency (NSE), correlation coefficients (R) and root mean square error (RMSE) were employed to verify the model's performance for calibration and validation. The correlation coefficient (R) measures how strongly two variables are related to each other.

Formulas to calculate these three parameters are shown in Equations (10)–(12), starting with the correlation coefficient (Equation (10)):

$$R = \frac{\frac{\sum_{i=1}^{n}\left(X_i-\overline{X}\right)\left(Y_i-\overline{Y}\right)}{(n-1)}}{\sqrt{\frac{1}{n-1}\sum_{i=1}^{n}\left(X_i-\overline{X}\right)^2}\sqrt{\frac{1}{n-1}\sum_{i=1}^{n}\left(Y_i-\overline{Y}\right)^2}} \quad (10)$$

The Nash–Sutcliffe efficiency (NSE) measure determines the magnitude of residual variation in comparison to recorded data variance:

$$NSE = \frac{\sum_{i=1}^{n}\left(X_i-\overline{X}\right)^2 - \sum_{i=1}^{n}(X_i-Y_i)^2}{\sum_{i=1}^{n}\left(X_i-\overline{X}\right)^2} \quad (11)$$

The root mean square error (RMSE) is a measure of how different two datasets are, comparing one predicted value to a known or observed value:

$$RMSE = \sqrt{\frac{1}{n}\sum_{i=1}^{n}(X_i-Y_i)^2} \quad (12)$$

where X_i is the observed data at time i, Y_i is the simulated data at time i, $\overline{X}$ is the mean value of the observed data $\overline{X} = \frac{1}{n}\sum_{i=1}^{n} X_i$ and $\overline{Y}$ is the mean value of the simulated data $\overline{Y} = \frac{1}{n}\sum_{i=1}^{n} Y_i$.

Following that, using the best output data from the HD module, the AD and Eco-Lab modules were calibrated and verified. The calibration step continued until the best modeling output was obtained, while the validation step was used to test the calibrated parameters of the model. The rating of the selected efficiency criteria for R, NSE and RMSE follows the past work of Moriasi (2007) [49].

3. Results

3.1. Calibration and Validation Results of HD Modeling

During the calibration and validation procedure, the Manning's hydraulic roughness coefficient (n) was found to be 0.03 for the global value (and varied between 0.015 and 0.075 for the local values). The hydrographs of simulated and observed tidal amplitude and water levels in 2019 and 2020 at Long Xuyen station were plotted, and the results are shown in Figures 4 and A1, respectively, whilst, the accuracy of the hydrodynamic model is reported in Table 3. Overall, the model's NSE, R and RMSE values were in the range of 0.84–0.90, 0.94–0.96 and 0.15–0.22 for water level performance, respectively. This indicates the model's overall good performance. While the 2019 calibration performed better than the 2020 validation, the NSE value for 2019 is 0.89, compared to 0.85 in 2020, and the RMSE value for 2019 (0.17) is likewise lower than that of 2020 (0.21). During this time, the river morphology, river network, the number of dykes and the sluice systems all changed slightly. In 2019, the comparison of tidal amplitudes performed quite well, with 0.89, 0.73 and 0.04 for NSE, R and RMSE, respectively. The results for tidal amplitude tide verification in 2020 were 0.71, 0.79 and 0.05 for NSE, R and RMSE, respectively. The average tidal amplitude at the Long Xuyen station was 0.8, and varied from 0.52 m to 1 m in 2019. In 2020, the tidal average amplitude was 0.85, and varied from 0.49 m to 1 m. According to research by Phan (2019) [50], which revealed that the tidal amplitude in the East Sea of Vietnam fluctuates approximately less than 1.5 m., and that the tidal amplitude in the river tends to decrease compared that of the coast, this is also consistent with the study of Gagliano (1968) [51]. In addition, the correlation between the observed and simulated water levels and the tidal amplitudes showed strong agreement, as seen in Figure 5. This suggests that the developed HD module has a good performance and can be used for further simulation activities.

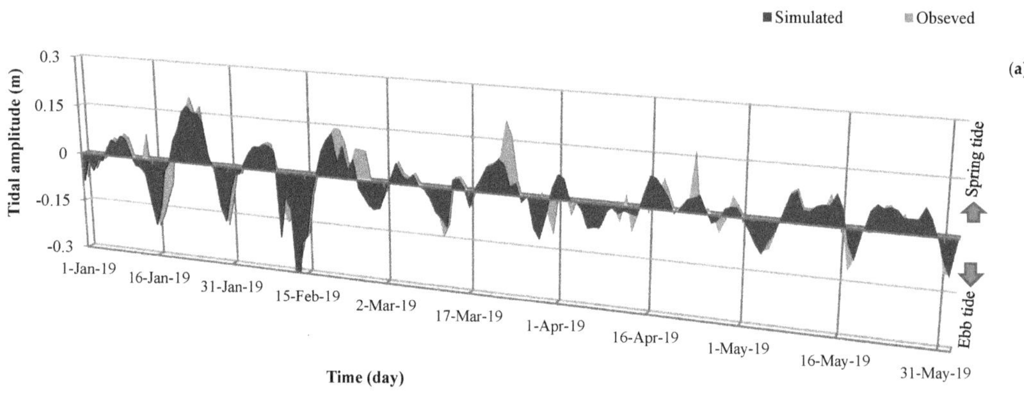

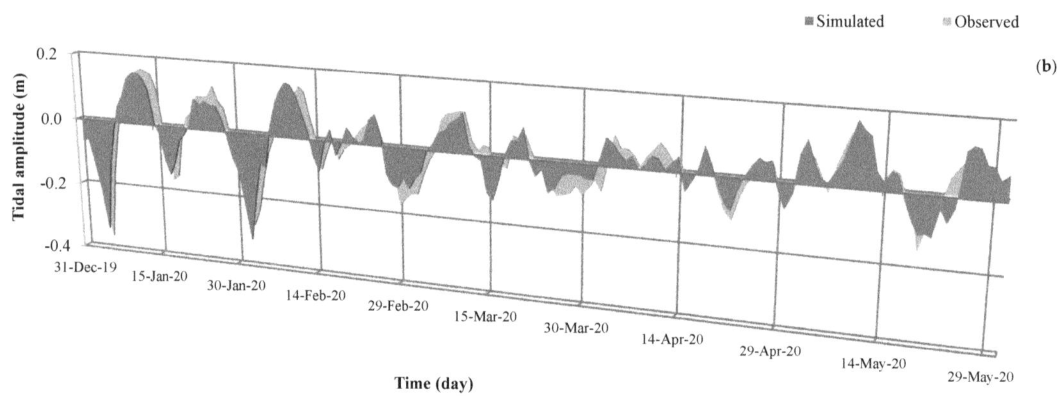

Figure 4. Time series of daily simulated and observed amplitudes in both 2019 (**a**) and 2020 (**b**) at Long Xuyen station.

Table 3. Summary of correlation coefficient (R), Nash–Sutcliffe efficiency (NSE) and root mean square error (RMSE) of water levels in the dry season (Jan–May) in 2019 and 2020 at Long Xuyen station.

	Time	2019			2020		
		NSE	**RMSE**	**R**	**NSE**	**RMSE**	**R**
Water level	January	0.90	0.15	0.96	0.85	0.20	0.94
	February	0.88	0.18	0.96	0.85	0.20	0.96
	March	0.88	0.18	0.95	0.84	0.21	0.96
	April	0.87	0.18	0.95	0.84	0.22	0.96
	May	0.89	0.17	0.95	0.86	0.21	0.95
	Jan–May	0.89	0.17	0.95	0.85	0.21	0.96
Tidal amplitude	Jan–May	0.89	0.04	0.73	0.71	0.05	0.79

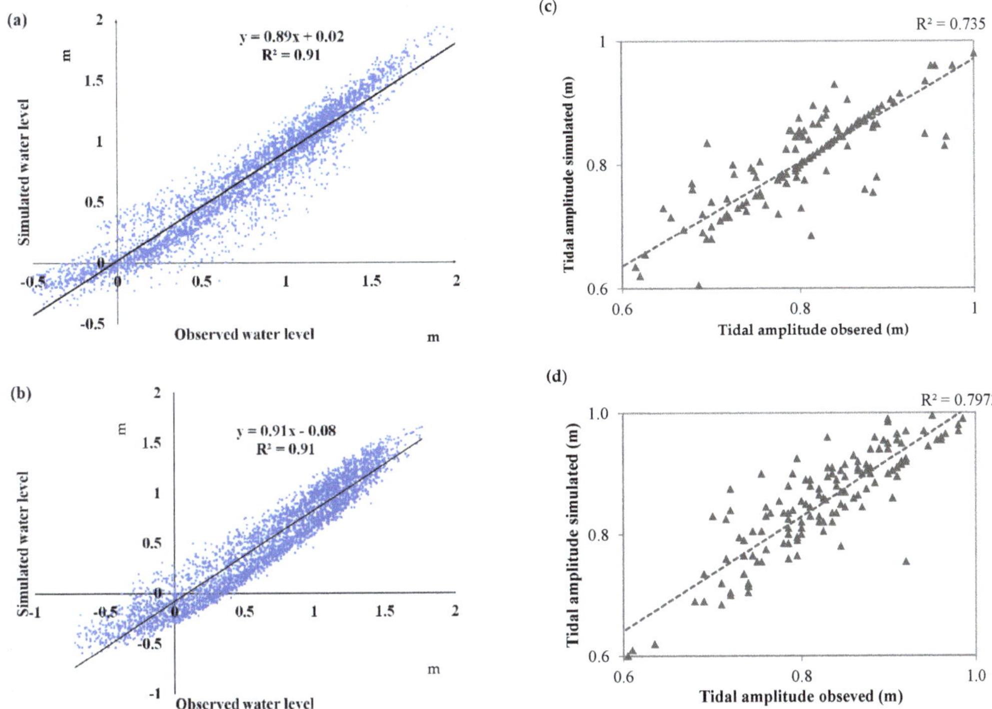

Figure 5. The correlation plots between observed and simulated water level data at an hourly scale for the dry seasons in (**a**) 2019 and (**b**) 2020. The correlation plots between observed and simulated tidal amplitude at a daily scale for the dry seasons in (**c**) 2019 and (**d**) 2020.

3.2. Calibration and Validation Results of Water Quality Modeling

The EcoLab module is linked to the advection–dispersion (AD) module, which describes both the transformation and transport processes of pollutants. Therefore, the relationship between BOD_5 and DO in different conditions can be used to describe the fluvial water quality. In the calibration step, the dispersion coefficients were seen to vary between 50 and 700, and BOD decay between 0.1 and 1.5. Figure 6 illustrates that the BOD_5 concentration difference between simulated and observed data varies by about 12 percent, 22 percent, 24 percent and 37 percent for first, second, third and fourth sites, respectively. The observed BOD_5 data of the first site were highest, while the accuracy of the fourth site was lowest. The lack of statistics on pollution load at site 4 is due to its proximately to Vinh Te Canal, which forms the boundary between Vietnam and Cambodia.

The simulation findings of water quality metrics demonstrate that the trend for temporal variation for the pollutants were similar between the main river channels and the smaller rivers and canals. The validation results of BOD_5 concentration are shown in Figure 7.

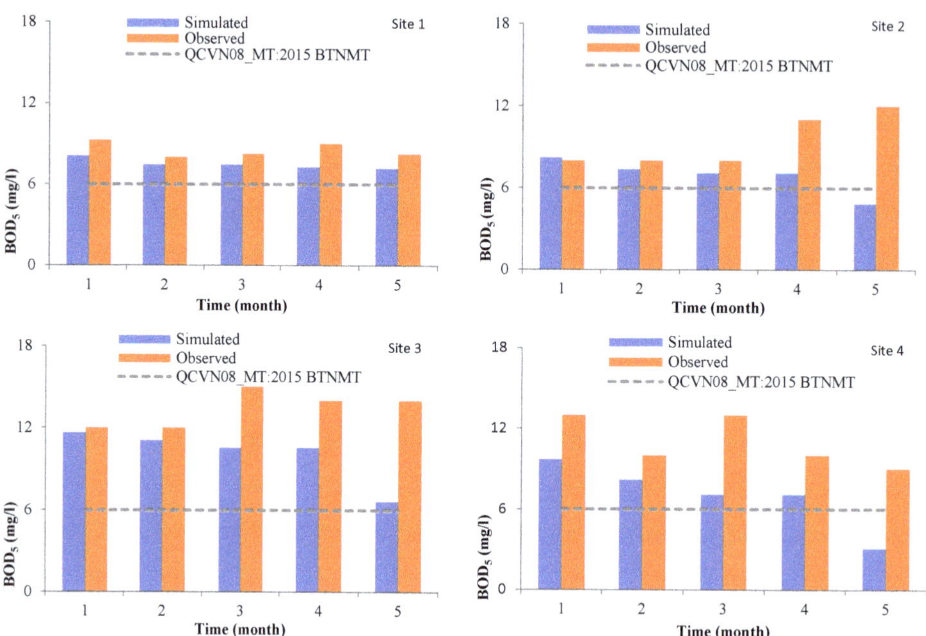

Figure 6. Calibrated results of BOD_5 concentration at the four sites within the study area for the year 2019. The National Technical Regulation on surface water quality (QCVN 08MT:2015) was approved by the Ministry of Environment and Natural Resources (MoNRE) in 2015.

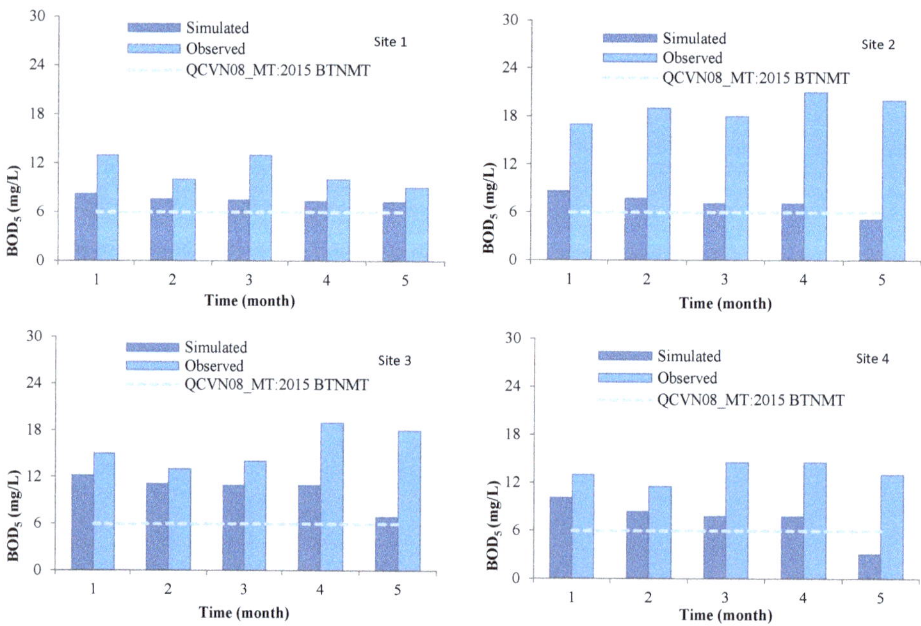

Figure 7. Validated results of BOD_5 concentration at the four locations within the study area for the year 2020.

For the first, second, third and fourth sites, the errors between the simulated and observed data were 30 percent, 62 percent, 32 percent and 44 percent, respectively. The accuracy of the first location was the highest, while the accuracy of the second locality was the lowest. Rapid riverbank and riverbed erosion and sedimentation in the research area has led to recent changes in the slope and cross-section of rivers and canals. Moreover, at site no. 2, as well as sites 3 and 4, located in the canal with low discharge volumes, the immediate pollution concentration is often strongly affected directly by the discharge source. In site no. 1, on the main river channel, a much higher discharge volume resulted in quicker pollution dilution, while pollutant concentrations fluctuated less in small canals.

The water quality model performed DO concentrations are in good agreement with the observed DO concentrations in 2019 and 2020 (Figure A2). For the first, second, third and fourth sites, the difference between simulated and observed DO values, respectively, varied by about 14 percent, 11 percent, 14 percent and 13 percent in 2019, and 7 percent, 13 percent, 12 percent and 10 percent in 2020.

An ANOVA was used to further analyze water quality data, i.e., BOD_5, DO and temperature, at stations along the Bassac River, infield canals and aquacultural site canals for the years 2019 and 2020. The aquacultural regions had the highest average BOD_5 (12.88 mg/L), followed by the infield canals (11.55 mg/L) and the Bassac River (9.07 mg/L) in 2019. The difference between BOD_5 in aquaculture and BOD_5 in the Bassac River, on the other hand, was only detected at a 5% significance level in the study. In the aquacultural area, the typical DO and temperature were around 4 mg/L and 30°C, respectively. At a 5% significance level, the aquacultural area's average DO (4.47 mg/L) and temperature (29.78°C) in 2020 differed from the river's average DO (5.04 mg/L) and temperature (29.08°C). The aquatic region had the lowest BOD_5 (13.65 mg/L), while the Bassac River (15.47 mg/L) and the infield canals (15.17 mg/L) displayed the highest. We found that water pollution, such as BOD_5 concentrations in the year of 2020, was lower than in the year of 2019. We suspect the reason is due to the impact of the COVID-19 pandemic, leading to decreased fish production and resulting in a decreased BOD_5 load.

However, according to QCVN 08-MT: 2015/BTNMT, column A2 of the National Technical Regulation on surface water quality, the surface water quality deterioration by aquaculture in 2019 and 2020 found in this study was far in exceedance of the desired water quality threshold. The DO concentration of water, for instance, was lower than the QCVN 08 threshold. Hence, when utilizing this water for domestic purposes, it is advised that people pretreat it carefully before use to ensure no long-term adverse health effects. This water quality model may be used by key stakeholders to forecast water quality in the aquacultural area so that appropriate in situ water treatment measures can be located and implemented. Furthermore, numerous studies have revealed transboundary environmental degradation caused by the Mekong River's flow through two major tributaries, the Mekong River and the Bassac River. The result highlights the need for surface water quality control checks in border areas.

Figure 8 shows the spatial distribution of BOD_5 max, BOD_5 min and BOD_5 max–BOD_5 min. The parameters are highest in the northwest and decrease gradually towards the southeast. High BOD_5 levels can be found in both Chau Doc and Tri Ton. Chau Doc is an urban area bordering Cambodia with moderate tourism activities. The main sources of wastewater in Chau Doc are from urban areas and tourism (average BOD_5 max and BOD_5 min concentrations are 25 mg/L and 17 mg/L, respectively). The population density in Chau Doc and the larger city of Long Xuyen was 963 and 2368 people per square kilometer in the year 2020, respectively. The remaining urban areas, particularly in Kien Giang province, have vastly lower population densities, whilst the impacts of urbanization on fluvial water quality remain minimal in coastal areas. Water quality in and around Long Xuyen is heavily affected not only by production activities, but also by waste from urban residential areas. Tri Ton, a semi-mountainous area in An Giang, is a popular tourist destination that is also undergoing rapid urbanization. Here, moderate levels of water quality were recorded, with the main source of pollution load being from surrounding aquacultural activities. Here, averages of BOD_5 max and BOD_5 min were 16 mg/L and 12 mg/L, respectively. However, on the main river (Bassac River) and inland canals, BOD_5 max was 17 mg/L and 13 mg/L, respectively. Aquacultural practices in the area to the north of Bassac River are typically in the form of in-river cage culture farming, while to the south, aquacultural production is typically practiced in ponds and lagoons. Overall, our results showed that the areas on or directly adjacent to the main river channels showed more variability due to the influence of higher flow volumes and exhibit, therefore, a better self-cleaning capacity [2,3]. Furthermore, according to Minh et al. [2,42], high pollution levels in the Vinh Te Canal have a degrading impact on the wider study area.

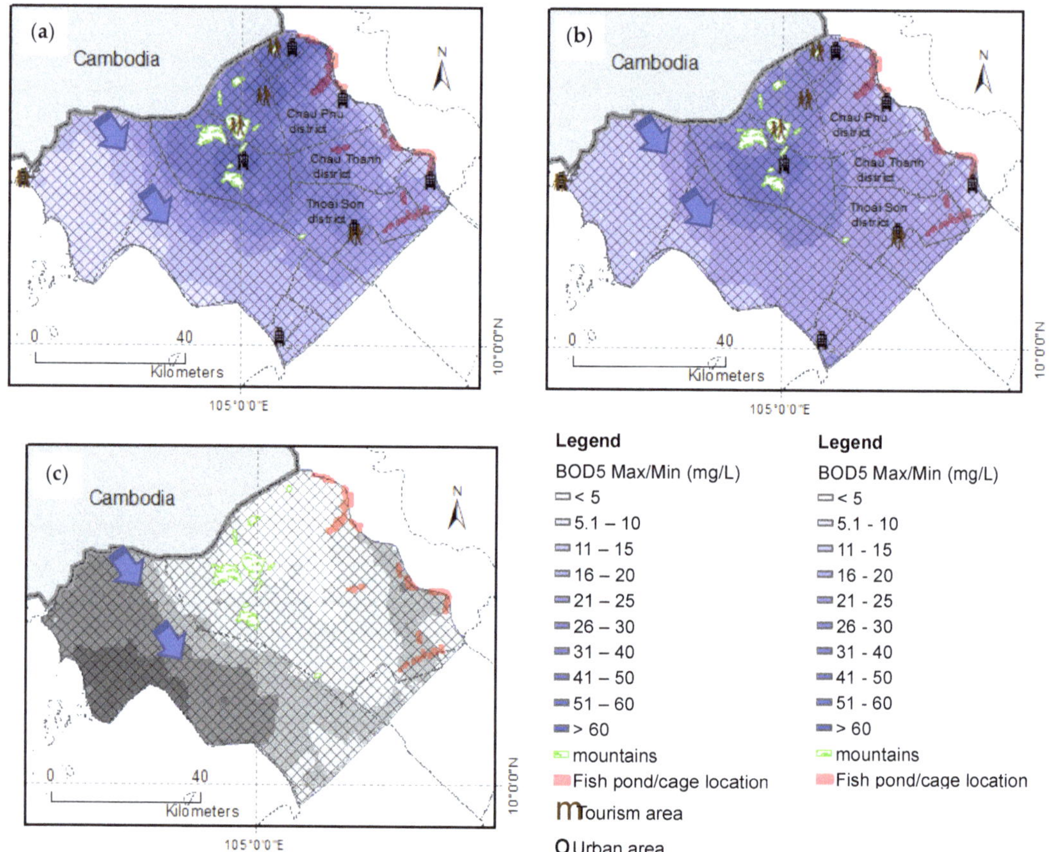

Figure 8. BOD_5 concentration in the dry season of 2020. Here, (**a**–**c**) represent the maximum, minimum and the difference between the maximum and minimum BOD_5 concentration values, respectively. Moreover, (**a**, **b**) also show the location of urban and tourism areas, and provincial boundaries. These arrows show the downtrends of BOD_5 concentration (**a**, **b**) and the downtrends of BOD_5 max–BOD_5 min.

High DO levels were found in the main river (Bassac) and main canals (Vinh Te Canal), while DO_5 max–DO_5 min, on the other hand, tends to follow the northwest direction (Figure 9).

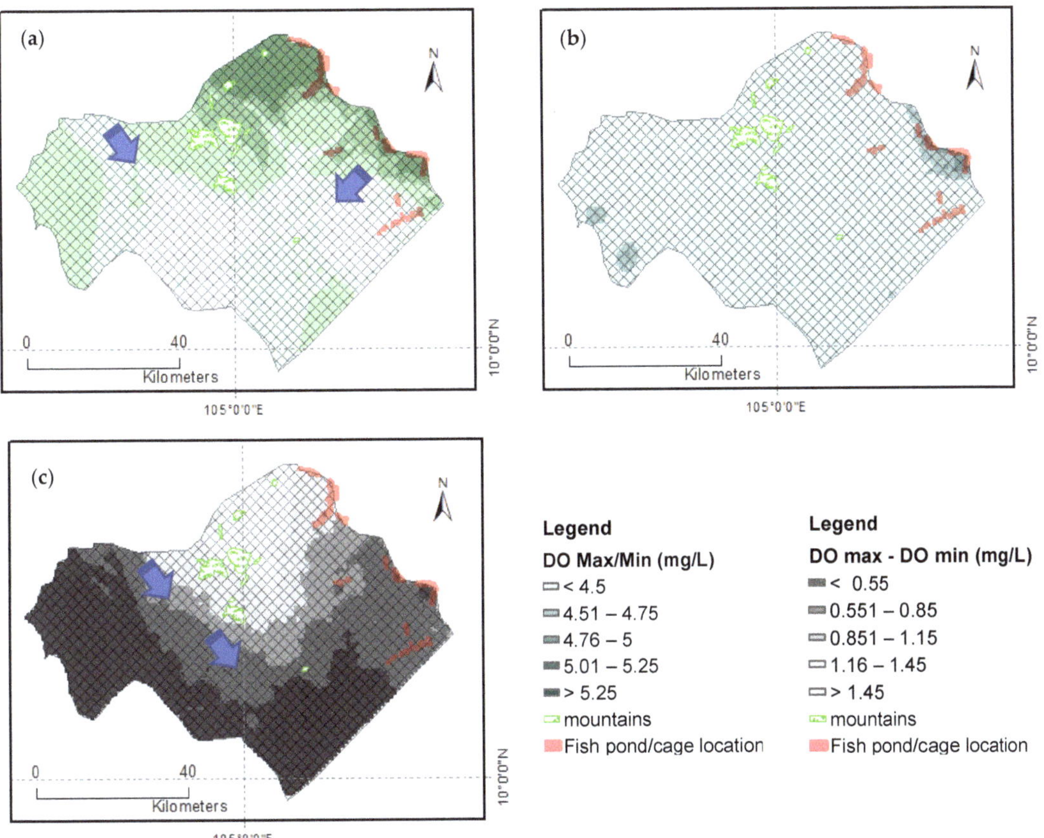

Figure 9. DO concentration in the dry season of 2020. Here, (**a**–**c**) represent the maximum, minimum and difference between the maximum and minimum DO concentration values, respectively. These arrows show the downward trend of DO concentration (9a,b) and the downward trend of DO max–DO min.

4. Discussion

Since the Streeter and Phelps model [52], more than a hundred further water quality models have been developed. The surface water quality models have progressed through many major stages, from assessing a single factor of water quality to multiple factors of water quality, from steady- to non-steady-state hydraulic models, from a point source model to a coupling-model of point and nonpoint sources and from one-dimensional to two-dimensional and three-dimensional models [53,54].

MIKE 11 is an advanced model with water quality management capabilities [55]. According to several studies, MIKE 11 necessitates a large amount of data and, in the case of using a small time step to ensure the stability of the model, simulation step takes a long time [56]. However, MIKE 11 also allows the user to select a simpler model that is appropriate for the target and dataset available. The EcoLab module, in particular, provides six different levels of conceptual water quality modeling complexity.

Moreover, some challenges remain, as models typically require a large amount of accurate data, an absence of which may result in a significant difference between the reality and the simulation results. Further, models do not include pollutant mechanisms, and are ambiguous about contaminant migration, meaning that we cannot predict how contaminants will migrate. Combination models are becoming popular for water quality modeling due to the complexity of water quality issues. In the case of water quality modeling, these models include combination models, artificial intelligence models and system integration [55].

In this study, MIKE 11 provided good results for the hydrodynamics simulation because the input data for the study area was of a high temporal resolution (hourly). However, the water quality monitoring data consisted of only monthly averages. As such, changes in water quality were not clearly visible. The use of an Artificial Neural Network (ANN) to overcome this limitation produced significantly better results than hydrodynamics and water quality modeling, as evidenced by goodness-of-fit indices. The ANN model, developed by McCulloch [57], consists of three layers, and the number of layer increases with the complexity. The ANN model has been used to predict some key water quality parameters in recent years, with results indicating that its accuracy is adequate for practical purposes.

Previous research also suggests that the ANN performs better in predicting water level, as well as nitrate and phosphate, compared to Support Vector Machine models [58,59]. Moreover, Rabindra et al. reveal that ANN does not require other physical parameters in the modeling process, which can reduce the complexities of modeling the system [60]. Most numerical models, on the other hand, have not been fully validated against field experimental data, which often requires large investments of time and capital to obtain. Data-based models have been used to replace the numerical model because water quality predictions can be made using only accumulated data. The ANN is one of many data-driven techniques that has been widely applied due to its efficiency in predicting and forecasting water quantity and quality variables in river systems [61].

5. Conclusions

The hydrodynamics module developed in this study was calibrated and validated to be in good agreement with the observed data. After successful flow simulation, the AD and EcoLab modules of MIKE 11 were used to simulate the surface water quality of the LXQ using discharge and water level data from existing the HD module in the dry season. The results of the BOD_5 transmission simulation meet the practical needs of the study area by assisting the appropriate management and planning of aquacultural areas in order to reduce pollution and the impact of harmful chemicals from other sources

High BOD_5 pollution was seen in most pond and lagoon aquacultural sites during the dry season in both 2019 and 2020. This is because ponds and lagoons have frequently slower growth stages, are less well ventilated than cage freshwater rearing areas, with limited water exchange and self-cleaning. However, the areas close to cage culture farming areas were found to have higher BOD_5 than areas surrounding pond/lagoon culture. The reason is due to the influences from other point pollution sources, such as upstream wastewater urban and tourism areas. Therefore, before locating and promoting aquacultural production, it is necessary to test site water quality. Furthermore, aquacultural areas should be located and zoned far from urban and tourist areas. Overall, the management and effective sustainable use of water resources needs to be strengthened to ensure the overall suitability of aquacultural locations.

Additionally, considering the data scarcity, several automatic measurement stations are proposed to aid the understanding of the spatio-temporal dynamics of water quality parameters, as well as data for improved model calibration and verification. Furthermore, diligent monitoring of all important hydrogeochemical parameters, such as salinity, isotopes/tracer elements, etc., which help in deciphering the seawater–freshwater mixing mechanisms in this dynamic hydrological system, should be considered as a future course of work.

Author Contributions: Conceptualization, V.N.U., V.P.D.T. and H.V.T.M.; methodology, V.N.U., V.P.D.T., T.T.T.D. and H.V.T.M.; software, T.T.T.D., A.V.H., V.P.D.T. and H.V.T.M.; writing—original draft preparation, V.N.U., V.P.D.T., R.A., P.K., T.V.T., T.T.T.D., N.K.D. and H.V.T.M.; writing—review and editing, V.N.U., V.P.D.T., R.A., P.K., T.V.T., N.K.D. and H.V.T.M. All authors have read and agreed to the published version of the manuscript.

Funding: This study is funded in part by the Can Tho University Improvement Project VN14-P6, supported by a Japanese ODA loan.

Institutional Review Board Statement: Not applicable.

Informed Consent Statement: Not applicable.

Data Availability Statement: Not applicable.

Conflicts of Interest: The authors declare no conflict of interest.

Appendix A

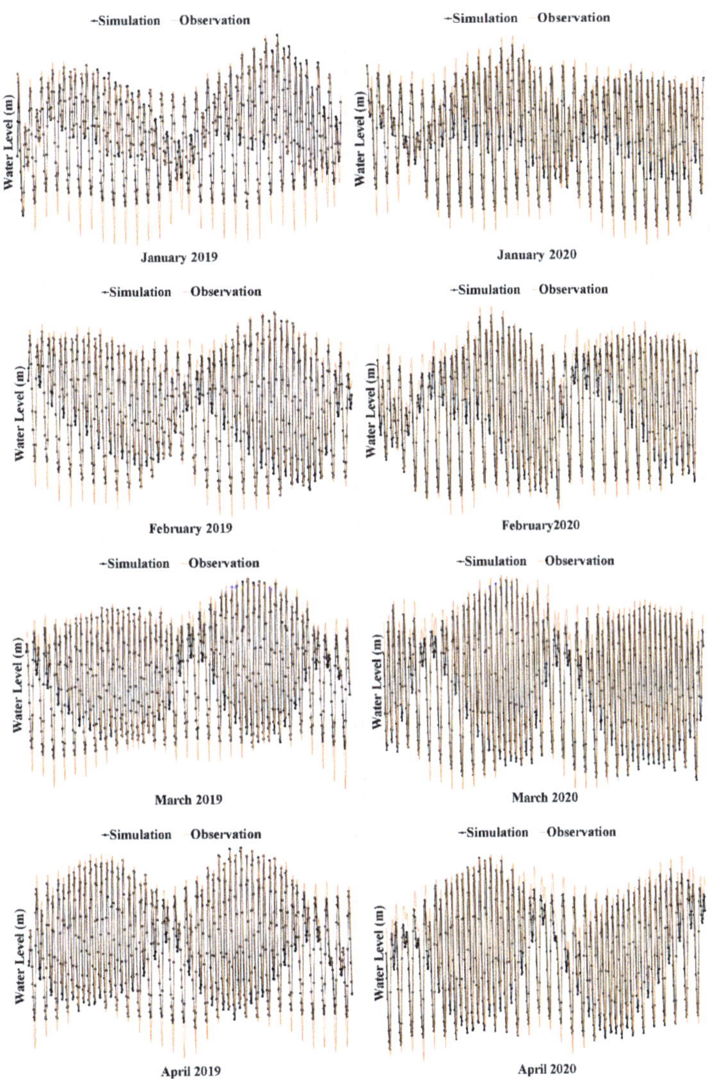

Figure A1. *Cont.*

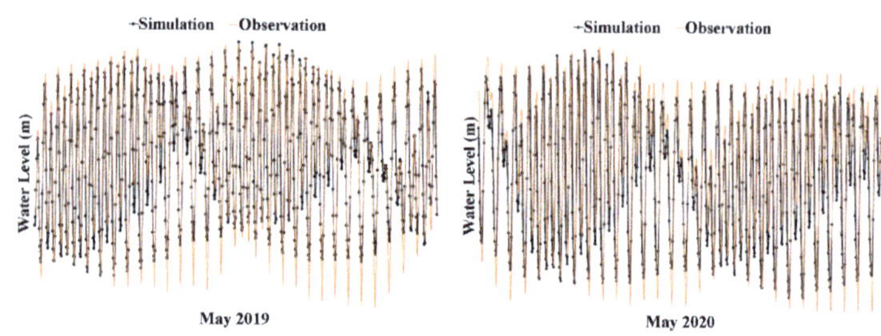

Figure A1. Comparison of simulated and observed water levels at the Long Xuyen station in the dry season 2019.

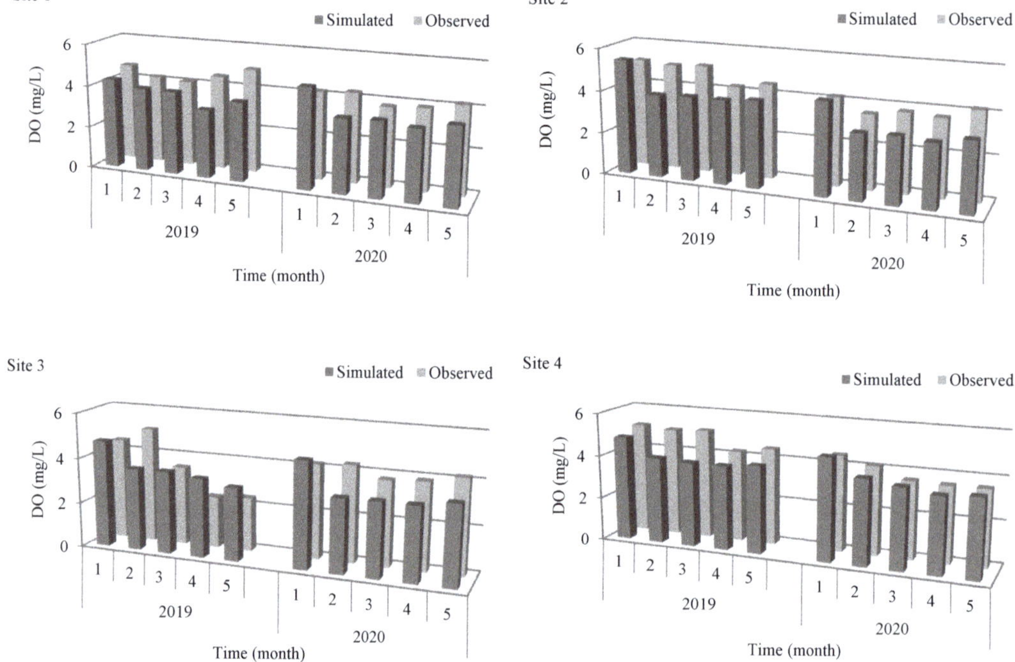

Figure A2. Calibrated and verified results of DO concentration at the four locations within the study area for the years 2019 and 2020.

Appendix B

Table A1. Location of aquaculture ponds and lagoons in the study area.

Location of Ponds and Lagoons	Locations of Water Quality Monitoring
Ponds and lagoons in Vinh Thanh Trung (Vinh Tre Canal)	TS5(TĐ)-CP
Ponds and lagoons Binh Thanh (Hau River)	TS6(TĐ)-CT
Ponds and lagoons My Hoa Hung (Hau River)	TS7(TĐ)-LX
Ponds and lagoons My Hoa Hung	TS8(TĐ)-LX
Ponds and lagoons Phu Thuan (Xa Doi Canal)	TS10(TĐ)-TS
Ponds and lagoons My Thoi (Cai Sao River)	TS11(TĐ)-LX
Ponds and lagoons Phú Thuận (Don Dong Canal)	TS12(TĐ)-TS
Ponds and lagoons Vinh Hanh (Nui Chac Canal)	TS13(TĐ)-CT
Ponds and lagoons Phu Thuan (Don Dong Canal)	TS14(TĐ)-TS
Ponds and lagoons Vinh Khanh (Cai Sao River)	TS15(TĐ)-TS

References

1. Duc, N.H.; Avtar, R.; Kumar, P.; Lan, P.P. Scenario-Based Numerical Simulation to Predict Future Water Quality for Developing Robust Water Management Plan: A Case Study from the Hau River, Vietnam. *Mitig. Adapt. Strateg. Glob. Chang.* **2021**, *26*, 33. [CrossRef]
2. Thu Minh, H.V.; Avtar, R.; Kumar, P.; Le, K.N.; Kurasaki, M.; Ty, T.V. Impact of Rice Intensification and Urbanization on Surface Water Quality in An Giang Using a Statistical Approach. *Water* **2020**, *12*, 1710. [CrossRef]
3. Minh, H.V.T.; Kurasaki, M.; Ty, T.V.; Tran, D.Q.; Le, K.N.; Avtar, R.; Rahman, M.M.; Osaki, M. Effects of Multi-Dike Protection Systems on Surface Water Quality in the Vietnamese Mekong Delta. *Water* **2019**, *11*, 1010. [CrossRef]
4. Vörösmarty, C.J.; McIntyre, P.B.; Gessner, M.O.; Dudgeon, D.; Prusevich, A.; Green, P.; Glidden, S.; Bunn, S.E.; Sullivan, C.A.; Liermann, C.R.; et al. Global Threats to Human Water Security and River Biodiversity. *Nature* **2010**, *467*, 555. [CrossRef] [PubMed]
5. Eng, C.T.; Paw, J.N.; Guarin, F.Y. The Environmental Impact of Aquaculture and the Effects of Pollution on Coastal Aquaculture Development in Southeast Asia. *Mar. Pollut. Bull.* **1989**, *20*, 335–343.
6. Schneider, P.; Asch, F. Rice Production and Food Security in Asian Mega Deltas—A Review on Characteristics, Vulnerabilities and Agricultural Adaptation Options to Cope with Climate Change. *J. Agron. Crop Sci.* **2020**, *206*, 491–503. [CrossRef]
7. Berg, H.; Tam, N.T. Use of Pesticides and Attitude to Pest Management Strategies among Rice and Rice-Fish Farmers IntheMekong Delta, Vietnam. *Int. J. Pest Manag.* **2012**, *58*, 153–164. [CrossRef]
8. Anh, P.T.; Kroeze, C.; Bush, S.R.; Mol, A.P. Water Pollution by Pangasius Production in the Mekong Delta, Vietnam: Causes and Options for Control. *Aquac. Res.* **2010**, *42*, 108–128. [CrossRef]
9. Mutea, F.G.; Nelson, H.K.; Au, H.V.; Huynh, T.G.; Vu, U.N. Assessment of Water Quality for Aquaculture in Hau River, Mekong Delta, Vietnam Using Multivariate Statistical Analysis. *Water* **2021**, *13*, 3307. [CrossRef]
10. Bouman, B.A.M.; Tuong, T.P. Field Water Management to Save Water and Increase Its Productivity in Irrigated Lowland Rice. *Agric. Water Manag.* **2001**, *49*, 11–30. [CrossRef]
11. Giao, N.T. Surface Water Quality in Aquacultural Areas in An Giang Province, Vietnam. *Int. J. Environ. Agric. Biotechnol.* **2020**, *5*, 1054–1061. [CrossRef]
12. Ndikumana, E.; Ho Tong Minh, D.; Dang Nguyen, T.H.; Baghdadi, N.; Courault, D.; Hossard, L.; El Moussawi, I. Estimation of Rice Height and Biomass Using Multitemporal SAR Sentinel-1 for Camargue, Southern France. *Remote Sens.* **2018**, *10*, 1394. [CrossRef]
13. Phung, D.; Huang, C.; Rutherford, S.; Dwirahmadi, F.; Chu, C.; Wang, X.; Nguyen, M.; Nguyen, N.H.; Do, C.M.; Nguyen, T.H. Temporal and Spatial Assessment of River Surface Water Quality Using Multivariate Statistical Techniques: A Study in Can Tho City, a Mekong Delta Area, Vietnam. *Environ. Monit. Assess.* **2015**, *187*, 229. [CrossRef] [PubMed]
14. Dunca, A.-M. Water Pollution and Water Quality Assessment of Major Transboundary Rivers from Banat (Romania). *J. Chem.* **2018**, *2018*, 9073763. [CrossRef]
15. Ighalo, J.O.; Adeniyi, A.G. A Comprehensive Review of Water Quality Monitoring and Assessment in Nigeria. *Chemosphere* **2020**, *260*, 127569. [CrossRef] [PubMed]
16. Avtar, R.; Kumar, P.; Singh, C.; Mukherjee, S. A Comparative Study on Hydrogeochemistry of Ken and Betwa Rivers of Bundelkhand Using Statistical Approach. *Water Qual. Expo. Health* **2011**, *2*, 169–179. [CrossRef]
17. Molekoa, M.D.; Avtar, R.; Kumar, P.; Minh, H.V.T.; Kurniawan, T.A. Hydrogeochemical Assessment of Groundwater Quality of Mokopane Area, Limpopo, South Africa Using Statistical Approach. *Water* **2019**, *11*, 1891. [CrossRef]

18. Singh, S.; Ghosh, N.; Gurjar, S.; Krishan, G.; Kumar, S.; Berwal, P. Index-Based Assessment of Suitability of Water Quality for Irrigation Purpose under Indian Conditions. *Environ. Monit. Assess.* **2018**, *190*, 29. [CrossRef] [PubMed]
19. Minh, H.V.T.; Avtar, R.; Kumar, P.; Tran, D.Q.; Ty, T.V.; Behera, H.C.; Kurasaki, M. Groundwater Quality Assessment Using Fuzzy-AHP in An Giang Province of Vietnam. *Geosciences* **2019**, *9*, 330. [CrossRef]
20. Kumar, P. Numerical Quantification of Current Status Quo and Future Prediction of Water Quality in Eight Asian Megacities: Challenges and Opportunities for Sustainable Water Management. *Environ. Monit. Assess.* **2019**, *191*, 319. [CrossRef]
21. Kumar, P.; Johnson, B.A.; Dasgupta, R.; Avtar, R.; Chakraborty, S.; Kawai, M.; Magcale-Macandog, D.B. Participatory Approach for More Robust Water Resource Management: Case Study of the Santa Rosa Sub-Watershed of the Philippines. *Water* **2020**, *12*, 1172. [CrossRef]
22. Kumar, P.; Avtar, R.; Dasgupta, R.; Johnson, B.A.; Mukherjee, A.; Ahsan, M.N.; Nguyen, D.C.H.; Nguyen, H.Q.; Shaw, R.; Mishra, B.K. Socio-Hydrology: A Key Approach for Adaptation to Water Scarcity and Achieving Human Well-Being in Large Riverine Islands. *Prog. Disaster Sci.* **2020**, *8*, 100134. [CrossRef]
23. Hanington, P.; To, Q.T.; Van, P.D.T.; Doan, N.A.V.; Kiem, A.S. A Hydrological Model for Interprovincial Water Resource Planning and Management: A Case Study in the Long Xuyen Quadrangle, Mekong Delta, Vietnam. *J. Hydrol.* **2017**, *547*, 1–9. [CrossRef]
24. Liang, J.; Yang, Q.; Sun, T.; Martin, J.; Sun, H.; Li, L. MIKE 11 Model-Based Water Quality Model as a Tool for the Evaluation of Water Quality Management Plans. *J. Water Supply Res. Technol.* **2015**, *64*, 708–718. [CrossRef]
25. Gedam, V.; Kelkar, P.; Jha, R.; Khadse, G.; Labhasetwar, P. Assessment of Assimilative Capacity of Kanhan River Stretch Using Mike-11 Modeling Tool Using Mike-11 Modeling Tool. *J Environ. Sci. Engg.* **2012**, *54*, 481–488.
26. Arnold, J.G.; Srinivasan, R.; Muttiah, R.S.; Williams, J.R. Large Area Hydrologic Modeling and Assessment Part I: Model Development. *JAWRA J. Am. Water Resour. Assoc.* **1998**, *34*, 73–89. [CrossRef]
27. Dang, T.D.; Cochrane, T.A.; Arias, M.E. Future Hydrological Alterations in the Mekong Delta under the Impact of Water Resources Development, Land Subsidence and Sea Level Rise. *J. Hydrol. Reg. Stud.* **2018**, *15*, 119–133. [CrossRef]
28. Smajgl, A.; Toan, T.Q.; Nhan, D.K.; Ward, J.; Trung, N.H.; Tri, L.; Tri, V.; Vu, P. Responding to Rising Sea Levels in the Mekong Delta. *Nat. Clim. Chang.* **2015**, *5*, 167–174. [CrossRef]
29. Chinh, P.V. Application of Mathematical Models to Assess Water Quality in the Downstream of Dong Nai River up to 2020. Master thesis. Da Nang University, Da Nang City, Vietnam. 2011. Master's Thesis, Da Nang University, Da Nang City, Vietnam, 2011.
30. Johnston, R.; Kummu, M. Water Resource Models in the Mekong Basin: A Review. *Water Resour. Manag.* **2012**, *26*, 429–455. [CrossRef]
31. Water Environment Partnership in Asia. Surface Water in Vietnam. *State of Water Environmental Issues.* Available online: http://www.wepa-db.net/policies/state/vietnam/surface.htm (accessed on 30 March 2021).
32. Kuenzer, C.; Guo, H.; Huth, J.; Leinenkugel, P.; Li, X.; Dech, S. Flood Mapping and Flood Dynamics of the Mekong Delta: ENVISAT-ASAR-WSM Based Time Series Analyses. *Remote Sens.* **2013**, *5*, 687–715. [CrossRef]
33. Vietnamese Government Resolution No. 120/NQ-CP-Resolution on Sustainable Climate-Resilient Development of the Vietnamese Mekong Delta, Vietnames Government, Vietnam. 2017; Available online: https://english.luatvietnam.vn/resolution-no120-nq-cp-dated-november-17-2017-of-the-government-on-sustainable-and-climate-resilient-development-of-the-mekong-river-delta-118378-Doc1.html (accessed on 18 November 2021).
34. Wilbers, G.-J.; Becker, M.; Nga, L.T.; Sebesvari, Z.; Renaud, F.G. Spatial and Temporal Variability of Surface Water Pollution in the Mekong Delta, Vietnam. *Sci. Total Environ.* **2014**, *485–486*, 653–665. [CrossRef] [PubMed]
35. Dinh, Q.; Balica, S.; Popescu, I.; Jonoski, A. Climate Change Impact on Flood Hazard, Vulnerability and Risk of the Long Xuyen Quadrangle in the Mekong Delta. *Int. J. River Basin Manag.* **2012**, *10*, 103–120. [CrossRef]
36. Duc Tran, D.; van Halsema, G.; Hellegers, P.J.G.J.; Phi Hoang, L.; Quang Tran, T.; Kummu, M.; Ludwig, F. Assessing Impacts of Dike Construction on the Flood Dynamics of the Mekong Delta. *Hydrol. Earth Syst. Sci.* **2018**, *22*, 1875–1896. [CrossRef]
37. Manh, N.V.; Dung, N.V.; Hung, N.N.; Merz, B.; Apel, H. Large-Scale Suspended Sediment Transport and Sediment Deposition in the Mekong Delta. *Hydrol. Earth Syst. Sci.* **2014**, *18*, 3033–3053. [CrossRef]
38. Le Anh Tuan, C.T.H.; Miller, F.; Sinh, B.T. Flood and Salinity Management in the Mekong Delta, Vietnam. The Sustainable Mekong Research Network (Sumernet): Bangkok, Thailand, 2007; pp. 15–68.
39. Balica, S.; Dinh, Q.; Popescu, I.; Vo, T.Q.; Pham, D.Q. Flood Impact in the Mekong Delta, Vietnam. *J. Maps* **2014**, *10*, 257–268. [CrossRef]
40. DoNRE. An Giang. *Summary Report on Environmental Monitoring Results in An Giang Province in 2020*; DoNRE: Long Xuyen, Vietnam, 2020; p. 361.
41. San Diego-McGlone, M.L.; Smith, S.V.; Nicolas, V.F. Stoichiometric Interpretations of C:N:P Ratios in Organic Waste Materials. *Mar. Pollut. Bull.* **2000**, *40*, 325–330. [CrossRef]
42. Minh, H.V.T.; Tam, N.T.; Nhu, Đ.T.; Ty, T.V. Assessment of the Surface Water Quality and Effectiveness of Triple-Glutinous Rice Cropping System in the Full–Dike Protected Area of Bac Vam Nao, An Giang Province. *Vietnam J. Hydrometeorol.* **2021**, *732*, 38–48. [CrossRef]
43. UNEP. *Pollutants from Land-Based Sources in the Mediterranean*; UNEP Regional Seas Reports and Studies No. 3; UNEP (United Nations Environment Programme): Washingtone DC, USA, 1984; p. 104.
44. Le, X.S.; Le, V.N. Load of Pollutions onto Da Nang Bay. *J. Mar. Sci. Technol.* **2015**, *15*, 165–175. [CrossRef]

45. Mai, T.H.; Ngô, X.N.; Trần, V.T.; Mai, T.H. Research to Determine Pollutant Load into Truong Giang River, Quang Nam Province. *J. Clim. Change Science* **2018**, *34*, 71–79.
46. Dung, N.V.; Merz, B.; Bárdossy, A.; Thang, T.D.; Apel, H. Multi-Objective Automatic Calibration of Hydrodynamic Models Utilizing Inundation Maps and Gauge Data. *Hydrol. Earth Syst. Sci.* **2011**, *15*, 1339–1354. [CrossRef]
47. Saito, Y.; Nguyen, V.L.; Ta, T.K.O.; Tamura, T.; Kanai, Y.; Nakashima, R. *Tide and River Influences on Distributary Channels of the Mekong River Delta*; American Geophysical Union: Washington, DC, USA, 2015; Volume 2015, p. GC41F-1148.
48. Wolanski, E.; Huan, N.N.; Nhan, N.H.; Thuy, N.N. Fine-Sediment Dynamics in the Mekong River Estuary, Vietnam. *Estuar. Coast. Shelf Sci.* **1996**, *43*, 565–582. [CrossRef]
49. Moriasi, D.N.; Arnold, J.G.; Van Liew, M.W.; Bingner, R.L.; Harmel, R.D.; Veith, T.L. Model Evaluation Guidelines for Systematic Quantification of Accuracy in Watershed Simulations. *Trans. ASABE* **2007**, *50*, 885–900. [CrossRef]
50. Phan, H.M.; Ye, Q.; Reniers, A.J.; Stive, M.J. Tidal Wave Propagation along The Mekong Deltaic Coast. *Estuar. Coast. Shelf Sci.* **2019**, *220*, 73–98. [CrossRef]
51. Gagliano, S.; McIntire, W. *Reports on the Mekong Delta*; Technical Report; Louisiana State University Coastal Studies Institute: Baton Rouge, LA, USA, 1968; Volume 57.
52. Wang, Q.; Li, S.; Jia, P.; Qi, C.; Ding, F. A Review of Surface Water Quality Models. *Sci. World J.* **2013**, *2013*, 231768. [CrossRef]
53. Wang, Q.; Zhao, X.; Yang, M.; Zhao, Y.; Liu, K.; Ma, Q. Water Quality Model Establishment for Middle and Lower Reaches of Hanshui River, China. *Chin. Geogr. Sci.* **2011**, *21*, 646–655. [CrossRef]
54. Burn, D.H.; McBean, E.A. Optimization Modeling of Water Quality in an Uncertain Environment. *Water Resour. Res.* **1985**, *21*, 934–940. [CrossRef]
55. Gao, L.; Li, D. A Review of Hydrological/Water-Quality Models. *Front. Agric. Sci. Eng.* **2015**, *1*, 267–276. [CrossRef]
56. Cox, B. A Review of Currently Available In-Stream Water-Quality Models and Their Applicability for Simulating Dissolved Oxygen in Lowland Rivers. *Sci. Total Environ.* **2003**, *314*, 335–377. [CrossRef]
57. McCulloch, W.S.; Pitts, W. A Logical Calculus of the Ideas Immanent in Nervous Activity. *Bull. Math. Biophys.* **1943**, *5*, 115–133. [CrossRef]
58. Stamenković, L.J. Application of ANN and SVM for Prediction Nutrients in Rivers. *J. Environ. Sci. Health Part A* **2021**, *56*, 867–873. [CrossRef]
59. Cuong, N.P.; Ty, T.V.; An, T.V.; Minh, H.V.T. Application of Artificial Neural Network (ANN) to Predict Water Levels for Urban Inundation Prediction in Can Tho City. *Agric. Rural Dev. J.* **2019**, *1*, 53–60.
60. Panda, R.K.; Pramanik, N.; Bala, B. Simulation of River Stage Using Artificial Neural Network and MIKE 11 Hydrodynamic Model. *Comput. Geosci.* **2010**, *36*, 735–745. [CrossRef]
61. Maier, H.R.; Dandy, G.C. The Use of Artificial Neural Networks for the Prediction of Water Quality Parameters. *Water Resour. Res.* **1996**, *32*, 1013–1022. [CrossRef]

Article

Assessing the Groundwater Reserves of the Udaipur District, Aravalli Range, India, Using Geospatial Techniques

Megha Shyam [1], Gowhar Meraj [1,2], Shruti Kanga [1], Sudhanshu [1], Majid Farooq [1,2], Suraj Kumar Singh [3,*], Netrananda Sahu [4] and Pankaj Kumar [5,*]

1 Centre for Climate Change & Water Research, Suresh Gyan Vihar University, Jaipur 302017, India; shyam528541@gmail.com (M.S.); gowharmeraj@gmail.com (G.M.); shruti.kanga@mygyanvihar.com (S.K.); cm@mygyanvihar.com (S.); majid_rsgis@yahoo.com (M.F.)
2 Department of Ecology, Environment and Remote Sensing, Government of Jammu and Kashmir, Srinagar 190018, India
3 Centre for Sustainable Development, Suresh Gyan Vihar University, Jaipur 302017, India
4 Department of Geography, Delhi School of Economics, University of Delhi, Delhi 110007, India; nsahu@geography.du.ac.in
5 Institute for Global Environmental Strategies, Hayama 240-0115, Kanagawa, Japan
* Correspondence: suraj.kumar@mygyanvihar.com (S.K.S.); kumar@iges.or.jp (P.K.)

Abstract: Population increase has placed ever-increasing demands on the available groundwater (GW) resources, particularly for intensive agricultural activities. In India, groundwater is the backbone of agriculture and drinking purposes. In the present study, an assessment of groundwater reserves was carried out in the Udaipur district, Aravalli range, India. It was observed that the principal aquifer for the availability of groundwater in the studied area is quartzite, phyllite, gneisses, schist, and dolomitic marble, which occur in unconfined to semi-confined zones. Furthermore, all primary chemical ingredients were found within the permissible limit, including granum. We also found that the average annual rainfall days in a year in the study area was 30 from 1957 to 2020, and it has been found that there are chances to receive surplus rainfall once in every five deficit rainfall years. Using integrated remote sensing, GIS, and a field-based spatial modeling approach, it was found that the dynamic GW reserves of the area are 637.42 mcm/annum, and the total groundwater draft is 639.67 mcm/annum. The deficit GW reserves are 2.25 mcm/annum from an average rainfall of 627 mm, hence the stage of groundwater development is 100.67% and categorized as over-exploited. However, as per the relationship between reserves and rainfall events, surplus reserves are available when rainfall exceeds 700 mm. We conclude that enough static GW reserves are available in the studied area to sustain the requirements of the drought period. For the long-term sustainability of groundwater use, controlling groundwater abstraction by optimizing its use, managing it properly through techniques such as sprinkler and drip irrigation, and achieving more crop-per-drop schemes, will go a long way to conserving this essential reserve, and create maximum groundwater recharge structures.

Keywords: groundwater hydrology; groundwater resource evaluation; groundwater management; groundwater reserves; sustainable water resource management

Citation: Shyam, M.; Meraj, G.; Kanga, S.; Sudhanshu; Farooq, M.; Singh, S.K.; Sahu, N.; Kumar, P. Assessing the Groundwater Reserves of the Udaipur District, Aravalli Range, India, Using Geospatial Techniques. *Water* **2022**, *14*, 648. https://doi.org/10.3390/w14040648

Academic Editor: Adam Milewski

Received: 16 January 2022
Accepted: 17 February 2022
Published: 19 February 2022

1. Introduction

On Earth, water is an essential resource for the existence of life. Among the various components of the hydrological cycle, groundwater is an essential reserve of freshwater, particularly in regions that do not have any other freshwater sources. As rainfall is the source of groundwater, an area's geological setting governs its existence and determines the stocks or reserves of groundwater in any region [1,2]. At a global scale, 71% of Earth's surface is covered with 326 million cubic miles of water [3]. Around 97% of it lies in the oceans, i.e., around 320 million cubic miles, which is too mineralized to be useful for consumptive uses in sustaining life such as drinking, agriculture, and other activities. Only 3% of water is freshwater suitable for consumptive use. Of this, 2.5% is locked in ice caps, glaciers, soil, atmosphere, and hence unavailable [4]. The remaining 0.5% of freshwater is available for direct consumptive use when sourced from lakes, ponds, streams, rivers, and groundwater [5].

Groundwater is sometimes the solely available water supply in desert areas that supports or grows agricultural production. Increased groundwater extraction (groundwater draft) for irrigation has significantly contributed to the agricultural revolution and an enhanced global food supply, since irrigated agriculture accounts for around 40% of world food production [6]. However, in many places, this has resulted in a permanent drop in storage (the volume of water stored in aquifers), known as groundwater depletion [7]. Although the consequences of groundwater extraction are most acute and visible at local scales, due to worldwide distribution, possible ramifications for water and food security, and sea-level rise, groundwater decline is considered to be a global problem [8]. However, there is a paucity of scientific literature regarding the severity of this problem [9–11]. Given that worldwide groundwater extractions are minimal relative to global recharge, the problem of global groundwater quantity has recently been addressed by water conservationists [12,13].

Locally, aquifer depletion is an established fact in many areas, as demonstrated by significant lowering in the groundwater table measured in wells and, more recently, through gravity observations from the GRACE satellites at the basin or watershed scale [14,15]. Groundwater depletion has a variety of repercussions that vary depending on the aquifer and its water-holding capacity [16,17]. As stated, one of the most apparent effects is a lowering of water tables. This results in the drying up of wells, and higher pumping costs that ultimately affect users. It also results in lower groundwater flow to streams, springs, and wetlands, affecting ecosystem services [18]. This can lead to land subsidence, reducing storage irreversibly and potentially damaging infrastructure [19]. Lower water tables cause groundwater movement, which can cause salinization in coastal areas due to saltwater intrusion or leakage from neighboring layers containing saline water [20]. Therefore, there is a need to assess and evaluate groundwater reserves to help conserve and efficiently manage this essential source of freshwater [21].

Globally, various studies have been carried out to assess groundwater reserves. Rehmati et al. (2016) investigated the groundwater potential in the Mehran region of Iran using the maximum entropy (ME) and random forest (RF) models. The study used various groundwater conditioning parameters to determine potential sites for groundwater, namely altitude, slope aspect, slope percentage, drainage density, topographic wetness index (TWI), distance from rivers, land use, topographic wetness index (TWI), plan curvature, lithology, and soil texture, all of which affect groundwater storage. The analysis discovered several zones with extremely high groundwater reservoirs [22]. Lezzaik and Milewski (2018) used a distributed ArcGIS-based model to estimate groundwater reserves in the Middle East and North Africa (MENA), based on derived aquifer saturation thickness and effective porosity estimates. The authors calculated changes in groundwater storage between 2003 and 2014 using monthly gravimetric datasets (GRACE) and land-surface parameters (GLDAS). They found that groundwater reserves in the region were estimated at 1.28 × 106 cukm, with an uncertainty range between 816,000 and 1.93 × 106 cukm [23]. Based on an exhaustive study of available maps, publications, and data, MacDonald et al. (2012) demonstrated continental-scale aquifer reserves and possible borehole yields in Africa. According to their calculations, total groundwater storage in Africa was estimated to be 0.66 million cukm. They demonstrated that boreholes located and constructed properly in numerous African countries would support handpump abstraction and contain enough storage to support abstraction over inter-annual recharge changes. Their maps also demonstrated that the possibility for higher-yielding boreholes is significantly reduced. This study indicated that plans based on extensive drilling of high-yielding boreholes that aim to enhance irrigation or supply water to rapidly urbanizing cities are likely to fail [24]. In India, Singh et al. (2017) used the Gravity Recovery and Climate Experiment (GRACE) to examine the water budget by monitoring gravity anomalies to predict changes in total water storage (TWS) content over India's north-west. From 2003 to 2012, the surface and groundwater estimates indicated a loss of 86.43 km^3/y on average over a ten-year period [15].

Due to an increase in population, urbanization, industrialization, and agricultural activities, India has encountered an extraordinary demand for groundwater in recent decades [25–29]. Therefore, following global trends, the need to regulate the use of groundwater for all activities in India is of utmost importance. Groundwater management is a challenge in a country such as India where the demand for water is greater than its replenishing rates [30]. Moreover, due to the loss of potential groundwater recharge zones to urbanization, the long-term sustainability of this essential ecosystem is in jeopardy [31,32]. Therefore, quantifying the groundwater resource in India is significant to understand the storage of groundwater and its projected life [25,33]. This will help set up new efficient systems for GW allocation for all activities, along with techniques for the management of groundwater reuse and recycling for long-term sustainability [34,35].

The present work evaluated the current and projected groundwater reserves in the Udaipur region, India, and assessed its use for human consumption using chemical assessment. Udaipur is in Rajasthan's agro-climatic zone IV-A and has a tropical, semi-arid, and hot environment. May is the warmest month of the year, with daily maximum and minimum temperatures of 38 °C and 24 °C, respectively. January is the coldest month, with typical daily maximum and minimum temperatures of 24 °C and 7.8 °C, respectively. The average annual rainfall is 624 mm. As it is arid, there is huge demand for groundwater in this region. There are various methodologies involved in evaluating groundwater reserves, as described in previous sections. GRACE data are mainly used for this purpose; however, due to issues of local scale uncertainties in the estimations, various authors have preferred water-balance equation-based approaches [36,37]. The water equation is based on the assessment of groundwater hydrology equations, and involves the assessment of the topography of the area, geomorphological conditions, climate variations, rainfall distributions, drainage characteristics, and hydrogeological characteristics [38–40]. The complete method involves assessing the geological formation of the area, the type of the aquifers with its hydraulic parameters, water levels, water-level fluctuation, water-level trends, groundwater flow direction, and all major chemical ingredient distribution and its concentration in groundwater [41–43]. Together, all this information is essential for assessing the availability of the groundwater and its usage characteristics. Overall, an integrated methodology involving the hydrological, hydrogeological, and hydrochemical characterization of the Udaipur region was adopted to evaluate groundwater reserves and assess their fitness for human consumptive use [44,45]. Using water-balance equations and statistical analysis in a spatial modeling framework, we assessed the groundwater reserves of Udaipur, Rajasthan.

2. Materials and Methods

Udaipur (the Lake city of Rajasthan) falls between 23°48′05.79″ to 25°06′16.75″ North and 73°01′23.10″ to 74°26′20.87″ East (11,773 km^2) in southern Rajasthan (Figure 1). The Precambrian-age Aravalli range circumscribes the entire district [42,43]. The elevation of the study area falls in the range 155–1313 m above mean sea level (AMSL). The overall physiographic gradient is towards the south and south-east of the Udaipur district. Rocky hills mainly cover the most north-west to central portion of the district belonging to Aravalli range, with elevation ranging from 1313 m to 155 m AMSL, and are considered to be good runoff zones [46,47].

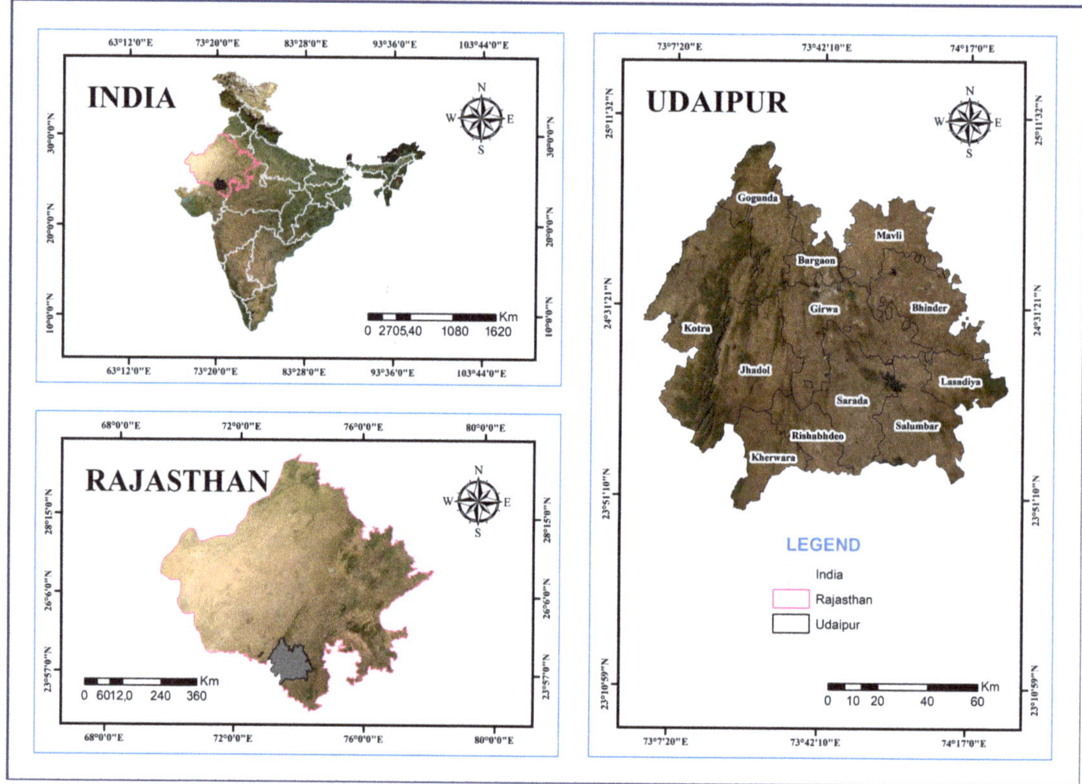

Figure 1. Location of Udaipur with respect to the State of Rajasthan and, overall, India.

Primary data, i.e., observation of physical conditions, vegetation growth, water level, groundwater yield, water quality in terms of TDS, and type of aquifer, using a hydro-inventory for the studied area, was collected during a field visit. Secondary data about rainfall, geology, geomorphology, groundwater level, groundwater quality, aquifer parameters, and groundwater draft was gathered from different sources such as the Water Resource Department (WRD) Govt. of Rajasthan India, Central Groundwater Board (CGWB), and Indian Metrological Department (IMD) [48]. Primary GIS layers were prepared in a vector format using ArcGIS 10.8.

We also performed a chemical analysis of the groundwater to establish its suitability for consumptive use. Electric conductivity (EC), pH, carbonate (CO_3), chloride (Cl), sulfate (SO_4), nitrate (NO_3), phosphate (PO_4), total hardness (TH), calcium (Ca), magnesium (Mg), sodium (Na), potassium (K), fluoride (F), iron (Fe), silicon dioxide (SiO_2), total alkalinity, total dissolved solids (TDS) and uranium (U) of groundwater were interpolated, and their limits were assessed for quality purposes. In the studied area, about 32 groundwater samples from 2016 to 2020 have been collected from existing representative wells/bore wells, and analyzed for various chemical ingredients.

The point locations of the observation stations (water depth and chemical analysis samples) were used for interpolation using the inverse distance weighted (IDW) technique in ArcGIS. Inverse distance weighted (IDW) is a probabilistic estimating interpolator that uses a linear set of attributes at known places to compute unknown values [49]. IDW produces surfaces by generating a neighborhood search of points and weighting these points by a power function, assuming that every input point has a local influence that reduces with distance [50]. Since the observation points were almost equally distributed,

IDW was considered to be the appropriate interpolation technique as reported by various other workers [49–55]. Interpolation of the water table estimates was used to evaluate groundwater flow direction in pre- and post-monsoon seasons and determine the hydraulic gradient for estimating groundwater reserves.

Analysis of average rainfall distribution, number of rainy days, peak daily rainfall, and drought years was carried out using historical rainfall data. Physiographic studies related to topography, drainage, and geomorphology were carried out using SRTM DEM (90 m) [56]. The water-level fluctuation and groundwater-level trends were analyzed using hydrograph analysis techniques and aquifer distribution.

To assess the age of groundwater reserves and sustainability of available reserves for long-term use, we evaluated the total groundwater resource. The methodology adopted in the current study is shown in Figure 2. It involves the use of the water-balance equation and statistical analysis and is a standard method laid down by the Groundwater Estimation Committee (2015), India [57–60].

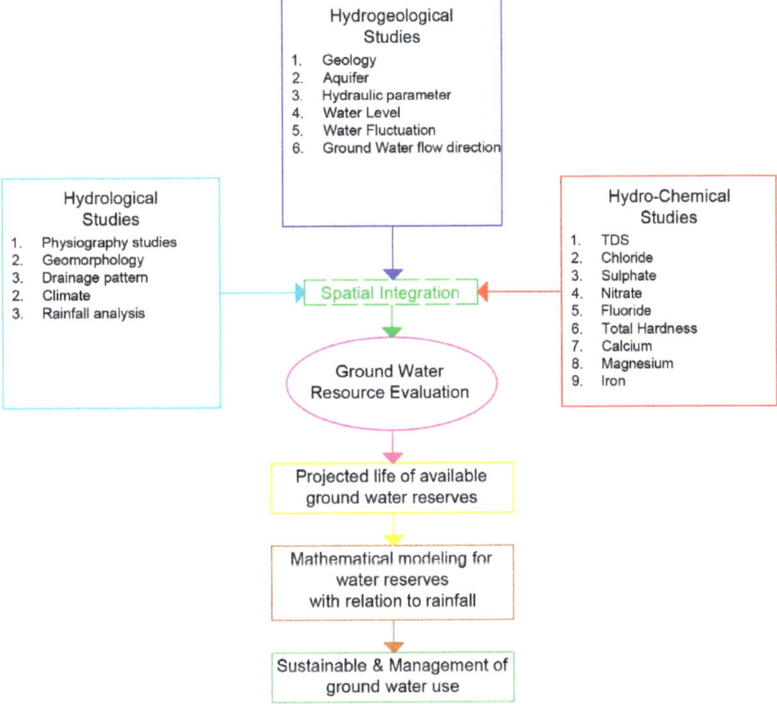

Figure 2. The overall methodology employed in the present study.

The methodology for groundwater resource assessment is based on the principal water-balance equation as given below [47–50]:

$$\text{Inflow} - \text{Outflow} = \text{Change in Storage (of an aquifer)} \tag{1}$$

The equations for estimation of total dynamic reserves (RT), groundwater draft (DT), surplus/deficit reserves, stage of groundwater development, and static reserves are given in Table 1.

Table 1. The equations for estimation of Groundwater Resource Evaluation.

	Dynamic Reserves (RT)	**Rr + RR + Rp + Ri**
Where	Rr = Recharge due to rainfall	Rr = A × S.F. × Sy Rr = Recharge due to Rainfall A = Total rechargeable area S.F. = Average Seasonal Fluctuation in the studied area Sy = Specific Yield
	RR = Recharge due to river	T × $\Delta H/\Delta I$ × L × no. of days T = Transmissivity (As per APT results) $\Delta H/\Delta I$ = Hydraulic Gradient (As per Water-Level Contour Map) L = length of river section, No. of days of river flow as reported in field = 30 days
	Rp = recharge due to ponds	Spread area of pond × Seepage factor × No. of days of water storage Seepage rate = 1.4 mm/day = 0.0014 m/day (As per GEC)
	Ri = recharge due to applied irrigation	Irrigated area (As per collected data from Revenue Department of Jaitaran and Raipur) × Recharge factor for Paddy/Non-Paddy
	Groundwater Draft (DT)	Dd + Di + DI + De
Where	Draft due to domestic consumption (Dd)	Population × Water requirement per day in m^3 × no. of days
	Draft due to applied irrigation (Di)	Average irrigated area × Average crop factor for general mixed crops
	Draft due to Industrial consumption (DI)	Water requirement per day in m^3 × no. of days
	Draft due to natural outflow (Do)	Do = T × $\Delta H/\Delta I$ × L × No. of days in a year T = Average Transmissivity of all aquifers $\Delta H/\Delta I$ = Average Hydraulic Gradient L = Length of out flow boundary
	Draft due to Evapotranspiration (De)	Replenishable reserves × Evapotranspiration Factor
	Surplus/Deficit Reserves	Total dynamic groundwater reserves–Total present groundwater draft
	Stage of Groundwater Development	Total groundwater Draft × 100 Total groundwater reserves
	Static Reserves (Sr)	A × S.T. × Sy where A = Area of different aquifers S.T. = Average saturated thickness

3. Results and Discussion

3.1. Geomorphological Characterization

The study area's north-east, east, and south-east zones have plain, gentler slopes with an elevation between 700–155 m AMSL and are considered to be good recharge zones because they facilitate the percolation of the rainfall events (Figure 3a). These zones help with the movement, transportation, and deposition of erosion of soils/sediments using streams in the studied area. The Sabarmati, Mahe, Banas, and Luni are the principal rivers to carry rainfall-runoff water in the studied area. These rivers are seasonal, with dendritic to sub-dendritic drainage (stream order between 5 and 6) (Figure 3b). The area has been classified into three parts as per the flow direction of the surface water during rainfall events, i.e., from central to south and south-east, north and north-west to east, and north-central to west of the district, and each zone with stream order between 5 and 6.

The area is subtropical and subhumid, with semi-arid climatic conditions. The average annual rainfall in the study from 1957 to 2020 was 627.77 mm, with the annual lowest and highest rainfall being 234.04 mm (1969) and 1282.15 mm (1973), respectively (Figure 4a). As per rainfall analysis, 57.17% of overall time series of annual rainfall years have a below-average rainfall (627.78 mm), whereas the remaining 42.86% have surplus rainfall. This suggests surplus rainfall following 5 successive deficit rainfall years. The average number of rainy days in a year is 30 (Figure 4b), with maximum daily rainfall being 299 mm (2015) (Figure 4c).

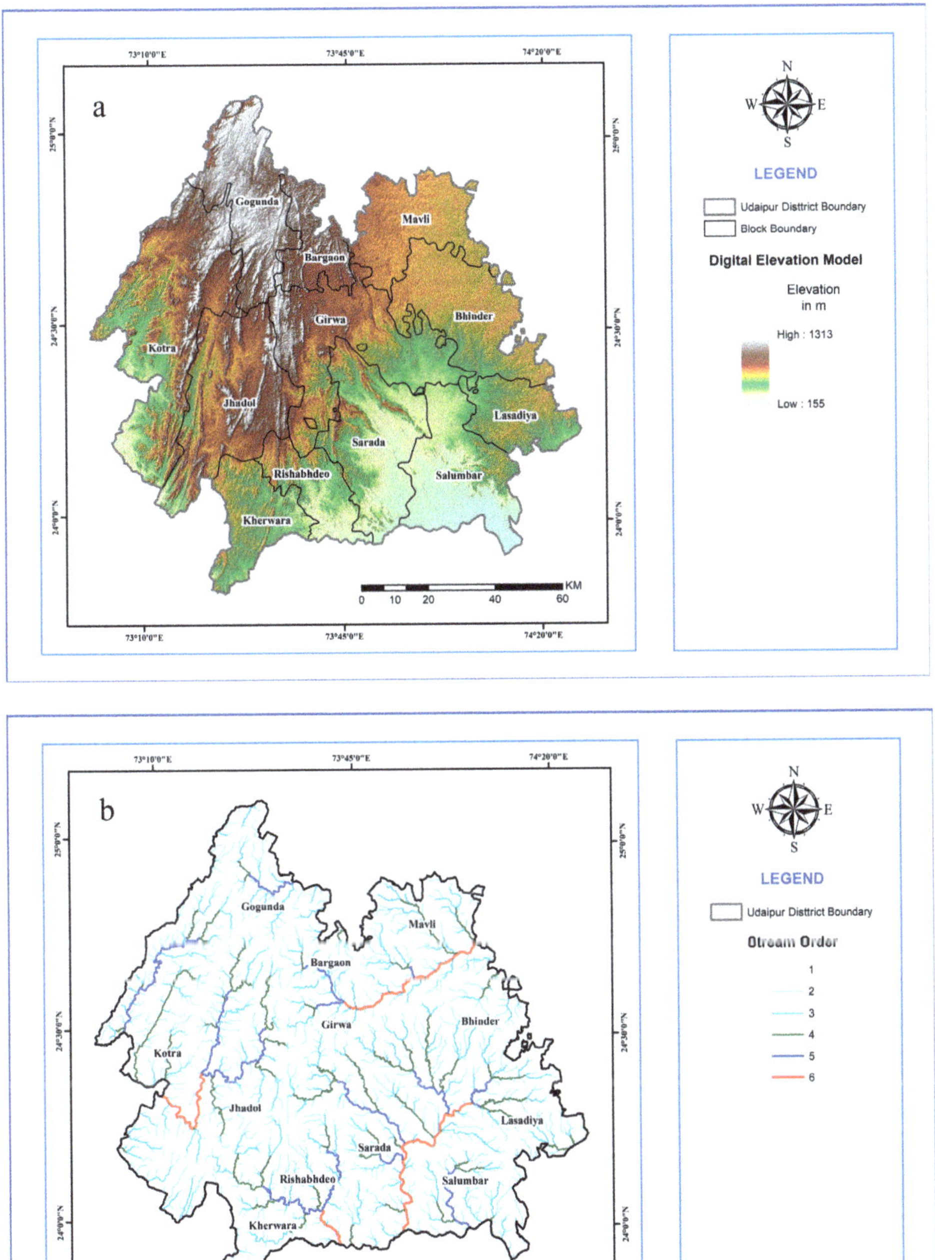

Figure 3. (**a**) Elevation, and (**b**) drainage characteristics of the study area.

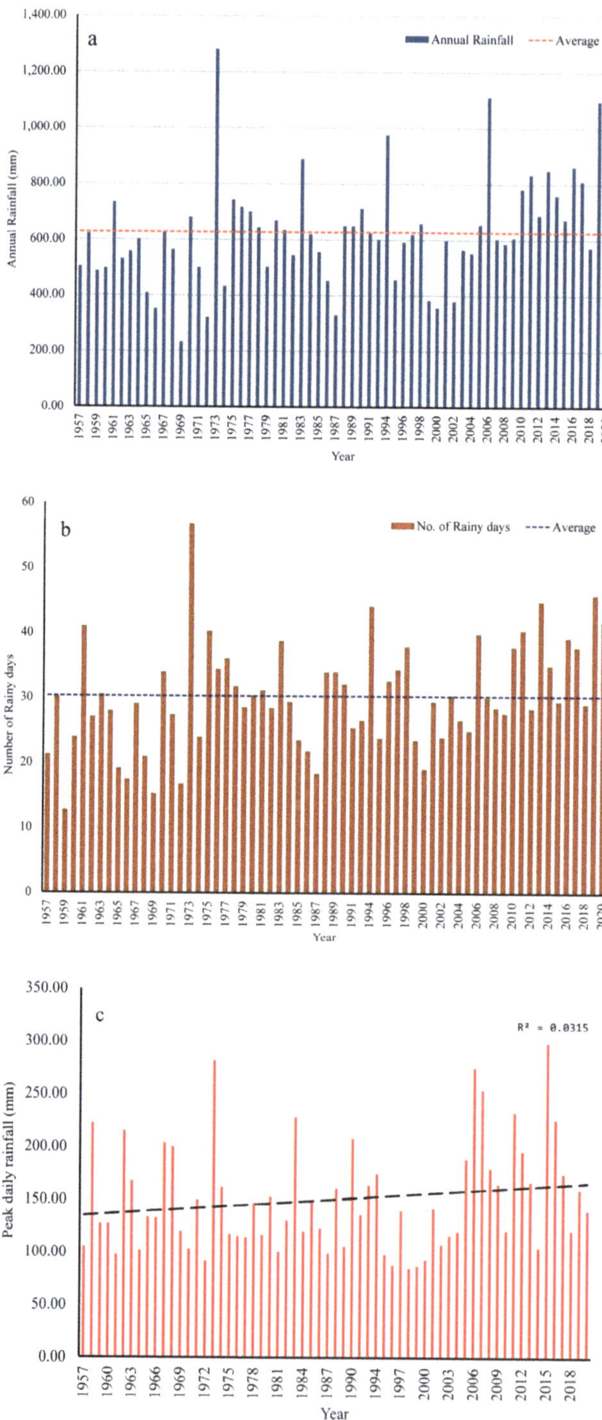

Figure 4. (**a**) Average rainfall, (**b**) number of rainy days, and (**c**) peak daily rainfall of the study area.

Geomorphologically, the area can be sub-divided into three major geomorphological units, i.e., hills (structural/linear/denudational), denudational origin (pediment/buried pediment), and fluvial origin (valley fill) (Figure 5). Most of the area is covered by hills, mostly runoff zones; the north-east and south of the district are covered by denudational origin, which is formed by erosion, stripping, and leaching, and serves as good recharge zones. Nearby the water bodies, the area is covered by fluvial origin, which is formed by the mass movement, transportation, and deposition and erosion of soil/sediment by streams, and serves as good recharge zones [25].

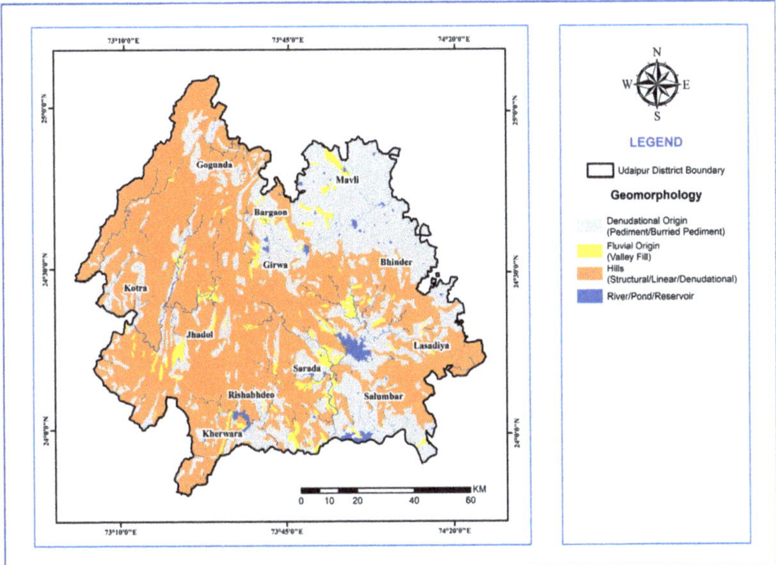

Figure 5. Geomorphological classes of the study area.

3.2. Hydrogeological Characterization

As per field observations and the available literature in the studied area, the northern to southern portion of the studied area belongs to the younger formation of the Aravalli supergroup of the Palaeoproterozoic age. The north-east and east of the area belongs to the oldest formation of the Bhilwara supergroup of the Palaeoproterozoic and Archaean age. The western and small part of the central zone belongs to the younger formation Delhi supergroup of the Palaeoproterozoic–Mesoproterozoic age. Isolated pockets in western, central, and eastern portions of the area belong to the extrusive/intrusive formation of the Palaeoproterozoic, Palaeoproterozoic–Mesoproterozoic, and Archaean age [47]. Table 2 summarizes the stratigraphic geological succession of the area.

Table 2. Summary of the classes of the stratigraphic geological succession of the study area.

Age	Super Group	Group	Lithology
Palaeoproterozoic	Aravalli	Barilake	Meta volcanics, chlorite schists, amphibolite, quartzite, and conglomerate
		Debari	Meta arkose, quartzite, phyllite, dolomitic marble, and dolomite
		Jharol	Chlorite-mica schist, calc shist, and quartzite
		Nathdwara	Banded gneissic complex (BGC)
		Udaipur	Phyllite, mica schists, meta siltstone, quartzite, dolomite, gneisses and migmatites
Palaeoproterozoic	Bhiwara	Rajpura-Dariba	Meta-volcano-sedimentory rocks of banded gneissic complex (BGC)
Palaeoproterozoic–Mesoproterozoic	Delhi	Gogunda	Calc schist, gneisses, mica shists, garnetiferous biotite-schists, quartzites, and migmatites
		Kumbhalgarh	Carbonate, mafic volcanic, and argillaceous rocks
Palaeoproterozoic–Mesoproterozoic	Extrusive/Intrusive	Phulad Ophiolite Suite	Banded gneissic complex (BGC)
Palaeoproterozoic		Rakhabdev Ultramafic Suite	Serpentinite, talc-chlorite-schist, actinolite-tremolite schist, and asbestos
Mesoproterozoic		Sendra-Ambaji Granite and Gneiss	Schists, gneisses, and composite gneiss Quartzites
Palaeoproterozoic		Udaipur/Salumbar/Udaisagar/Darwal Granite	
-		Undifferentiated Granite	
Unconformity			
Archaean	Bhiwara	Hindoli	-
		Mangalwar Complex	Migmatites, gneisses, quartzite, felspathic granite ferrous mica shists and para-amphibolites
-	Extrusive/Intrusive	Untala and Gingla Granites	Politic gneiss, quartzite, marble, calc-silicates

The groundwater availability in the district is generally maintained by topographic and structural units existing in the geological formation, i.e., quartzite, phyllite, gneisses, schist, and dolomitic marble, which are the principle aquifers in the district. The availability and movement of groundwater creates pore spaces between grains, fractures, and bedding plains in the geological formation. The distribution of aquifer in the Udaipur district is given in Figure 6.

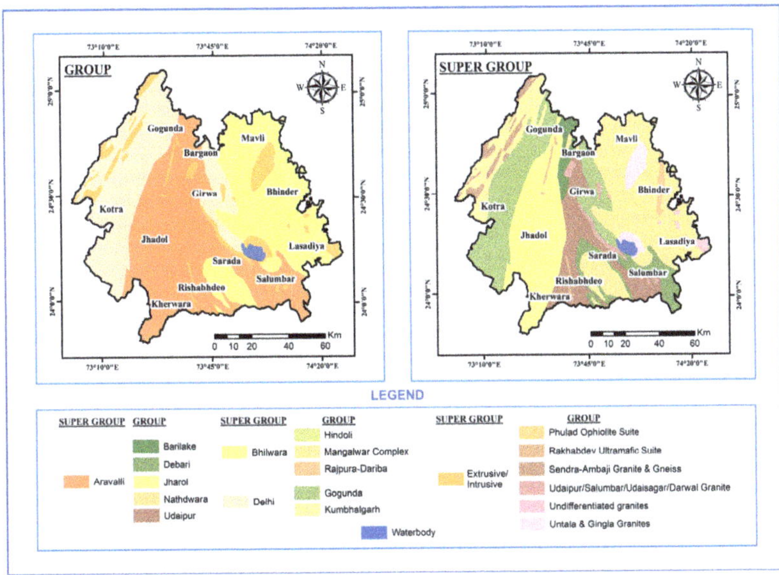

Figure 6. The distribution of aquifer in different groups in the Udaipur district.

The average groundwater yield from all aquifers through groundwater abstraction structures, such as tube wells/bore wells/dug-cum-bore wells at different locations is, per the reported information, of the order of 47 m^3/day, and the data are given in Table 3. The combined hydraulic parameters of all aquifers, i.e., transmissivity (15.63 m^2/day) and specific yield (1.5%) are also presented.

Table 3. The Average Yield of Groundwater at different locations.

Type of Aquifer	Name of the Location	Yield Range in m^3/day	Depth Range of Groundwater Abstraction Structure in m
Calc schist and gneiss	Gogunda	40–60	15–20
	Kotra	40–50	15–20
Granite	Kotra	35–50	15–20
Quartzite	Jhadol	25–35	20–25
Phyllite and schist	Bargaon	40–60	15–20
	Girwa	50–80	25–30
	Gogunda	50–80	20–25
	Jhadol	40–60	25–30
	Kherwara	40–60	20–25
	Kotra	40–60	20–25
	Mavli	40–60	25–30
	Salumbar	40–60	15–20
	Sarada	40–60	15–20
	Bhinder	35–50	15–20
Granites and gneiss	Sarada	35–45	15–20
	Salumbar	35–45	20–25
	Mavli	35–45	20–30
	Girwa	35–45	20–25

The locations of hydrograph stations for 2016–2017, 2017–2018, 2018–2019, and 2019–2020 are given in Figure 7, and the corresponding lithology classes are shown in Figure 8.

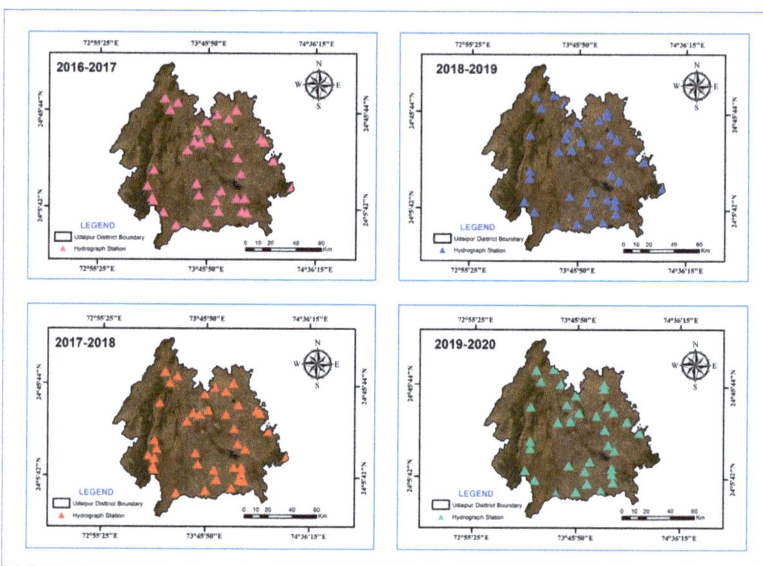

Figure 7. The distribution of hydrograph stations in the Udaipur district.

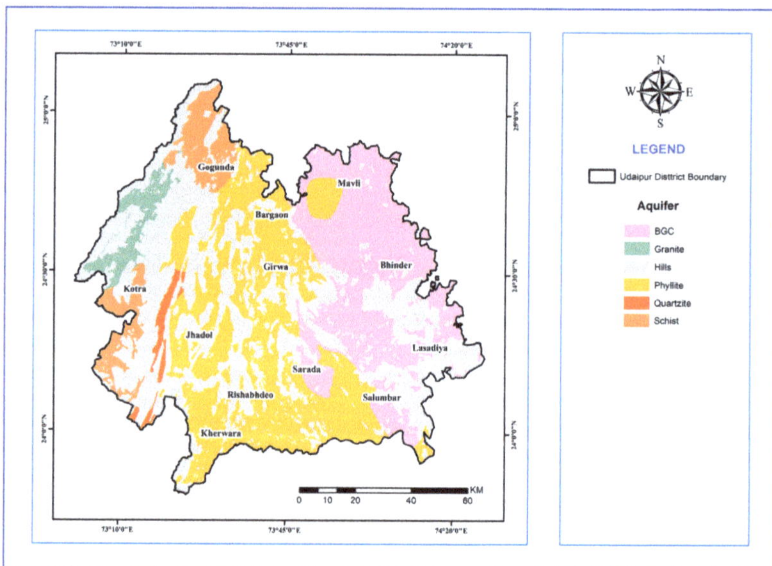

Figure 8. The distribution of different lithologies in the Udaipur district.

As per the available data, the distribution of water level below groundwater level, and water-level contour map (AMSL) for pre- and post-monsoon has been prepared to show the water zones and the groundwater flow direction in the area (Figures 9 and 10).

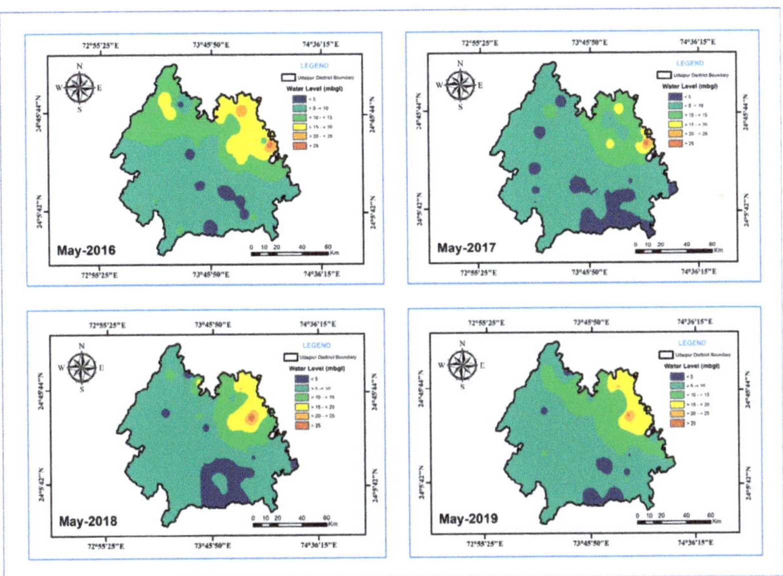

Figure 9. The water-level distribution during pre-monsoon season in the Udaipur district.

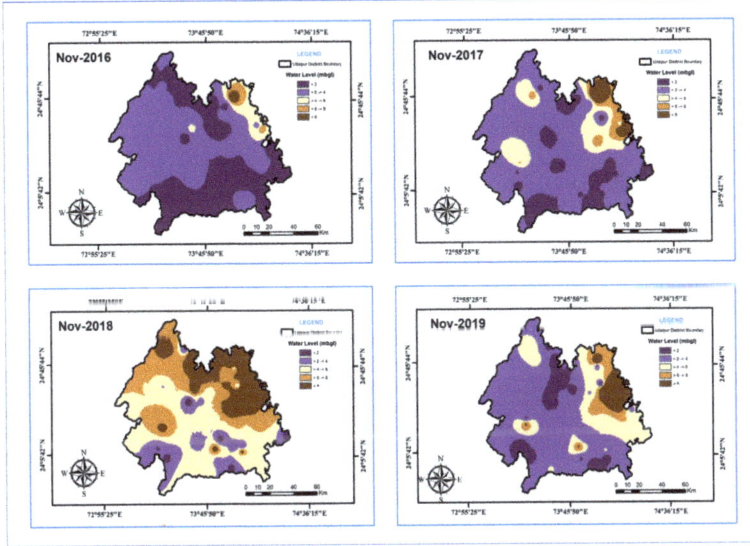

Figure 10. The water-level distribution of during post-monsoon season in the Udaipur district.

In pre-monsoon (May 2016 to 2019), most of the depths of water levels are less than 25 m. However, in the northern and north-eastern parts of the district, the water-level zones are slightly decreased. Similarly, for post-monsoon, the water levels in the area are shallow, which is less than 10 m, as per data collected from the hydrograph station. However, the general groundwater flow direction in the studied area is south-eastwards (Figures 11 and 12). According to the water-level contour map, the hydraulic gradient is 1/127.5, equivalent to 1/130. Water-level fluctuation in the area is around 3 m.

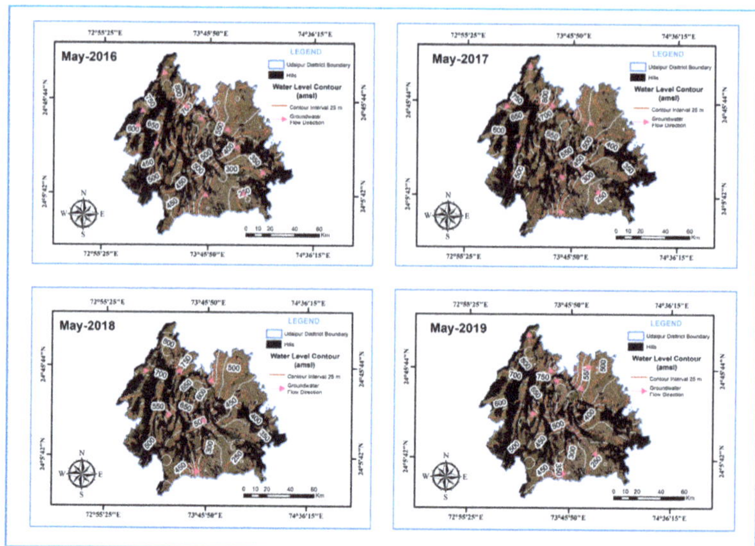

Figure 11. The water-level distribution, specifically contour information during pre-monsoon season in the Udaipur district.

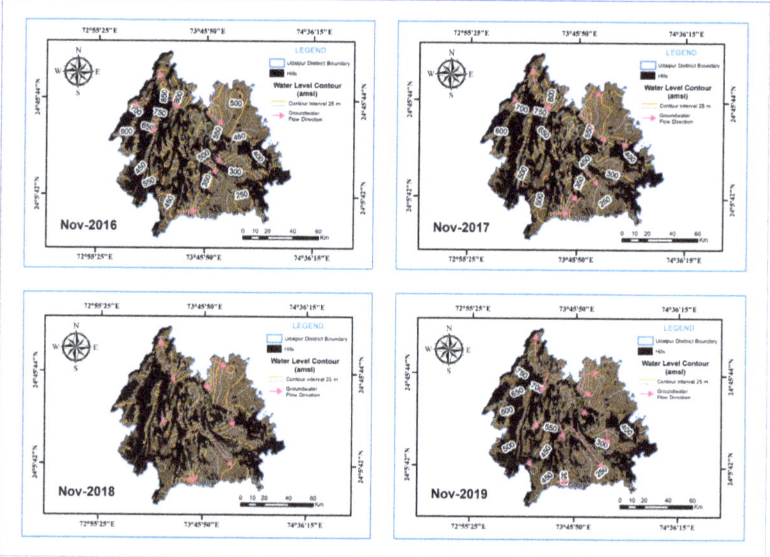

Figure 12. The water-level distribution, specifically contour information during post-monsoon season in the Udaipur district.

In the Aravalli formation, the central part consists of phyllite, quartzite, and dolomite, and are the principal aquifer lithologies for groundwater availability have low to medium permeability. Under unconfined zones, the availability and movement of groundwater is limited to weathered zones such as schistosity, joints, fissures, fractures, and bedding plains. The yield from these aquifers ranges from 20 to 200 cum/day [47]. In the Bhilwara formation, the eastern part of the studied area is characterized by schist, gneisses, and granite rocks. In a few places, extrusive/intrusive formations also exist with low permeability. Groundwater in this zone is in weathered joints and foliation planes under unconfined to semi-confined zones. The yield from these formations is 20 to 60 cum/day [61]. In the Delhi formation, the western-most zones consist of quartzite, biotite schist, calc schist, and calc gneiss with medium permeability. The groundwater occurs in joints and fractures with yields ranging from 12 to 250 cum/day under a semi-confined nature [62]. In the Alluvium formation, water occurs under unconfined zones and is highly permeable. However, due to overexploitation, these zones are dried out in the studied area. In these unconsolidated formations, sand, gravel, cobbles, and boulders exist and are found close to rivers. Most of the study area is covered by hard pavements of rocks consisting of weathered portions, fractures, joints, and bedding plains. During rainfall events, the recharge of rainfall-runoff water is directly percolated into the ground by natural seepage and infiltration [63].

3.3. Hydrochemical Characterization

The location of collected samples for chemical analysis is shown in Figure 13. The groundwater quality in terms of TDS is under permissible limits as per drinking water norms IS 10500:2015, except for a few isolated pockets in the north-east for 2016 to 2018 [64]. However, in 2020, the total area was under the permissible limits, indicating the impact of groundwater recharge on its quality. Similarly, all other parameters improved in 2020 compared to previous years. The distribution of major chemical ingredients is shown in Figures 14–22. The measurement of uranium levels in the district is less than 30 µg/L, as per the prescribed norms of the WHO [65].

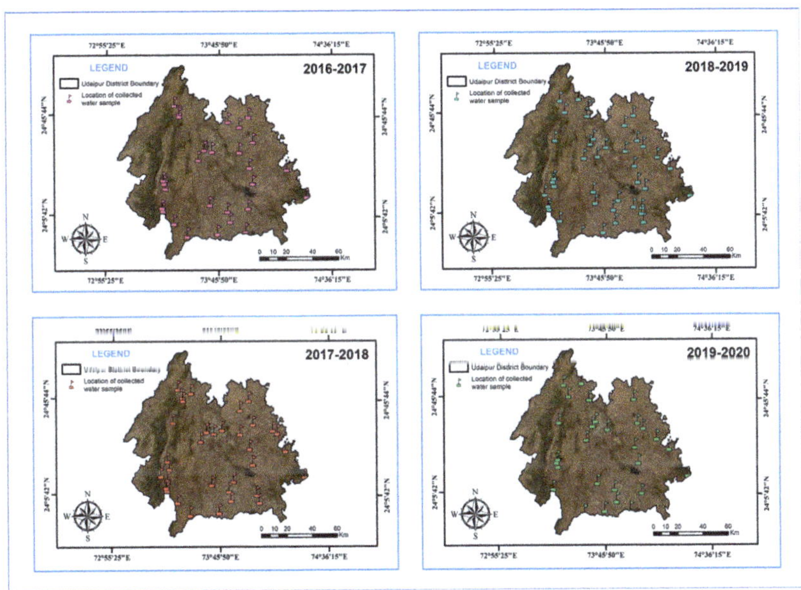

Figure 13. The location of collected samples for chemical analysis.

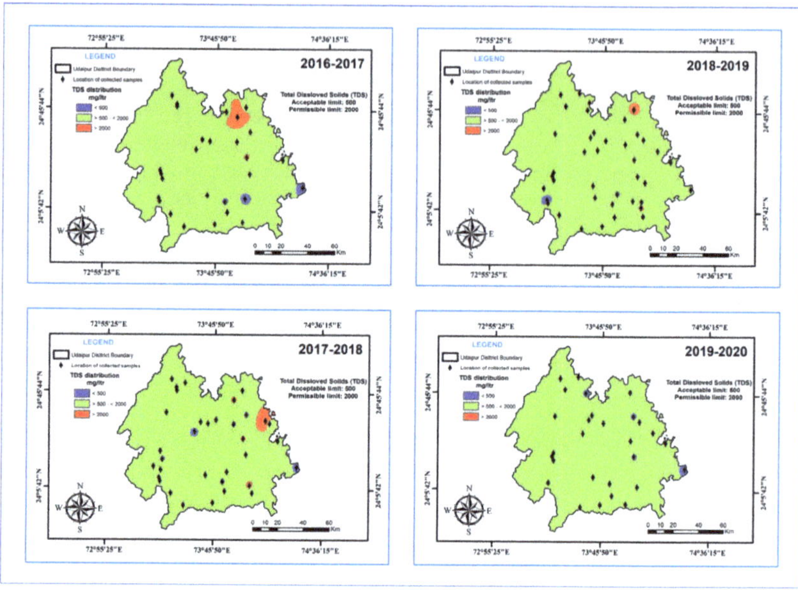

Figure 14. The changing distribution of TDS from 2016–2018 in the study area.

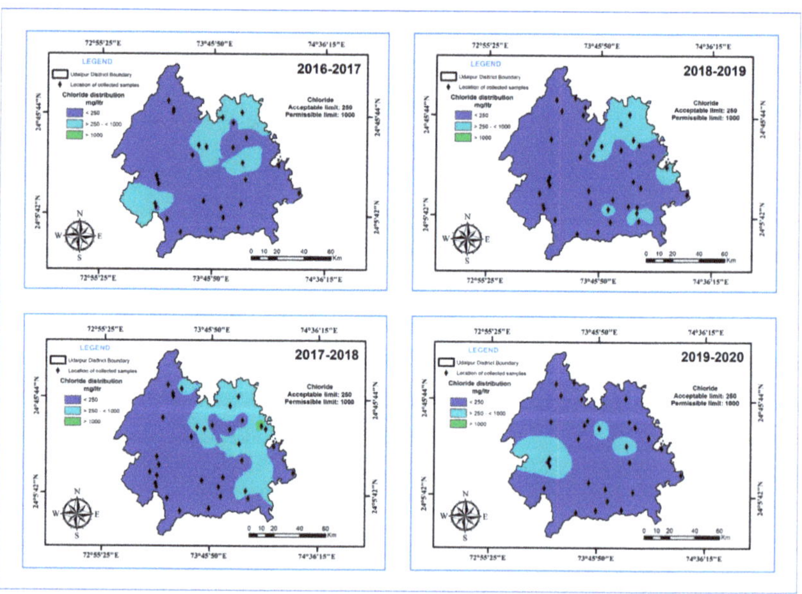

Figure 15. The changing distribution of chloride from 2016–2018 in the study area.

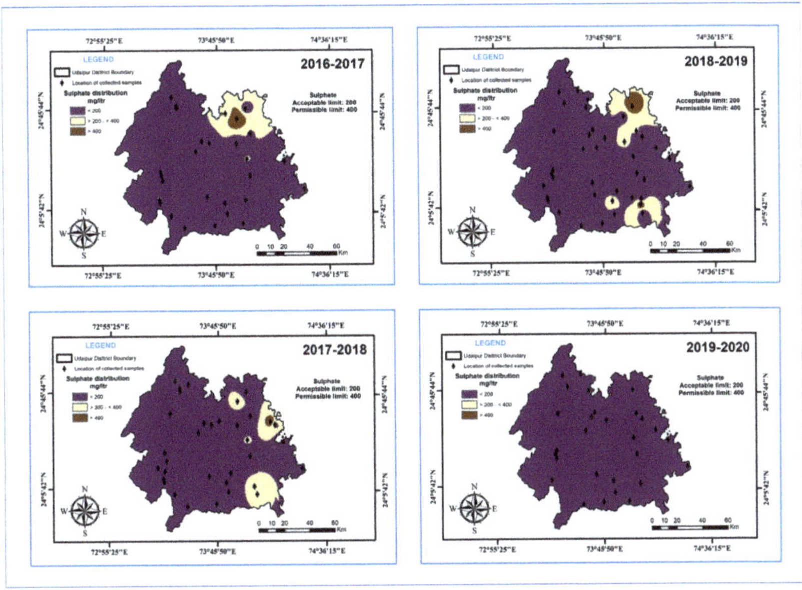

Figure 16. The changing distribution of sulfate from 2016–2018 in the study area.

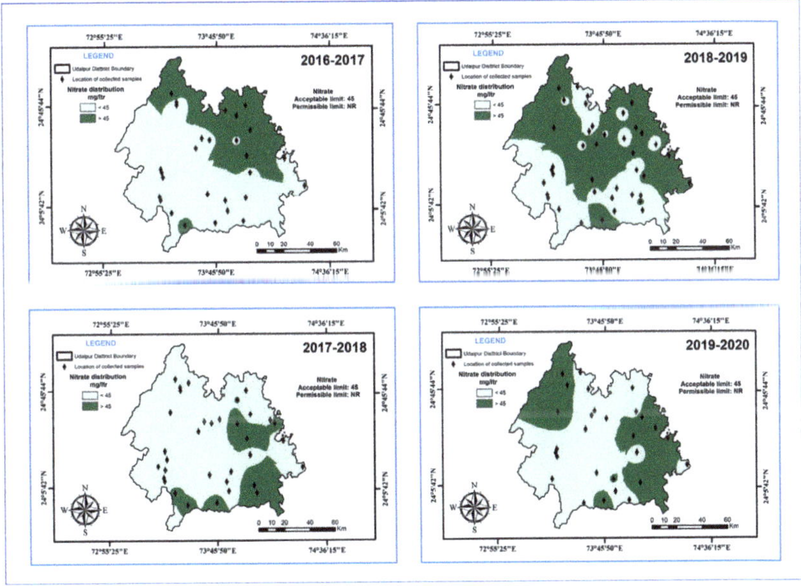

Figure 17. The changing distribution of nitrate from 2016 to 2018 in the study area.

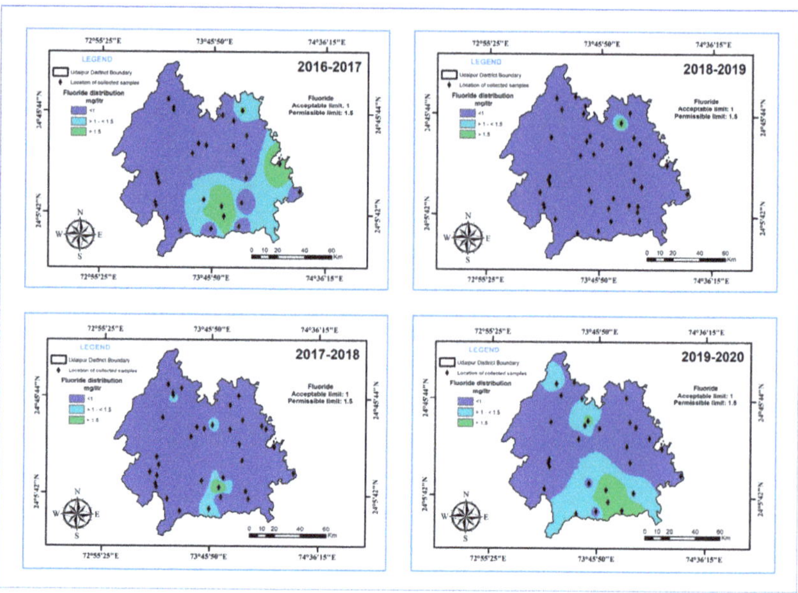

Figure 18. The changing distribution of fluoride from 2016–2018 in the study area.

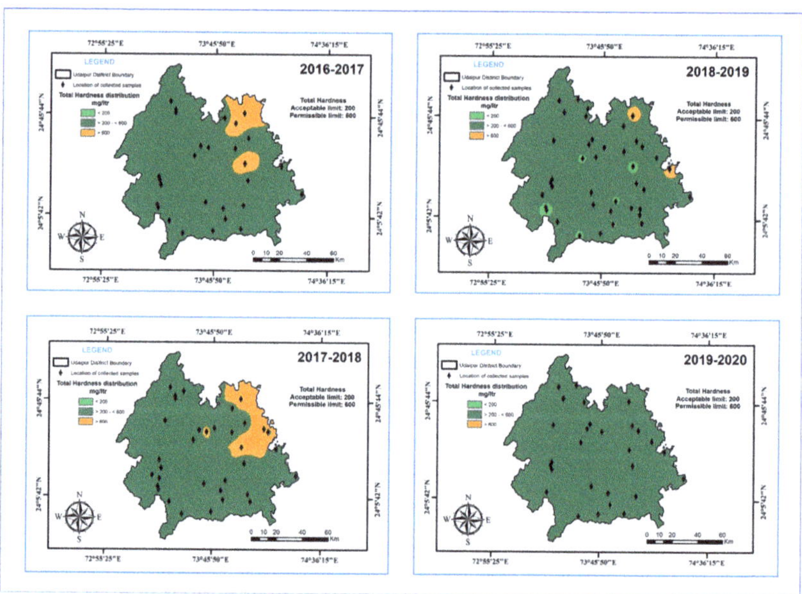

Figure 19. The changing distribution of total hardness from 2016–2018 in the study area.

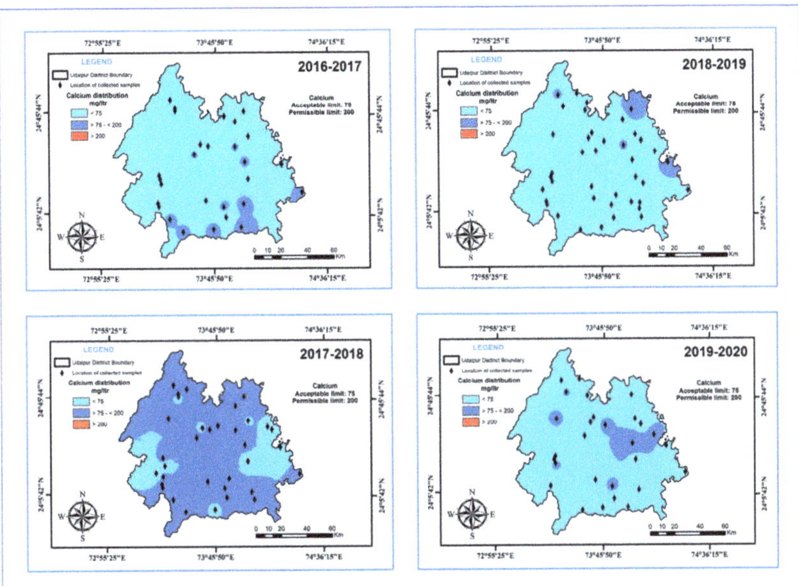

Figure 20. The changing distribution of calcium from 2016–2018 in the study area.

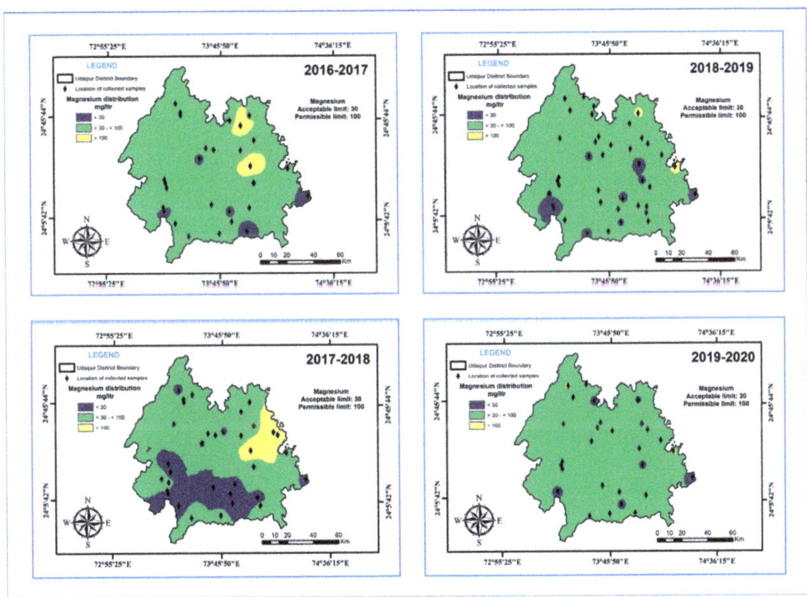

Figure 21. The changing distribution of magnesium from 2016–2018 in the study area.

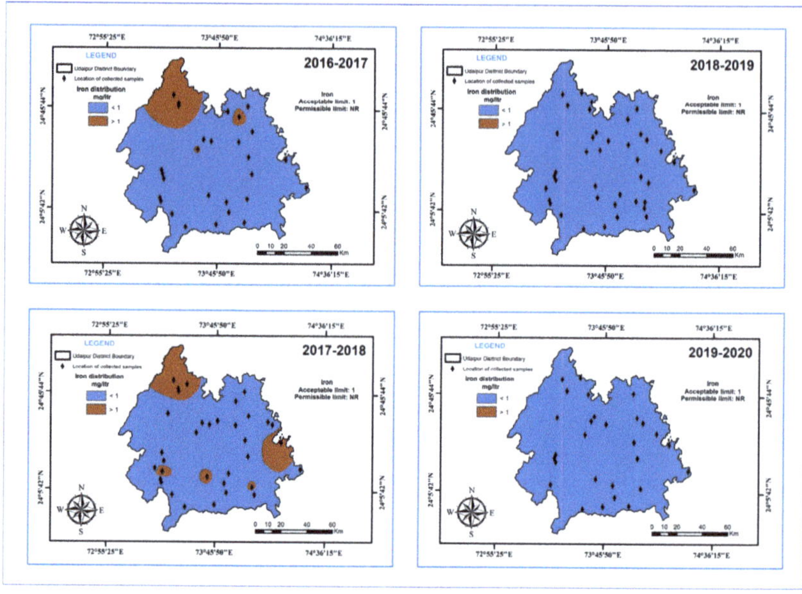

Figure 22. The changing distribution of iron from 2016–2018 in the study area.

3.4. Groundwater Resource Evaluation

As per the adopted methodology for the estimation of groundwater resource evaluation, the dynamic reserves of the area are 637.42 mcm/annum, consisting of recharge due to rainfall—353.19 mcm/annum, recharge due to river/stream—14.90 mcm/annum, recharge due to ponds—0.33 mcm/annum, and recharge due to applied irrigation—247.22 mcm/annum. The total groundwater draft is 639.67 mcm/annum, consisting of the draft due to domestic and other activities such as cattle—46.57 mcm/annum, draft due to industrial and mining projects—14.67 mcm/annum, draft due to applied irrigation—543.88 mcm/annum, draft due to evapotranspiration—0 mcm/annum, and draft due to natural outflow—34.55 mcm/annum.

The calculation reveals that there is a deficit of 2.25 mcm/annum. The stage of groundwater development is 100.67%, rendering the area in the over-exploited category, which is in line with categorization as a dynamic groundwater resource of India in 2020. However, there are enough static reserves to sustain consumptive groundwater use during the drought periods.

3.5. Projected Life of Reserves

The total deficit reserves are 2.25 mcm/annum based on average rainfall (627 mm). Using the linear equation model, the established relationship between rainfall and deficit/surplus reserves was used to project the availability of GW. The model is useful for predicting utilizable reserves in any nth year based on that year's rainfall. With the help of random number theory and correlation regression analysis, the following mathematical relationship has been calculated to estimate total water reserves in the region for a minimum to maximum rainfall [66–71]. The equation governing the above relationship is

$$Y = 1.10274\ X - 638.84 \quad (2)$$

where Y = water reserves in mcm/annum and X = rainfall in mm/annum. With the help of the above analysis, total water reserves are predicted for various rainfall values, as given in Table 4. Figure 23 reveals that if rainfall is below average, there would exist deficit reserves, and if the rainfall is above 700 mm, there will be surplus reserves available on the present groundwater draft (10% growth rate of groundwater draft on every year) [72–74].

Table 4. Deficit/Surplus reserves at different rainfall events.

Rainfall in mm/annum (X)	Dynamic Reserves in mcm/annum	Groundwater Draft in mcm/annum	Total Deficit/Surplus Reserves in mcm/annum (Y)
100	101.66	639.67	−538.01
200	203.32	639.67	−436.35
300	304.99	639.67	−334.68
400	406.65	639.67	−233.02
500	508.31	639.67	−131.36
600	609.97	639.67	−29.70
627	637.42	639.67	−2.25
700	711.63	639.67	71.96
800	914.96	639.67	275.29
900	914.96	639.67	275.29
1000	1016.62	639.67	376.95
1100	1118.28	639.67	478.61
1200	1219.94	639.67	580.27

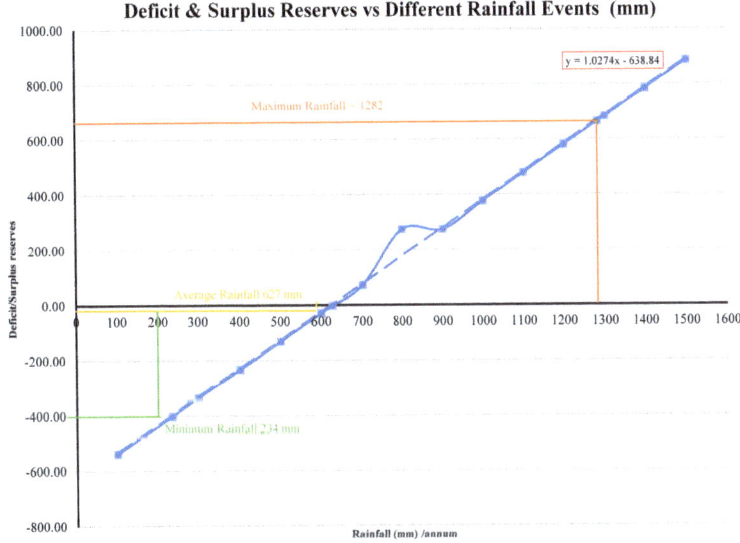

Figure 23. Regression model between deficit/surplus reserves and the rainfall.

3.6. Sustainability Management of Groundwater Reserves

The total deficit reserves are 2.25 mcm/annum based on average rainfall (627 mm). In such a situation, for sustainable groundwater development, groundwater recharge measures equivalent to deficit reserves are required from large rainwater-harvesting structures, water conservation, reuse–recycle measures, and regulation of existing groundwater draft [75–78]. The draft may increase in future scenarios due to growth in population, industrial development/expansion of existing industrials, mining/expansion of mining, and agriculture sectors. Therefore, the net groundwater draft will be more than what is required at present [78,79].

The available dynamic reserves are 637.42 mcm/annum, and the deficit reserves are drawn from static reserves. Hence, it is essential to control groundwater abstraction and optimize groundwater use by modernizing the existing irrigation practices using a

22. Rahmati, O.; Pourghasemi, H.R.; Melesse, A.M. Application of GIS-based data driven random forest and maximum entropy models for groundwater potential mapping: A case study at Mehran Region, Iran. *Catena* **2016**, *137*, 360–372. [CrossRef]
23. Lezzaik, K.; Milewski, A. A quantitative assessment of groundwater resources in the Middle East and North Africa region. *Hydrogeol. J.* **2018**, *26*, 251–266. [CrossRef]
24. MacDonald, A.M.; Bonsor, H.C.; Dochartaigh, B.É.Ó.; Taylor, R.G. Quantitative maps of groundwater resources in Africa. *Environ. Res. Lett.* **2012**, *7*, 024009. [CrossRef]
25. Singh, S.K.; Pandey, A.C. Geomorphology and the controls of geohydrology on waterlogging in Gangetic plains, North Bihar, India. *Environ. Earth Sci.* **2014**, *71*, 1561–1579. [CrossRef]
26. Farooq, M.; Singh, S.K.; Kanga, S. Inherent vulnerability profiles of agriculture sector in temperate Himalayan region: A preliminary assessment. *Indian J. Ecol.* **2021**, *48*, 434–441.
27. Kanga, S.; Meraj, G.; Das, B.; Farooq, M.; Chaudhuri, S.; Singh, S.K. Modeling the spatial pattern of sediment flow in lower Hugli estuary, West Bengal, India by quantifying suspended sediment concentration (SSC) and depth conditions using geoinformatics. *Appl. Comput. Geosci.* **2020**, *8*, 100043. [CrossRef]
28. Narasimha Prasad, N.B.; Shivraj, P.V.; Jagatheesan, M.S. Evaluation of Groundwater Development Prospects in Kadalundi River Basin. *J. Geol. Soc. India* **2007**, *69*, 1103–1110.
29. Jeelani, G. Aquifer response to regional climate variability in a part of Kashmir Himalaya in India. *Hydrogeol. J.* **2008**, *16*, 1625–1633. [CrossRef]
30. Shah, T. Towards a Managed Aquifer Recharge strategy for Gujarat, India: An economist's dialogue with hydro-geologists. *J. Hydrol.* **2014**, *518*, 94–107. [CrossRef]
31. Meraj, G.; Singh, S.K.; Kanga, S.; Islam, M.N. Modeling on comparison of ecosystem services concepts, tools, methods and their ecological-economic implications: A review. *Model. Earth Syst. Environ.* **2021**. [CrossRef]
32. Meraj, G.; Farooq, M.; Singh, S.K.; Islam, M.; Kanga, S. Modeling the sediment retention and ecosystem provisioning services in the Kashmir valley, India, Western Himalayas. *Model. Earth Syst. Environ.* **2021**. [CrossRef]
33. Pandey, A.C.; Singh, S.K.; Nathawat, M.S. Analysing the impact of anthropogenic activities on waterlogging dynamics in Indo-Gangetic plains, Northern Bihar, India. *Int. J. Rem. Sens.* **2012**, *33*, 135–149. [CrossRef]
34. Zarghami, M.; Akbariyeh, S. System dynamics modeling for complex urban water systems: Application to the city of Tabriz, Iran. *Resour. Conserv. Recycl.* **2012**, *60*, 99–106. [CrossRef]
35. Kumar, R.; Singh, R.D.; Sharma, K.D. Water resources of India. *Curr. Sci.* **2005**, *5*, 794–811.
36. Hu, X.; Shi, L.; Zeng, J.; Yang, J.; Zha, Y.; Yao, Y.; Cao, G. Estimation of actual irrigation amount and its impact on groundwater depletion: A case study in the Hebei Plain, China. *J. Hydrol.* **2016**, *543*, 433–449. [CrossRef]
37. Dangar, S.; Asoka, A.; Mishra, V. Causes and implications of groundwater depletion in India: A review. *J. Hydrol.* **2021**, *596*, 126103. [CrossRef]
38. Moore, I.D.; Grayson, R.B.; Ladson, A.R. Digital terrain modelling: A review of hydrological, geomorphological, and biological applications. *Hydrol. Processes* **1991**, *5*, 3–30. [CrossRef]
39. Fenta, A.A.; Kifle, A.; Gebreyohannes, T.; Hailu, G. Spatial analysis of groundwater potential using remote sensing and GIS-based multi-criteria evaluation in Raya Valley, northern Ethiopia. *Hydrogeol. J.* **2015**, *23*, 195–206. [CrossRef]
40. Portoghese, I.; Uricchio, V.; Vurro, M. A GIS tool for hydrogeological water balance evaluation on a regional scale in semi-arid environments. *Comput. Geosci.* **2005**, *31*, 15–27. [CrossRef]
41. Wałęga, A.; Młyński, D.; Wojkowski, J.; Radecki-Pawlik, A.; Lepeška, T. New empirical model using landscape hydric potential method to estimate median peak discharges in mountain ungauged catchments. *Water* **2020**, *12*, 983. [CrossRef]
42. Espinha Marques, J.; Samper, J.; Pisani, B.; Alvares, D.; Carvalho, J.M.; Chaminé, H.I.; Marques, J.M.; Vieira, G.T.; Mora, C.; Sodré Borges, F. Evaluation of water resources in a high-mountain basin in Serra da Estrela, Central Portugal, using a semi-distributed hydrological model. *Environ. Earth Sci.* **2011**, *62*, 1219–1234. [CrossRef]
43. Mussa, K.R.; Mjemah, I.C.; Machunda, R.L. Open-source software application for hydrogeological delineation of potential groundwater recharge zones in the singida semi-arid, fractured aquifer, Central Tanzania. *Hydrology* **2020**, *7*, 28. [CrossRef]
44. Asadi, H.; Shahedi, K.; Jarihani, B.; Sidle, R.C. Rainfall-runoff modelling using hydrological connectivity index and artificial neural network approach. *Water* **2019**, *11*, 212. [CrossRef]
45. Arabi, M.; Govindaraju, R.S.; Hantush, M.M. A probabilistic approach for analysis of uncertainty in the evaluation of watershed management practice. *J. Hydrol.* **2007**, *333*, 459–471. [CrossRef]
46. Singh, P.K.; Dahiphale, P.; Yadav, K.K.; Singh, M. Delineation of groundwater potential zones in Jaisamand basin of Udaipur district. In *Groundwater*; Springer: Singapore, 2018; pp. 3–20.
47. Roy, A.B.; Paliwal, B.S. Evolution of lower Proterozoic epicontinental deposits: Stromatolite-bearing Aravalli rocks of Udaipur, Rajasthan, India. *Precambrian Res.* **1981**, *14*, 49–74. [CrossRef]
48. Central Ground Water Board (CGWB). *Manual on Artificial Recharge of Groundwater*; Ministry of Water Resources, Government of India: New Delhi, India, 2007.
49. Gong, G.; Mattevada, S.; O'Bryant, S.E. Comparison of the accuracy of kriging and IDW interpolations in estimating groundwater arsenic concentrations in Texas. *Environ. Res.* **2014**, *130*, 59–69. [CrossRef]
50. Luo, W.; Taylor, M.C.; Parker, S.R. A comparison of spatial interpolation methods to estimate continuous wind speed surfaces using irregularly distributed data from England and Wales. *Int. J. Climatol. J. R. Meteorol. Soc.* **2008**, *28*, 947–959. [CrossRef]

51. Nawar, S.; Corstanje, R.; Halcro, G.; Mulla, D.; Mouazen, A.M. Delineation of soil management zones for variable-rate fertilization: A review. *Adv. Agron.* **2017**, *143*, 175–245.
52. Kanga, S.; Sudhanshu Meraj, G.; Farooq, M.; Nathawat, M.S.; Singh, S.K. Reporting the management of COVID-19 threat in India using remote sensing and GIS based approach. *Geocarto Int.* **2020**, 1–8. [CrossRef]
53. Kanga, S.; Meraj, G.; Farooq, M.; Nathawat, M.S.; Singh, S.K. Analyzing the risk to COVID-19 infection using remote sensing and GIS. *Risk Anal.* **2021**, *41*, 801–813. [CrossRef] [PubMed]
54. Tomar, P.; Singh, S.K.; Kanga, S.; Meraj, G.; Kranjčić, N.; Đurin, B.; Pattanaik, A. GIS-Based Urban Flood Risk Assessment and Management—A Case Study of Delhi National Capital Territory (NCT), India. *Sustainability* **2021**, *13*, 12850. [CrossRef]
55. Meraj, G. Ecosystem service provisioning–underlying principles and techniques. *SGVU J. Clim. Chang. Water* **2020**, *7*, 56–64.
56. Pall, I.A.; Meraj, G.; Romshoo, S.A. Applying integrated remote sensing and field- based approach to map glacial landform features of the Machoi Glacier valley, NW Himalaya. *SN Appl. Sci.* **2019**, *1*, 488. [CrossRef]
57. Singh, S.; Raju, N.J.; Ramakrishna, C. Evaluation of groundwater quality and its suitability for domestic and irrigation use in parts of the Chandauli-Varanasi region, Uttar Pradesh, India. *J. Water Resour. Prot.* **2015**, *7*, 572. [CrossRef]
58. Chinnasamy, P.; Maheshwari, B.; Prathapar, S. Understanding groundwater storage changes and recharge in Rajasthan, India through remote sensing. *Water* **2015**, *7*, 5547–5565. [CrossRef]
59. Groundwater Resource Estimation Committee (GEC). *Ground Water Resource Estimation Methodology*; India Ministry of Water Resources Report; India Ministry of Water Resources: Delhi, India, 2009; p. 107.
60. Maheshwari, B.; Varua, M.; Ward, J.; Packham, R.; Chinnasamy, P.; Dashora, Y.; Dave, S.; Soni, P.; Dillon, P.; Purohit, R.; et al. The Role of Transdisciplinary Approach and Community Participation in Village Scale Groundwater Management: Insights from Gujarat and Rajasthan, India. *Water* **2014**, *6*, 3386–3408. [CrossRef]
61. Vikas, C.; Kushwaha, R.; Ahmad, W.; Prasannakumar, V.; Reghunath, R. Genesis and geochemistry of high fluoride bearing groundwater from a semi-arid terrain of NW India. *Environ. Earth Sci.* **2013**, *68*, 289–305. [CrossRef]
62. Bhattacharjee, S.; Nandy, S. Geology of the western Arunachal Himalaya in parts of Tawang and West Kameng districts, Arunachal Pradesh. *J. Geol. Soc. India* **2008**, *72*, 199–207. [CrossRef]
63. Bhattacharya, A.K. Artificial ground water recharge with a special reference to India. *Int. J. Res. Rev. Appl. Sci.* **2010**, *4*, 214–221.
64. Kumari, P.; Kumar, G.; Prasher, S.; Kaur, S.; Mehra, R.; Kumar, P.; Kumar, M. Evaluation of uranium and other toxic heavy metals in drinking water of Chamba district, Himachal Pradesh, India for possible health hazards. *Environ. Earth Sci.* **2021**, *80*, 271. [CrossRef]
65. Ibrahim, S.A.; Al-Tawash, B.S.; Abed, M.F. Environmental assessment of heavy metals in surface and groundwater at Samarra City, Central Iraq. *Iraqi J. Science* **2018**, *59*, 1277–1284.
66. Takeuchi, K. Regional Water Exchange for Drought Alleviation. Doctoral Dissertation, Colorado State University, Fort Collins, CO, USA, 1974.
67. Zhang, Y.; Chiew, F.H.S.; Zhang, L.; Li, H. Use of remotely sensed actual evapotranspiration to improve rainfall–runoff modelling in southeast Australia. *J. Hydrometeorol.* **2009**, *10*, 969–980. [CrossRef]
68. Kwon, M.; Kwon, H.H.; Han, D. A hybrid approach combining conceptual hydrological models, support vector machines and remote sensing data for rainfall-runoff modeling. *Rem. Sens.* **2020**, *12*, 1801. [CrossRef]
69. Hamdan, A.N.A.; Almuktar, S.; Scholz, M. Rainfall-runoff modeling using the HEC-HMS model for the Al-Adhaim River catchment, Northern Iraq. *Hydrology* **2021**, *8*, 58. [CrossRef]
70. ASCE Manual. *Groundwater Management*; American Society of Civil Engineers: New York, NY, USA, 1987.
71. Ibrahim, A.; Zakaria, N.; Harun, N.; Muzamil, M.; Hashim, M. Rainfall runoff modeling for the basin in Bukit Kledang, perak. In *IOP Conference Series: Materials Science and Engineering, Proceedings of the International Nuclear Science Technology and Engineering Conference 2020 (iNuSTEC 2020), Selangor, Malaysia, 17–19 November 2020*; IOP Publishing Ltd.: Bristol, UK, 2021; Volume 1106, p. 012033. [CrossRef]
72. Shah, T.; Molden, D.; Sakthivadivel, R.; Seckler, D. Global groundwater situation: Opportunities and challenges. *Econ. Political Wkly.* **2001**, *36*, 4142–4150.
73. Narasimha Prasad, N.B.; Sharma, K.K. Geophysical Investigations for Groundwater Exploration in Amini Island. *Bull. Pure Appl. Sci.* **2003**, *22*, 31–41.
74. Narasimha Prasad, N.B. Assessment of Groundwater Resources in Nileshwar Basin. *J. Appl. Hydrol.* **2003**, *XVI*, 52–60.
75. Wali, S.U.; Abubakar, I.M.; Umar, M.A.; Shera, A.A. Reassessing Groundwater Potentials and Subsurface water Hydro-chemistry in a Tropical Anambra Basin, Southeastern Nigeria. *J. Geol. Res.* **2020**, *2*, 1–24. [CrossRef]
76. Boryczko, K.; Rak, J. Method for assessment of water supply diversification. *Resources* **2020**, *9*, 87. [CrossRef]
77. India Meteorological Department (IMD). Meteorological monograph: Hydrology N0. In *Rainfall Profile of Udaipur*; 15/2013; Meteorological Centre, Jaipur & Meteorological Department: New Delhi, India, 2013; pp. 1–113.
78. Chatterjee, R.; Gupta, B.K.; Mohiddin, S.K.; Singh, P.N.; Shekhar, S.; Purohit, R. Dynamic groundwater resources of National Capital Territory, Delhi: Assessment, development and management options. *Environ. Earth Sci.* **2009**, *59*, 669–686. [CrossRef]
79. Shin, S.; Park, H. Achieving cost-efficient diversification of water infrastructure system against uncertainty using modern portfolio theory. *J. Hydroinform.* **2018**, *20*, 739–750. [CrossRef]

80. Shekhar, S. An approach to interpretation of step drawdown tests. *Hydrogeol. J.* **2006**, *14*, 1018–1027. [CrossRef]
81. Shekhar, S.; Purohit, R.; Kaushik, Y.B. Groundwater management in NCT Delhi, Technical paper included in the special session on Ground water. In Proceedings of the 5th Asian Regional Conference of INCID, Vegan Bhawan, New Delhi, India, 9–11 December 2009; Available online: https://www.researchgate.net/publication/238708371_Groundwater_Management_in_NCT_Delhi (accessed on 31 December 2021).

Article

Enhancing Water Supply Resilience in a Tropical Island via a Socio-Hydrological Approach: A Case Study in Con Dao Island, Vietnam

Duc Cong Hiep Nguyen [1,*], Duc Canh Nguyen [2,3], Thi Tang Luu [4], Tan Cuong Le [5], Pankaj Kumar [6,*], Rajarshi Dasgupta [6] and Hong Quan Nguyen [4,7]

1 Southern Institute for Water Resources Planning, Ho Chi Minh City 72710, Vietnam
2 Institute of Fundamental and Applied Sciences, Duy Tan University, Ho Chi Minh City 70000, Vietnam; nguyenduccanh3@duytan.edu.vn
3 Faculty of Environmental and Chemical Engineering, Duy Tan University, Da Nang 50000, Vietnam
4 Center of Water Management and Climate Change, Institute for Environment and Resources, Vietnam National University, Ho Chi Minh City 71308, Vietnam; luuthitang@gmail.com (T.T.L.); nh.quan@iced.org.vn (H.Q.N.)
5 Institute for Environment and Resources, Vietnam National University, Ho Chi Minh City 72409, Vietnam; letancuongmttn@gmail.com
6 Institute for Global Environmental Strategies, Hayama 240-0115, Japan; dasgupta@iges.or.jp
7 Institute for Circular Economy Development, Vietnam National University, 01 Marie Curie, VNU Campus, Ho Chi Minh City 71308, Vietnam
* Correspondence: hiepndc@yahoo.com (D.C.H.N.); kumar@iges.or.jp (P.K.); Tel.: +81-046-855-3858 (P.K.)

Citation: Nguyen, D.C.H.; Nguyen, D.C.; Luu, T.T.; Le, T.C.; Kumar, P.; Dasgupta, R.; Nguyen, H.Q. Enhancing Water Supply Resilience in a Tropical Island via a Socio-Hydrological Approach: A Case Study in Con Dao Island, Vietnam. *Water* **2021**, *13*, 2573. https://doi.org/10.3390/w13182573

Academic Editor: Maria Mimikou

Received: 20 August 2021
Accepted: 15 September 2021
Published: 18 September 2021

Abstract: Socio-hydrological approaches are gaining momentum due to the importance of understanding the dynamics and co-evolution of water and human systems. Various socio-hydrological approaches have been developed to improve the adaptive capacity of local people to deal with water-related issues. In this study, a social-hydrological approach was developed to enhance the water supply resilience in Con Dao Island, Vietnam. We used a water-balance model, involving the Water Evaluation and Planning (WEAP) tool, to conduct a scenario-based evaluation of water demands. In doing so, we assessed the impacts of socio-economic development, such as population growth and climate change, on increasing water demand. The modelling results showed that the existing reservoirs—the main sources to recharge the groundwater (accounting for 56.92% in 2018 and 65.59% in 2030)—play a critical role in enhancing water supply resilience in the island, particularly during the dry season. In addition, future water shortages can be solved by investment in water supply infrastructures in combination with the use of alternative water sources, such as rainwater and desalinated seawater. The findings further indicate that while the local actors have a high awareness of the role of natural resources, they seem to neglect climate change. To meet the future water demands, we argue that upgrading and constructing new reservoirs, mobilizing resources for freshwater alternatives and investing in water supply facilities are among the most suitable roadmaps for the island. In addition, strengthening adaptive capacity, raising awareness and building professional capacity for both local people and officials are strongly recommended. The research concludes with a roadmap that envisages the integration of social capacity to address the complex interaction and co-evolution of the human–water system to foster water-supply resilience in the study area.

Keywords: socio-hydrology; Con Dao Island; water resilience; WEAP

1. Introduction

Swift global changes, frequent extreme weather conditions along with increasing water demand has significantly impacted the socio-economic development through our co-evolution with water [1]. Keeping in mind the limited availability of freshwater resources as well as the fact that one-third of the global population is living in water stress, sustainable water management is a global challenge of high priority [2]. It is reported that out of this available little freshwater deposit, half of the global population depends on groundwater to meet their potable water demand [3]. For the Asian continent, despite having a significant share of global freshwater resources, the average per capita water availability is much less than that of the rest of the world owing to water pollution, water conflicts, poor governance and management strategies and socio-cultural practices [3–5].

Because the interaction between the hydrological cycle and its interaction with the biophysical environment is very complex, different holistic approaches, such as integrated water resource management (IWRM) models [2], socio-hydrological approaches [6], etc., have been used by both scientific communities as well as decision-makers. Among various IWRM tools, some of the models, such as WEAP (Water Evaluation and Planning), MIKE, RIBASIM (river basin simulation model), and WBalMo (water balance model), have been widely used across the globe [2,7–9].

Socio-hydrological science is gaining interest due to the importance of understanding the dynamics and co-evolution of water and human systems, and hence there is a greater impact on decision making [10]. In practice, the socio-hydrological approach requires participatory models where the community has a great influence at every step of the analysis, and hence the outcome guarantees a result for common good [11]. In the literature, a lot of studies on the relationship between social and hydrological systems have been carried out in the field of socio-hydrological resilience [12], prediction in a socio-hydrological world [13], system understanding [14], risk management [15], and land-use management [16]. These studies generally used socio-hydrological models to support decision-makers analyze the people–water interaction and receive feedback before offering effective decisions [1].

Recently, a socio-hydrological approach has been developed by [6] to improve the adaptive capacity of residents in isolated riverine islands in the context of water scarcity due to rapid global changes and climate change (Figure 1). This approach built a resilient water environment for achieving human wellbeing in three large riverine islands in Asia, namely Fraserganj, South 24 Parganas (Ganges River, India), Dakshin Bedkashi (Padma River, Bangladesh), and Con Dao Island (Mekong River, Vietnam). The authors recommended the key steps to apply the socio-hydrological concept for project management, in which the use of numerical tools considering social aspects was highlighted to assist in the analysis and decision-making process. The study has opened a new way for sustainable water resource management in isolated riverine islands, particularly enhancing water supply resilience for Con Dao Island, based on the social-hydrological approach.

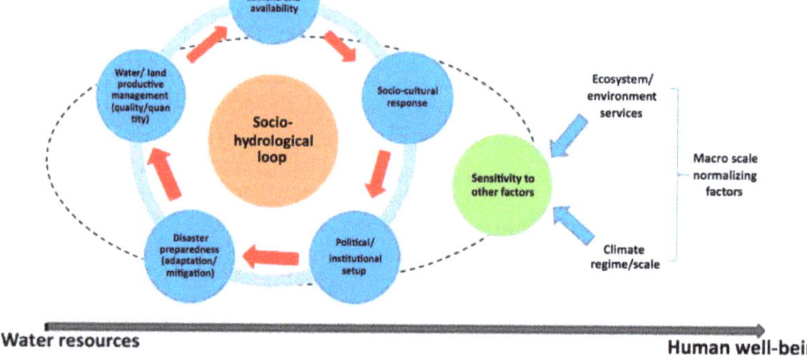

Figure 1. Conceptual diagram of socio-hydrological approach to bridge the gap between water resources and human wellbeing (Source: [6]).

Con Dao is an island in Southern Vietnam, a very popular tourist destination because of its natural sceneries [17]. However, in recent years, this island has been witnessing water scarcity because of various reasons, viz. a sharp increase in water demand because of rapid population growth and a decline in water availability [18].

Regarding numerical tools for water resource management in Con Dao, reference [19] developed a coupled hydrological model of HEC-HMS and MODFLOW to analyze the interaction between surface water and groundwater in Con Dao Island. The study proposed a drought management plan with solutions to improve water availability for coping with the increasing water demand in the future. However, the study did not consider the social/institutional analysis, i.e., water balance and water project implementation which are significant attributes in the socio-hydrological loop.

Several solutions have been proposed to resolve the problems, such as improving the current water supply in Con Dao by upgrading the storage capacity of the existing reservoirs, constructing new reservoirs, and desalinating water [20]. Other alternative water resources such as rainwater can be also a promising solution to resolve water problems in the island since it has plenty of precipitation annually (~2000 mm/year). However, there is no information available on the efficiency assessment of all proposed solutions/countermeasures in the island. To address the above gaps, the objectives of this paper are (i) to identify the interaction between people and water in Con Dao Island, Vietnam through the lens of a socio-hydrological approach, and (ii) to suggest a sustainable water management strategy for enhancing water supply resilience in the context of climate change and socio-economic development in the area.

2. Materials and Methods

2.1. Study Site

The study site is Con Dao, a typical island, located in Ba Ria-Vung Tau (BRVT) province, 230 km from Ho Chi Minh City, Vietnam (located at 8°40′57″ N and 106°36′26″ E) (Figure 2). It is an isolated historical island consisting of 16 small islands with a total area of 75.2 km^2. It has a tropical climate and receives approximately 2000 mm/year of precipitation. Despite the high precipitation rate and the relatively low population density (about 112 people/km^2), the island is facing a water shortage, with the situation expected to worsen in the future due to the limitations of the current water supply system, and the increasing demand due to the increase of residents and travelers. These rapid changes are casting extreme effects on the communities in the island due to their poor adaptive capacities (limited resources/infrastructure and institutional setup) as they are isolated geographically. Hence, achieving a healthy interaction between human and water systems in the island in the future is a critical concern of sustainability.

2.2. Current Water Supply Status in Con Dao Islands

There are three water resources in the island: surface water, groundwater, and rainwater.

Surface water is the main source of water, even though there is no large river in the island. The island consists of 45 short and small streams with a total length of 37.6 km (about 0.73 km/km^2). Most of the streams only have water in the rainy season, but no or little water in the dry season [19]. The rainfall is often not retained in the streams but flows directly into the sea due to the high and steep topography. To date, the Con Dao Islands have three large reservoirs, including two natural lakes (An Hai and Quang Trung 1 (QT1)) and the Quang Trung 2 (QT2) Reservoir, constructed in 2018, which are the main water sources to supply the decentralized water supply plant as intake resources. The storage capacities of the An Hai, QT1, and QT2 reservoirs are 540,000 m^3, 518,000 m^3, and 645,000 m^3, respectively. These reservoirs are linked to each other by the canals, and these are the main sources to recharge water into the aquifer. The decentralized water supply plant then treats the water and the supply for 3 residential sub-areas: Central Town, Co Ong Airport, and Ben Dam Port.

(Shallow) groundwater has been exploited for the water supply plant in Central Town through 25 wells around the QT1 reservoir. This water provides mostly domestic water to the water supply plant (with a capacity of 3440 m^3/day). The water not only supplies water to Central Town but also to Co Ong and Ben Dam through two water transfer stations with the capacity of 300 and 800 m^3/day, respectively. The shallow groundwater system is recharged from reservoirs seasonally.

Rainwater use is still not under current practice in the island, except in some remote areas where the water distribution system hardly reaches.

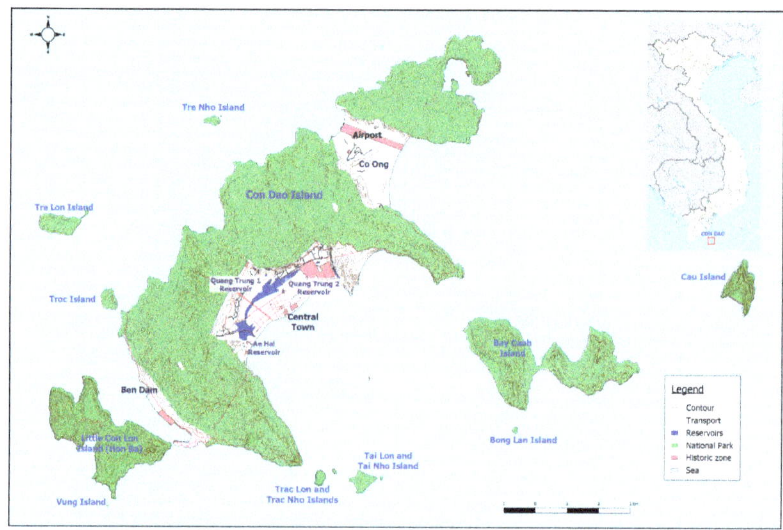

Figure 2. Map of Con Dao Islands (Modified from [18]).

2.3. *Socio Hydrology Approach*

To achieve the research objectives, an integrated socio-hydrological approach has been developed with a combination of quantitative assessment using numerical modeling and qualitative assessment using field surveys, focus group discussions (FGDs), and Key Informant Interviews (KIIs). The detailed methodology is shown in Figure 3.

2.3.1. Quantitative Analysis

The first step of quantitative analysis was to set up a hydrological numerical model. In this study, the Water Evaluation and Planning (WEAP) system, which is an integrated water resource management tool based on the basic principle of water balance accounting, was used as it has been widely applied for water resource problems all over the world [9,21–23].

2.3.2. Qualitative Analysis

Participatory Rural Appraisal (PRA) tools were applied in this study, which includes focus group discussion (FGD), in-depth interviews, and questionnaire surveys. The FGD was first conducted with 10 local government officials and technical staff for about three hours. Participants in the FGD included district officials from the People's Committee and different sectors that are related to water use/management in the island, e.g., the Office of Natural Resources and Environment (DNRE). Originally, criteria for the selection of the FGD participants included (1) working for the local authority at the district level, (2) leaders who were responsible for the development of their sectors such as chairmen or heads. In reality, the participants were selected following the arrangement of the Con Dao authority. Participants in the FGD were queried with open-ended questions related to water management and water use in Con Dao Island (Appendix C).

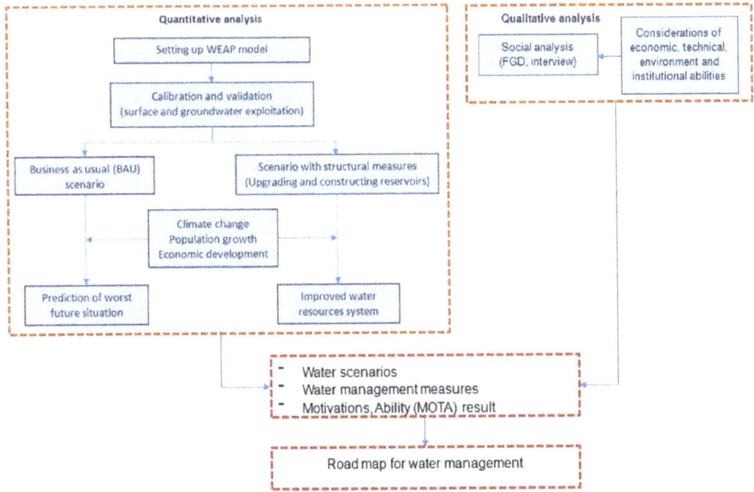

Figure 3. Study framework.

The in-depth interviews were then carried out with four local officials to assess their motivation and ability in water management based on the qualitative assessment aspect of the Motivation and Ability (MOTA) framework (Figure 4). Here, local officials include those who are working for the district authority and local state water companies. The motivation of the local officials is observed based on their perceived existing issues, solutions, and professional roles in solving the issues. The ability of the current governing system is described based on the perceptions of local officials on institutional capacity, financial capacity, and technical capacity [24,25].

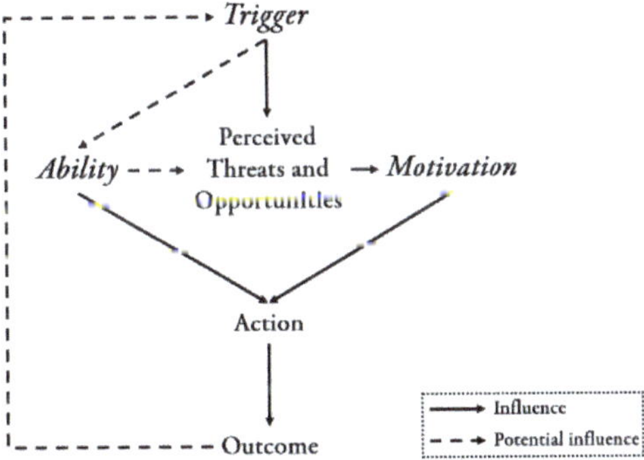

Figure 4. MOTA framework (Source: [25]).

Finally, surveys were conducted with 50 households. The households were selected based on 8 criteria including (1) geographic features; (2) hydrological features; (3) cultural features; (4) ecological features; (5) different administrative boundaries; (6) different sex; (7) different age; and (8) different jobs. The survey questionnaire was developed to cover the most possible factors responsible for water security issues in the study area.

The information obtained from the observations during the survey, FGD, and in-depth interviews was interpreted intersubjectively by the research team to examine the obstacles and enablers of sustainable water management in Con Dao Island.

3. Results

3.1. Hydrological Simulation

3.1.1. Model Setup

This step began with identifying water users and water sources in the study area, as well as developing a schematic (spatial layout) of the water supply and demand system for Con Dao (Figure 5). The main water users on the island are local residents, tourists, and agricultural and industrial activities, while the main water sources are rainwater, reservoirs, and groundwater (Table 1).

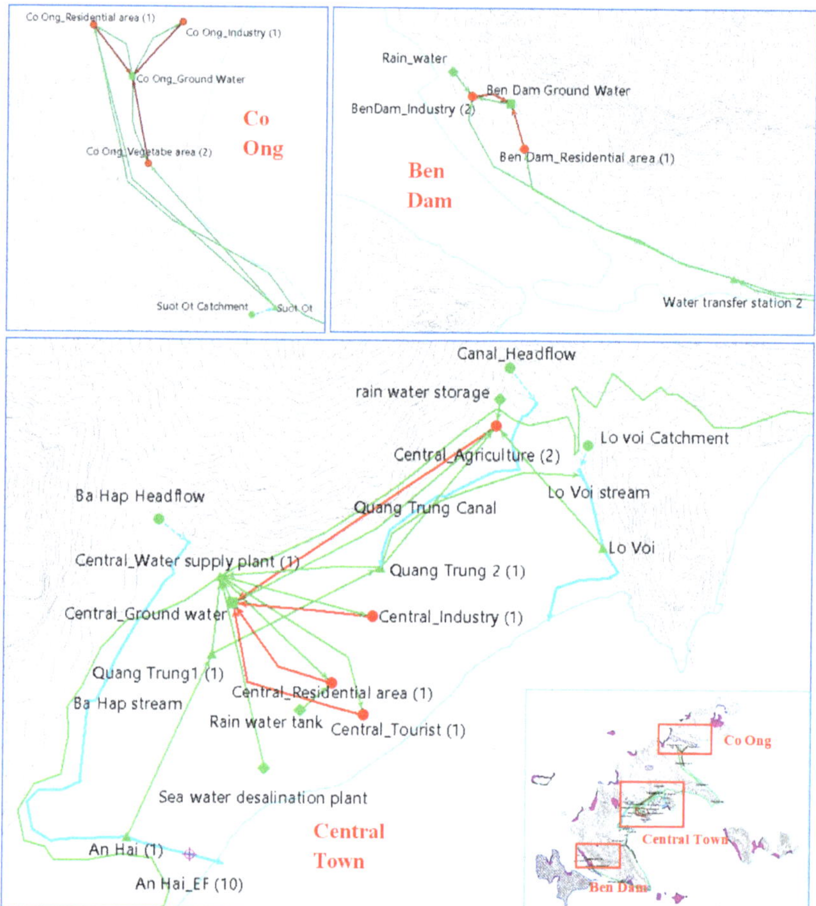

Figure 5. Schematic of the WEAP model for Con Dao Island.

Table 1. Water users and water supply sources for different areas in Con Dao.

Area	Water Users	Water Supply Sources	
		Current	Future
Co Ong	- Local residents - Tourists - Agriculture - Industry (airport)	- Groundwater - Rainwater - Water transferred from Central town	- Reservoir: Suoi Ot - Groundwater - Rainwater - Water transferred from Central town
Central town	- Local residents - Tourists - Agriculture - Industry	- Reservoirs: An Hai, Quang Trung 1, and Quang Trung 2 - Groundwater - Rainwater	- Reservoirs: An Hai, Quang Trung 1, Quang Trung 2, and Lo Voi - Groundwater - Rainwater
Ben Dam	- Local residents - Tourists - Industry	- Rainwater - Water transferred from Central town	- Reservoirs: Nui Mot - Groundwater - Rainwater - Water transferred from Central town

Next, water demand and water supply at present (2018) and in the future (2030) were calculated. For water demand, the input data in WEAP included population, the number of tourists, type of crops (vegetables, annual crop, fruits), crop calendar, livestock (the number of buffaloes, cattle, pigs, chickens, ducks, and goats), industrial area, type of industry (ice production). These data were taken from the 2018 Statistical Yearbook of Con Dao [26] and other related reports, such as the master plan on socio-economic development of Con Dao and the agricultural development planning for Con Dao to 2020, vision to 2030 [20] (as shown in Appendix A). The water demand for agricultural production was calculated by the crop coefficient approach, whereas other demands (i.e., domestic use, tourism, industry, and agriculture) were determined by the following formula:

$$Q_{day} = Nxq/1000\ (m^3/day) \tag{1}$$

in which:

q is the water use standard regulated in TCXDVN 33:2006 of the Ministry of Construction on the water supply–distribution system and facility–design standard;

N is the number of people or the area of the industrial zone (ha).

Meanwhile, the input data for water supply included meteorological data (monthly rainfall and evaporation), the storage of reservoirs, groundwater storage and recharge, the capacity of the water supply plant, and water transfer stations. These data were provided by the Division for Water Resources Planning and Investigation for the South of Vietnam (DWRPIS), Con Dao Meteorological station, and the Con Dao local government (shown in Appendix B).

In this model, the groundwater–surface water interactions in Con Dao were simulated using the method in WEAP, i.e., the amount of groundwater inflow from or outflow to reservoirs was specified.

In the second step, calibration and validation were carried out to assess the reliability of WEAP for Con Dao. The model was run to simulate monthly surface and ground water exploitation with the input data from step 1, including water demand for different sectors (e.g., domestic use, tourism, agriculture, industry) and water supply sources (e.g., reservoirs and groundwater). The calibration was achieved by the 2014 data when simulated and observed values of water exploitation were matched, i.e., the Nash–Sutcliffe efficiency (NSE) value will reach "1". The final parameters of the calibration process were then used to validate the model with the 2018 data. The better the validation model was, the closer the NSE value was to "1".

The third step was to set up scenarios for evaluating the current condition and future development of water resource management (Table 2). A reference scenario (BAU scenario—

S0) was set up to estimate future risks and challenges in water security considering key drivers, namely climate change, population growth, economic development, land-use change, and groundwater extraction rate. According to the master plan on socio-economic development of Con Dao in 2020, with vision to 2030 [20], population and tourists by 2030 were estimated to be about 30,000 and 300,000 people, respectively. Additionally, the industrial area will be expanded by 50 hectares. In this scenario, the forecast of temperature and rainfall in 2030 was based on the Vietnam climate change scenarios by [27].

Table 2. Summary of water resources management scenarios for Con Dao Island.

No.	Scenario	Description
1	Reference—BAU (S0)	- Increase in water demand due to socio-economic development and population growth - Climate change impacts (temperature increases 10% while rainfall decreases 10%) - No improvement in water supply capacity
2	Increasing water supply (S1)	Same as S0, but increasing in water supply capacity +Upgrading storage capacity of the existing reservoirs: An Hai (0.02×10^6 m^3), Quang Trung 1 (0.02×10^6 m^3) +Constructing new reservoirs: Nui Mot (25,000 ha, in 2020), Suoi Ot (171,000 ha, in 2021), Lo Voi (68,000 ha, in 2022) +Mitigating the loss from the water supply system (from 10% to 5%) +Investing on a surface water supply plant (3000 m^3/day)
3	Using low-cost alternative water resources (S2)	S1 + Using rainwater tanks at households in Central Town as a resource for domestic use
4	Using high-cost alternative water resources (S3)	S1 + Constructing a seawater desalination plant (capacity of 3000 m^3/day)
5	Combining (S4)	S2 + S3

Scenarios with countermeasures were similar to the reference scenario but took into account human intervention to improve water supply capacity. According to the master plan on the socio-economic development of Con Dao, the local government has planned to dredge the existing reservoirs (An Hai and Quang Trung 1), construct new reservoirs (including Nui Mot in 2020, Suoi Ot in 2021, Lo Voi in 2022), mitigate the losses from the water supply system, and invest in a surface-water supply plant (S1). Rainwater is also a promising water resource in such remote islands, which can be a sustainable way to obtain good-quality drinking water at a low cost and with little energy expenditure and was considered in this study (S2). In addition, desalination was considered a high-cost water resource alternative (S3) as proposed in the master plan on the socio-economic development of Con Dao [20]. All countermeasures were incorporated in Scenario S4.

3.1.2. Model Calibration and Validation

Only groundwater exploitation data was available for model calibration and validation. The WEAP model for Con Dao Island was calibrated using the groundwater exploited by the water supply plant in 2014 (Figure 6). This calibrated model was then validated with the data in 2018. The NSE values of the calibration and validation were 0.894 and 0.867, respectively. The results imply that the model was reliable to simulate the water scenarios for the Island.

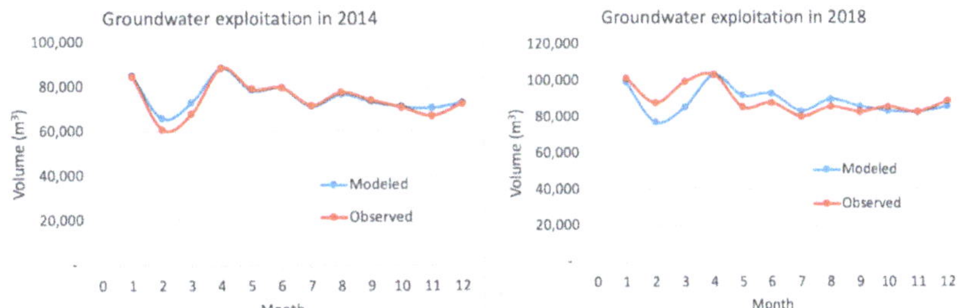

Figure 6. Calibration and validation results of the WEAP model for Con Dao Island.

3.1.3. Interaction between Surface and Ground Water

To simulate the flow from surface water to groundwater, the "Specify GW-SW Flows" method was used in this study. The outputs from the model showed that the volumes of groundwater replenished from the An Hai, QT1, and QT2 reservoirs in 2018 are about 497.7, 496.0, and 548.2 thousand m^3, respectively. The corresponding values will increase up to 574.5, 600.1, and 657.2 thousand m^3 in 2030. The increased groundwater recharge, in this case, can be explained by the increase in reservoir capacity after dredging. These outputs showed a close relationship between surface water and groundwater in Con Dao Island.

Compared to other sources of groundwater recharge, the rate of inflows from the three reservoirs of An Hai, QT1, and QT2 into an aquifer was 1.5 million m^3 (accounting for 56.92% in 2018) and 1.8 million m^3 (65.59% by 2030) (Figure 7). The WEAP model also presented that more than 80% of groundwater recharge in the period from January to April will come from these three reservoirs. This demonstrated the importance of the reservoirs on enhancing water supply resilience in Con Dao Island, particularly during the dry season. Therefore, the reservoirs need to be preserved and upgraded in terms of both quantity and quality for the future under the impact of climate change.

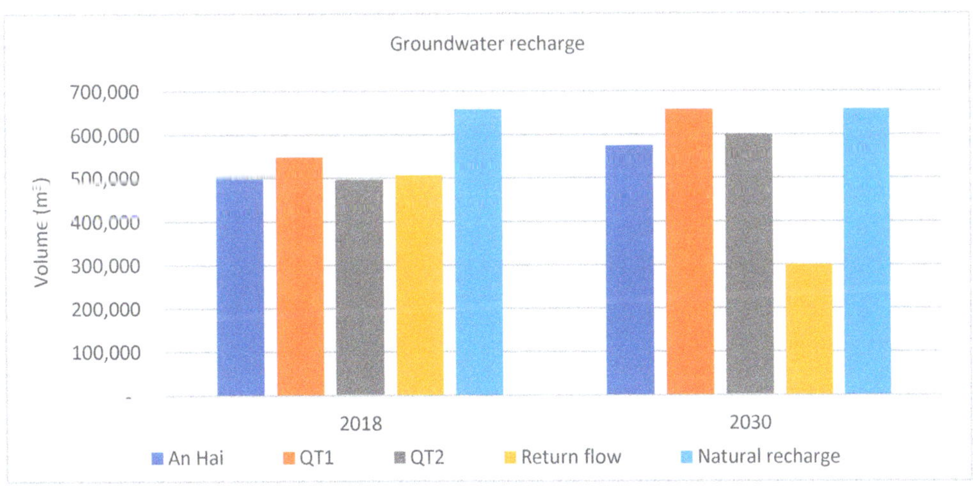

Figure 7. Groundwater recharge in Con Dao Island in 2018 and 2030.

3.1.4. Water Supply and Water Demand in Con Dao Island

As the population increases along with socio-economic development, the water demand will also increase (Appendix A). In 2018, the water demand was about 1.37 million m^3, in which industry, domestic use, tourists, and agricultural production occupied about 32.0%, 22.5%, 23.0%, and 18.0%, respectively. The water demand by 2030 will nearly triple to

3.83 million m^3, and domestic water use will be the largest, followed by water for tourists and industry. It is worth noting here that the number of tourists often spikes in the summer (peak tourist season), leading to a rapid increase in water demand. On the other hand, the impacts of climate change (i.e., temperature increases 10% while rainfall decreases 10%) could decrease water supply sources. Therefore, the water shortage will be more serious due to limited water supply capacity.

To deal with the increase in water demand, it is necessary to change the water supply. If people do not have any solutions to increase the capacity of water supply, the future demand will not be met, i.e., unmet water demand will be severe (about 2.56 million m^3 in 2030) (referring to scenario S0 in Figure 8a). For scenario S1, where the government would invest in the water supply infrastructure, the unmet demand will reduce by only about 0.48 million cubic meters in 2030, which will mainly occur in the dry season (Figure 8b). These results show that the policies proposed by local authorities are appropriate, but do not yet fully meet the water demand for the future.

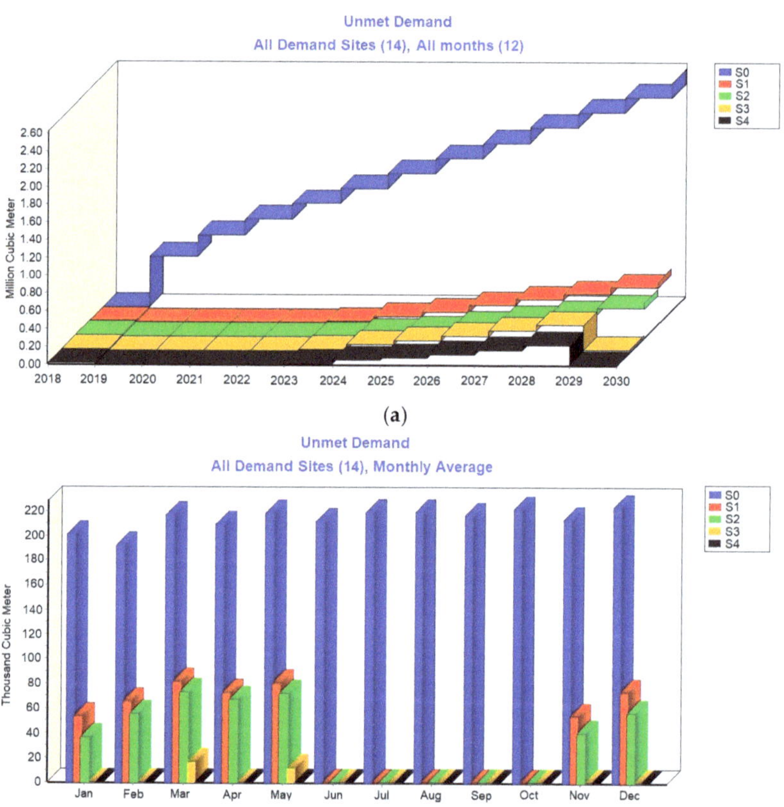

Figure 8. Unmet demand in Con Dao Island under S0, S1, S2, S3, S4 scenarios: (**a**) by year; (**b**) by monthly average.

While rainwater was considered as a resource for domestic use in the Central town (S2), the potential of rainwater harvesting quantity was calculated based on the number of households and the roof area of each household (estimated about 100 m^2). In this case, the water supply will increase by about 0.08 million m^3, resulting in a decrease in the unmet demand to 0.40 million m^3.

If a seawater desalination plant is invested in (S3), the water supply can be increased significantly and almost meet the future water demand. The unmet demand for this scenario will be only 0.03 million m^3. The future water demand will be fully met if all the methods of increasing the water supply are combined (S4). This can be considered a sustainable solution for the water supply system in the island.

3.2. *Social Observation*

3.2.1. Obstacles and Enablers in Sustainable Water Management in Con Dao Island

Perceptions of Local Actors in Natural Resources Management and Climate Change

Local people have shown a high awareness about the importance of natural resource protection such as the freshwater, ocean, and forest, partly thanks to the propaganda campaigns organized by local governments. However, only a limited number of surveyed households practice water reuse, reduction, and recycling, including collecting rainwater, due to less financial benefits. Most of them are not worried about or have never experienced a water shortage. Regarding climate change, a small portion of local people confirm that they heard about the term before, mostly those who used to work for the government or participated in social associations. Similarly, local officials have a high awareness of the protection of natural resources and the importance of the inclusiveness of locals' participation in the process. They have a good understanding of climate change in general, though raising awareness on this issue for local people is not mentioned properly. In general, both local people and officials might be aware of the importance of the protection of natural resources; nevertheless, they do not know how to apply this awareness in practice. Training organized by local governments is recommended to improve local household knowledge on specific practices.

Perceptions of Local Officials on Problems and Solutions Regarding Water Management

Several perceived risks are stated by local officials on water use and management in the Con Dao Islands. Firstly, the groundwater quantity is limited, especially during the dry season (April and May) when the number of tourists reaches a peak. Secondly, annual monitoring shows that the groundwater quality is decreasing due to pollution. A possible solution is dredging the existing reservoirs (interview of technical staff of DNRE).

3.2.2. Abilities to Implement Water Infrastructure Projects

Institutional Abilities

There are different institutions involved in the water management in the Con Dao Islands, including (1) the Department of Finance, which provides financial resources for relevant activities; (2) various associations such as the women's association, veteran association, etc., which play an important role in promoting and raising awareness of the local people; (3) the Economic Department, which is in charge of operating the dam systems; and (4) residential areas which play a supporting role in the local water management. The cooperation between the different institutions is reported to be effective and smooth.

Nevertheless, multiple institutional bottlenecks were raised by the respondents. Firstly, laws and policies on monitoring and controlling pollution for small-scale businesses or households are limited, especially in wastewater management for livestock. There are neither regulations nor mechanisms on the sanctions for causing environmental pollution that are applied to the smallholders. Secondly, some organizations are public non-business units; therefore, all the expenses need to be approved at higher managing levels through long and complicated procedures. Unclear and overlapping roles limit the effectiveness of the working process. An unclear position in the formal system also makes it difficult to coordinate with higher managing levels. Another bottleneck is the limited capacity of the officials. Not only is there a need for improving their professional skills, but the officials also expressed a need for additional training in other fields. Finally, there is a lack of information, especially on the groundwater capacity of the area; thus, it is difficult to expand the current capacity of the water supply.

Financial Abilities

Respondents stated that the financial capacity of the relevant agencies in water management is sufficient to carry out their tasks. Their budgets are provided by the central government following the current financial regulations of Vietnam. Priorities of the budget include paying tax to the government, paying salary to the staff, and contributing to a common fund for awarding and other social benefits. However, they do not have a financial budget for capacity building.

Technical Abilities

The technical abilities are varied in different agencies. In the case of the Water Supply Station, though most of the technicians have been working for a long time, since after the American War (1975), their capacity is nevertheless limited in operating the old water supply system. These technicians have lots of experience and they work manually very well; however, this might be a disadvantage if the Station wants to upgrade the water system using state-of-the-art technologies. This agency, therefore, has a high demand for capacity building. By contrast, the technical officials of the DNRE are confident in their professional skills. However, they raise concerns on concurrent tasks that they were not trained to work on. Local officials follow assigned tasks from higher managing levels. Besides working in their professional field, they are concurrently responsible for other tasks due to a lack of human resources. For example, the technical staff of the DNRE has the main responsibilities of checking the results and writing reports on the monitoring of water quality. They are additionally in charge of working on raising awareness of the local people on water protection.

4. Discussion

4.1. The Roadmaps for Sustainable Water Management in Con Dao Island

Assumptions on the socio-economic conditions as input for different scenarios are summarized in Table 3. This section elaborates on the possibility of the different scenarios in light of the enablers and obstacles analyzed from the previous section.

Table 3. Input socio-economic scenarios and output from the model.

Input of the Model: Scenarios	Output from the Model
S0 (current water supply)	Does not meet future demand
S1 (increasing centralized water supply capacity by upgrading and constructing new reservoirs and a surface water supply plant)	Reduces but still does not meet future demand
S2 (find a low-cost alternative water resource, e.g., rainwater)	Reduces but still does not meet future demand
S3 (find high-cost alternative water resources, e.g., seawater)	Almost meets future demand (99%)
S4 (combine S2 and S3)	Fully meet future demand, reduce risky if any part does not work well

The current awareness of officials on the possible increasing demand and the shortage of water supply due to, e.g., population increase and tourism, is an enabler to motivate them to achieve alternatives described in scenarios S1, S2, S3, and S4. Given the current dominance of the hierarchical management system and existing plans of the leading water suppliers, increasing centralized water supply capacity by upgrading and constructing new reservoirs seems to be the most foreseen scenario (S1) on the island [20]. Nevertheless, meeting S1 would require extra technical training for the employed technical staff and recruiting new ones. This is possible with reasonable budget allocation available as mentioned above.

Alternatives for low-cost freshwater resources such as rainwater (S2) or high-cost resources such as seawater treatment (S3) are well recognized by the top officials in Con Dao. However, these alternatives would require affordable and workable technologies (for seawater) [28–30] and raising awareness in local people's perceptions on saving freshwater (for rainwater and other facilities) [31–33]. While a considerable amount of finance and technical training to apply the new technologies is of concern, raising awareness is doable for local officials given their experience in campaigning and propaganda. Combining these measures would result in the most desirable scenario S4. With the current socio-economic and political situation of Con Dao, as well as enablers and obstacles as analyzed above, continuing on S1 at present while mobilizing resources for S2 and S3 would be the most suitable roadmap to achieve water demand of the island. Modelling well in advance and along the way to carefully assess the feasibility of each roadmap, considering possible changes will help to tackle uncertainties and support the Con Dao Island on the roadmap.

4.2. Further Recommendations for Water Management in the Con Dao Island

Due to the limits of water, land, and human resources, as well as the vulnerability to climate change, when it comes to water management in the island, an integrated approach should be considered. In integrated planning, all the different water resources and the uses and users of water resources must be considered together. Firstly, the available water resources and their sustainable yields should be assessed in the planning of water resource developments. The better-quality and cheaper water resources (rainwater, groundwater, surface water) need to be assessed initially. Other options, including desalination, may be required if the other sources are over-utilized or where the economy can afford them.

Conjunctive uses of different water should be considered. Better-quality water resources (i.e., rainwater) may be a suitable option for most basic demands that require high-quality standards, such as drinking and cooking with little or no treatment. While lower-quality water resources (surface water and seawater) require high treatment to supply high-quality standards, they can be suitable for meeting the non-drinking demands, such as toilet flushing and washing, with lower treatment. These approaches, which can enhance the sustainability of the water supply system and reduce the energy as well as the costs for treatment have also been widely proposed and practiced successfully in other studies [34–36].

While most of the large, centralized water developments (upgrading water reservoirs, water treatment and seawater treatment plants) consume a lot of time and resources, these methods only rely on human and economic resources from the government and do not involve the residents. A decentralized system, such as rooftop rainwater harvesting, can be much easier to implement. To promote a decentralized water supply system, the following strategies are recommended: first, more local pilot projects need to be implemented. Because water problems are site-specific, so should be the solutions. Technical criteria from other regions should be used only as a reference and should be customized to local conditions. Locally available materials and workers should be used where possible to minimize costs and increase the workmanship, as well as to enhance the awareness of residents. Operation and maintenance requirements should be minimized to enable the residents' operation and maintenance. Second, the promotion and education of water preservation and use should be carried out at all levels. Lastly, an innovative micro-funding system should be created in cooperation with Corporate Social Responsibility (CSR) or Environmental, Social, and Governance (ESG) activities of the public sector enterprises as a win–win tool. Together with seed money from the public sector, small islands can develop a localized business via spinoff effects.

4.3. Socio-Hydrological Approach from This Study

The new-science socio-hydrology has been promoted recently as a promising tool for water management to cope with contemporary challenges, including the uncertainty from climate change. Socio-hydrology has been attempting to build predictive models on the co-evolution of the coupled human–water system by integrating social aspects into existing hydrological models [10]. Despite the ongoing efforts, accommodating different social aspects into socio-hydrological models remains a challenge due to insufficient empirical data [37,38]. Attempting to overcome this challenge, social aspects, however, have been hardly integrated as inputs into our model due to several constraints including the limitations of the existing hydrological models and limited capacity of obtaining appropriate data for quantifying the social aspects for use in the selected models. Nevertheless, different attributes, such as the *socio-cultural responses*, *political/institutional setup*, and *water demand and availability*, in the socio-hydrological approach [6] have been addressed as inputs for the scenario development to identify the feasible roadmaps of Con Dao Island.

In this study, the importance of the protection of natural resources has been emphasized by local governments and well perceived by local people as a result of awareness-raising propaganda and campaigning. However, understanding climate change and its consequences has been neglected. Though having a good awareness of natural resources, local people are highly dependent on local governments for water use and management. The link between climate change and environmental issues, especially in the water sector, are not well perceived by local people, leading to a gap in perceptions of the existing issues. This is perceived as a consequence of the unclear and overlapping of the hierarchical management system in Vietnam, as well as a limited capacity of local officials. This research encounters several limitations, such as the small survey sample size and non-randomly selected samples. Better research design to adapt the questionnaires to local contexts and model preparation is recommended to continue the efforts in future research.

5. Conclusions

The proposed social-hydrological approach combining modeling and social methods such as focus group discussion, in-depth interviews, and a survey was applied to address water resource management in Con Dao Island. The analysis of the interaction between surface water and groundwater showed that the An Hai, QT1, and QT2 reservoirs play an important role in enhancing water supply resilience in the island, particularly during the dry season, as they are the main sources to recharge into the groundwater. The WEAP results also showed that the water demand in the island tends to increase in the context of socio-economic development, population growth, and climate change, making the water shortage more serious due to the limited water supply capacity. Although the water supply capacity in the island has been improved by the investment of the local government in water supply infrastructures, the water demand may not be fully met in the near future. The sustainable solution for the water supply system in Con Dao needs to be combined with the use of alternative water sources, such as rainwater and desalinated seawater.

Despite having a high awareness of the role of natural resources, local actors seem to neglect climate change and its consequences due to the hierarchical management system and the dependence of local people on the government. Given these enablers and obstacles, upgrading and constructing new reservoirs while mobilizing resources for freshwater alternatives and saving facilities is the most desirable roadmap for the island. To achieve the roadmap, strengthening adaptive capacity, raising awareness, and building a professional capacity for both local people and officials are important to address existing issues in water and relevant sectors in Con Dao Islands.

Further research could explore the possibility of integrating the information on social aspects into the existing models to understand better the interaction and co-evolution of the human–water system in the case of islands.

Author Contributions: Conceptualization and methodology, H.Q.N. and D.C.H.N.; data collection, H.Q.N., D.C.H.N., T.C.L. and T.T.L.; model set-up, D.C.H.N.; social analysis, T.T.L. and H.Q.N.; writing—original draft preparation, D.C.H.N., T.T.L. and H.Q.N.; writing—review and editing, D.C.H.N., T.T.L., H.Q.N., P.K., D.C.N. and R.D. All authors have read and agreed to the published version of the manuscript.

Funding: This work is supported by Asia Pacific Network for Global Change Research (APN) under the Collaborative Regional Research Programme (CRRP) with project reference number CRRP2019-01MY-Kumar.

Institutional Review Board Statement: Not applicable.

Informed Consent Statement: Not applicable.

Data Availability Statement: The data developed in this study will be made available on request to the corresponding authors.

Conflicts of Interest: The authors declare no conflict of interest.

Appendix A

Table A1. Water demand in Con Dao (Con Dao Statistical Yearbook 2018).

Area	Water users	Quantity		Water Use Standard (m^3/day)		Water Demand (m^3/year)	
		2018	2030	2018	2030	2018	2030
Central Town	Local residents	5625	13,500	0.1	0.2	205,313	985,500
	Tourists	4000	6500	0.2	0.3	292,000	711,750
	Annual crop (ha)	12	5			58,696	24,457
	Vegetables (ha)	6	33	Crop coefficient		29,348	161,415
	Fruit trees (ha)	14	11			111,992	87,994
	Cattles (head)	207	500	0.05	0.05	3778	9125
	Pigs (head)	1072	2000	0.03	0.03	11,738	21,900
	Chickens, ducks (head)	11,000	20,000	0.01	0.01	20,075	36,500
	Goat (head)	157	300	0.03	0.03	1433	2738
	Industry (ha)	7	11	22.0	22.0	54,750	91,250
Ben Dam	Local residents	1300	3000	0.1	0.2	47,450	19,000
	Tourists	-	500	0.2	0.3	-	54,750
	Industry (ha)	45	62	22.0	22.0	361,350	97,860
Co Ong	Local residents	1500	4000	0.1	0.2	54,750	292,000
	Tourists	300	2500	0.2	0.3	21,900	273,750
	Vegetables(ha)	2		Crop coefficient		9783	24,457
	Industry (ha)	3	5	22.0	100.0	21,900	36,500

Appendix B

Table A2. The meteorological data—monthly rainfall and evaporation in Con Dao (unit: mm).

2018												
	Jan.	Feb.	Mar.	Apr.	May	Jun.	Jul.	Aug.	Sep.	Oct.	Nov.	Dec.
Rainfall	7.8	4.6	3.2	4.8	32.2	301.9	526.8	436.5	223.2	256.4	102.1	66.6
Evaporation	117.1	101.3	103.1	97.6	93.1	88.5	92.9	97.2	86	75.9	95.1	113.3
2014												
	Jan.	Feb.	Mar.	Apr.	May	Jun.	Jul.	Aug.	Sep.	Oct.	Nov.	Dec.
Rainfall	3.5	0	0	1.5	72.6	163.1	415.9	153.2	174.2	282.4	126.6	61.7
Evaporation	105.5	108.5	112.2	101.8	83.8	91	93.4	102	80	87.2	87.4	94.8

Appendix C. Questions for the Focus Group Discussion

1. What kind of activities do you need to use water for? For example, agriculture, aquaculture, industry, etc.
2. What kind of issues related to water use are you currently facing in your area?
3. Do you experience any water shortages for domestic use?
4. Do you think the water quality has an effect on your mental and physical health?
5. Do you think water shortage has an effect on your social relationships?
6. Do you feel annoyed when you experience water shortages or low water quality?
7. Are you discriminated against due to water shortages or low water quality?
8. Do you know about policies that are related to water in your area? If yes, please specify.

References

1. Blair, P.; Buytaert, W. Socio-hydrological modelling: A review asking "why, what and how?". *Hydrol. Earth Syst. Sci.* **2016**, *20*, 443–478. [CrossRef]
2. Kumar, P. Numerical quantification of current status quo and future prediction of water quality in eight Asian megacities: Challenges and opportunities for sustainable water management. *Environ. Monit. Assess.* **2019**, *191*, 319. [CrossRef] [PubMed]
3. United Nations World Water Assessment Programme (WWAP). *The United Nations World Water Development Report 2017*; Wastewater: The Untapped Resource; UNESCO: Paris, France, 2017; p. 198.
4. Jalilov, S.-M.; Kefi, M.; Kumar, P.; Masago, Y.; Mishra, B.K. Sustainable Urban Water Management: Application for Integrated Assessment in Southeast Asia. *Sustainability* **2018**, *10*, 122. [CrossRef]
5. Peña-Ramos, J.; Bagus, P.; Fursova, D. Water Conflicts in Central Asia: Some Recommendations on the Non-Conflictual Use of Water. *Sustainability* **2021**, *13*, 3479. [CrossRef]
6. Kumar, P.; Avtar, R.; Dasgupta, R.; Johnson, B.A.; Mukherjee, A.; Ahsan, N.; Nguyen, D.C.H.; Nguyen, H.Q.; Shaw, R.; Mishra, B.K. Socio-hydrology: A key approach for adaptation to water scarcity and achieving human well-being in large riverine islands. *Prog. Disaster Sci.* **2020**, *8*, 100134. [CrossRef]
7. Ingol-Blanco, E.; McKinney, D.C. Development of a Hydrological Model for the Rio Conchos Basin. *J. Hydrol. Eng.* **2013**, *18*, 340–351. [CrossRef]
8. Slaughter, A.R.; Mantel, S.K.; Hughes, D.A. Investigating possible climate change and development effects on water quality within an arid catchment in South Africa: A comparison of two models. In Proceedings of the 7th International Congress on Environmental Modelling and Software, San Diego, CA, USA, 15–19 June 2014; Volume 3, pp. 1568–1575.
9. Yang, L.; Bai, X.; Khanna, N.Z.; Yi, S.; Hu, Y.; Deng, J.; Gao, H.; Tuo, L.; Xiang, S.; Zhou, N. Water evaluation and planning (WEAP) model application for exploring the water deficit at catchment level in Beijing. *Desalination Water Treat.* **2018**, *118*, 12–25. [CrossRef]
10. Sivapalan, M.; Savenije, H.H.G.; Blöschl, G. Socio-hydrology: A new science of people and water. *Hydrol. Process.* **2011**, *26*, 1270–1276. [CrossRef]
11. Herrera-Franco, G.; Montalván-Burbano, N.; Carrión-Mero, P.; Bravo-Montero, L. Worldwide Research on Socio-Hydrology: A Bibliometric Analysis. *Water* **2021**, *13*, 1283. [CrossRef]
12. Eslamian, S.; Reyhani, M.N.; Syme, G. Building socio-hydrological resilience: From theory to practice. *J. Hydrol.* **2019**, *575*, 930–932. [CrossRef]
13. Srinivasan, V.; Sanderson, M.; Garcia, M.; Konar, M.; Bloschl, G.; Sivapalan, M. Prediction in a socio-hydrological world. *Hydrol. Sci. J.* **2017**, *62*, 338–345. [CrossRef]
14. Viglione, A.; Di Baldassarre, G.; Brandimarte, L.; Kuil, L.; Carr, G.; Salinas, J.L.; Scolobig, A.; Blöschl, G. Insights from socio-hydrology modelling on dealing with flood risk—Roles of collective memory, risk-taking attitude and trust. *J. Hydrol.* **2014**, *518*, 71–82. [CrossRef]
15. Falter, D.; Schröter, K.; Dung, N.V.; Vorogushyn, S.; Kreibich, H.; Hundecha, Y.; Apel, H.; Merz, B. Spatially coherent flood risk assessment based on long-term continuous simulation with a coupled model chain. *J. Hydrol.* **2015**, *524*, 182–193. [CrossRef]
16. Fish, R.D.; Ioris, A.A.; Watson, N.M. Integrating water and agricultural management: Collaborative governance for a complex policy problem. *Sci. Total. Environ.* **2010**, *408*, 5623–5630. [CrossRef]
17. Dang, T.K.P. Tourism imaginaries and the selective perception of visitors: Postcolonial heritage in Con Dao Islands, Vietnam. *Isl. Stud. J.* **2021**, *16*, 249–270. [CrossRef]
18. National Institute of Agricultural Planning and Projection (NIAPP). *Agricultural Development Planning of Con Dao District to 2020, Vision to 2030*; National Institute of Agricultural Planning and Projection (NIAPP): Hanoi, Vietnam, 2017.
19. Long, T.T.; Koontanakulvong, S. SW-GW Interaction Analysis for Drought Management in Con Son Valley, Con Dao Island, Ba Ria-Vung Tau Province, Vietnam. In Proceedings of the 1st AUN/SEED-Net Regional Conference on Natural Disaster, Yogyakarta, Indonesia, 22–23 January 2014; Volume 22, p. 23.
20. Minister of Planning and Investment (MPI). *Decision No. 1742/QD-BKHDT of the on Approving the Master Plan on Socio-economic Development of Con Dao District, Ba Ria-Vung Tau Province, Up to 2020, with a Vision toward 2030*; Minister of Planning and Investment (MPI): Hanoi, Vietnam, 2011.
21. Mounir, Z.M.; Ma, C.M.; Amadou, I. Application of Water Evaluation and Planning (WEAP): A Model to Assess Future Water Demands in the Niger River (In Niger Republic). *Mod. Appl. Sci.* **2011**, *5*, 38. [CrossRef]
22. Dang, D.K.; Tran, N.A.; Mai, T.N. Application of WEAP model for integrated water balance in Lam River Basin. *Vietnam J. Sci. Technol.* **2015**, *31*, 186–194.
23. Agarwal, S.; Patil, J.P.; Goyal, V.C.; Singh, A. Assessment of Water Supply–Demand Using Water Evaluation and Planning (WEAP) Model for Ur River Watershed, Madhya Pradesh, India. *J. Inst. Eng. Ser. A* **2019**, *100*, 21–32. [CrossRef]
24. Nguyen, H.Q.; Korbee, D.; Ho, L.; Weger, J.; Hoa, P.T.T.; Duyen, N.T.T.; Luan, P.D.M.H.; Luu, T.T.; Thao, D.H.P.; Trang, N.T.T.; et al. Farmer adoptability for livelihood transformations in the Mekong Delta: A case in Ben Tre province. *J. Environ. Plan. Manag.* **2019**, *62*, 1603–1618. [CrossRef]
25. Phi, H.L.; Hermans, L.M.; Douven, W.J.; Van Halsema, G.E.; Khan, M.F. A framework to assess plan implementation maturity with an application to flood management in Vietnam. *Water Int.* **2015**, *40*, 984–1003. [CrossRef]

26. Ba Ria—Vung Tau (BRVT) Statistics Office. *BRVT Statistical Yearbook 2018*; Statistical Publishing House of BRVT: BRVT, Vietnam, 2019.
27. Minister of Natural Resources and Environment (MONRE). Climate Change and Sea Level Rise Scenarios for Vietnam. 2016. Available online: https://vihema.gov.vn/wp-content/uploads/2015/12/02.-Tom-tat-Kich-ban-BDKH-va-NBD-cho-VN_2016-Tieng-Anh.pdf (accessed on 25 September 2020).
28. Urrea, S.A.; Reyes, F.D.; Suárez, B.P.; Bencomo, J.A.D.L.F. Technical review, evaluation and efficiency of energy recovery devices installed in the Canary Islands desalination plants. *Desalination* **2018**, *450*, 54–63. [CrossRef]
29. Borge-Diez, D.; García-Moya, F.J.; Cabrera-Santana, P.; Rosales-Asensio, E. Feasibility analysis of wind and solar powered desalination plants: An application to islands. *Sci. Total. Environ.* **2020**, *764*, 142878. [CrossRef]
30. Duong, H.C.; Tran, L.T.T.; Truong, H.T.; Nelemans, B. Seawater membrane distillation desalination for potable water provision on remote islands – A case study in Vietnam. *Case Stud. Chem. Environ. Eng.* **2021**, *4*, 100110. [CrossRef]
31. Thuy, B.T.; Dao, A.D.; Han, M.; Nguyen, D.C.; Nguyen, V.-A.; Park, H.; Luan, P.D.M.H.; Duyen, N.T.T.; Nguyen, H.Q. Rainwater for drinking in Vietnam: Barriers and strategies. *J. Water Supply Res. Technol.* **2019**, *68*, 585–594. [CrossRef]
32. Kim, Y.; Han, M.; Kabubi, J.; Sohn, H.-G.; Nguyen, D.-C. Community-based rainwater harvesting (CB-RWH) to supply drinking water in developing countries: Lessons learned from case studies in Africa and Asia. *Water Supply* **2016**, *16*, 1110–1121. [CrossRef]
33. Temesgen, T.; Han, M.; Park, H.; Kim, T.-I. Policies and Strategies to Overcome Barriers to Rainwater Harvesting for Urban Use in Ethiopia. *Water Resour. Manag.* **2016**, *30*, 5205–5215. [CrossRef]
34. Nguyen, D.C.; Han, M.Y. Design of dual water supply system using rainwater and groundwater at arsenic contaminated area in Vietnam. *J. Water Supply Res. Technol.* **2014**, *63*, 578–585. [CrossRef]
35. Kourtis, I.M.; Kotsifakis, K.G.; Feloni, E.G.; Baltas, E.A. Sustainable Water Resources Management in Small Greek Islands under Changing Climate. *Water* **2019**, *11*, 1694. [CrossRef]
36. Lautze, J.; Holmatov, B.; Saruchera, D.; Villholth, K.G. Conjunctive management of surface and groundwater in transboundary watercourses: A first assessment. *Hydrol. Res.* **2018**, *20*, 1–20. [CrossRef]
37. Aerts, J.C.J.H. Integrating agent-based approaches with flood risk models: A review and perspective Integrating agent-based approaches with flood risk models: A review and perspective. *J. Water Secur.* **2020**, *11*, 100076. [CrossRef]
38. Baldassarre, G.D.; Sivapalan, M.; Rusca, M.; Cudennec, C. Socio-hydrology: Scientific Challenges in Addressing a Societal Grand Challenge. *Water Resour. Res.* **2019**, *55*, 6327–6355. [PubMed]

 *water*

Article

Vulnerability and Risk Assessment to Climate Change in Sagar Island, India

Aparna Bera [1], Gowhar Meraj [1,2], Shruti Kanga [1], Majid Farooq [1,2], Suraj Kumar Singh [3,*], Netrananda Sahu [4] and Pankaj Kumar [5,*]

1 Centre for Climate Change and Water Research, Suresh Gyan Vihar University, Jaipur 302017, India; aparna.62232@mygyanvihar.com (A.B.); gowharmeraj@gmail.com (G.M.); shruti.kanga@mygyanvihar.com (S.K.); majid_rsgis@yahoo.com (M.F.)
2 Department of Ecology, Environment and Remote Sensing, Government of Jammu and Kashmir, Srinagar 190018, India
3 Centre for Sustainable Development, Suresh Gyan Vihar University, Jaipur 302017, India
4 Department of Geography, Delhi School of Economics, University of Delhi, Delhi 110007, India; nsahu@geography.du.ac.in
5 Institute for Global Environmental Strategies, Hayama 240-0115, Kanagawa, Japan
* Correspondence: suraj.kumar@mygyanvihar.com (S.K.S.); kumar@iges.or.jp (P.K.)

Abstract: Inhabitants of low-lying islands face increased threats due to climate change as a result of their higher exposure and lesser adaptive capacity. Sagar Island, the largest inhabited estuarine island of Sundarbans, is experiencing severe coastal erosion, frequent cyclones, flooding, storm surges, and breaching of embankments, resulting in land, livelihood, and property loss, and the displacement of people at a huge scale. The present study assessed climate change-induced vulnerability and risk for Sagar Island, India, using an integrated geostatistical and geoinformatics-based approach. Based on the IPCC AR5 framework, the proportion of variance of 26 exposure, hazard, sensitivity, and adaptive capacity parameters was measured and analyzed. The results showed that 19.5% of mouzas (administrative units of the island), with 15.33% of the population at the southern part of the island, i.e., Sibpur–Dhablat, Bankimnagar–Sumatinagar, and Beguakhali–Mahismari, are at high risk (0.70–0.80). It has been concluded that the island has undergone tremendous land system transformations and changes in climatic patterns. Therefore, there is a need to formulate comprehensive adaptation strategies at the policy- and decision-making levels to help the communities of this island deal with the adverse impacts of climate change. The findings of this study will help adaptation strategies based on site-specific information and sustainable management for the marginalized populations living in similar islands worldwide.

Keywords: risk; vulnerability; climate change; principal component analysis; low-lying delta; IPCC AR 4 and AR 5

Citation: Bera, A.; Meraj, G.; Kanga, S.; Farooq, M.; Singh, S.K.; Sahu, N.; Kumar, P. Vulnerability and Risk Assessment to Climate Change in Sagar Island, India. *Water* **2022**, *14*, 823. https://doi.org/10.3390/w14050823

Academic Editor: Rafael J. Bergillos

Received: 21 January 2022
Accepted: 2 March 2022
Published: 6 March 2022

Publisher's Note: MDPI stays neutral with regard to jurisdictional claims in published maps and institutional affiliations.

1. Introduction

Climate change is a major concern that has increased the rapid and slow onset of climate events globally [1,2]. Rising ocean and air temperatures, increasing occurrence and intensity of tidal surges, violent stormy cyclones, severe flooding, and extreme precipitation events are some of the manifestations of climate change [3]. The low-lying coastal regions are witnessing adverse impacts, such as inland flooding, submergence, and coastal erosion, due to rising sea levels [4]. According to an estimation, by 2050, almost a million people living in three significant deltas, namely, the Mekong Delta, Nile delta, Ganges–Brahmaputra–Meghna delta, will be adversely impacted by rising sea levels [5]. For the Indian Bengal Delta, such an increase could be as high as 70% [6]. Climate variability greatly influences the environment and socioeconomic aspects, such as agriculture, livelihood, health, and biodiversity [7]. Biophysical vulnerability manifests in communities' exposure to climate change; hence, due to the greater social vulnerability, they are more exposed to adverse impacts [8]. Climate change primarily affects the poor, disabled, aged, and marginalized populations, increasing social vulnerabilities [9].

Apart from broad-scale increases, local factors influence household-level vulnerability [10]. Population pressure, changes in land use, and intensive agriculture can exacerbate risks and exposure [11]. These catalyze the displacement of endangered people and increase the number of population traps, which can cause internal and external population movement [12,13]. Sundarbans is a very good instance of the manifestations of climate change, wherein underdevelopment and over-reliance on climate-dependent subsistence have rendered the whole ecosystem vulnerable [14].

Sagar Island has encountered the impacts of climate change in the form of rising sea levels, tidal surges, increased soil salinity, violent cyclones, and severe coastal erosion [15,16]. Part of an archipelago of 102 islands in the Sundarban coastal region, Sagar Island is the most significant. Its inhabitants are losing their land under their feet day by day. The surrounding four islands, named Bedford, Lohachara, Khasimara, and Suparivanga, were diluviated by coastal erosion in the last few decades. Bishalakkhipur mouza of Sagar Island was submerged, and Sagar mouza has become uninhabitable due to the excessive erosion. Ghoramara island will soon be submerged by the rising sea and accelerating erosion [17]. The range of apparent sea level increase varies between 3–8 mm/year in the Sundarbans, beyond the global average of 3 mm/year [18]. This present rate can result in a 20 percent enhanced flooding risk, over 1.520 mm by 2070 [19]. According to [18], the Indian Sundarban has lost approximately 4% of forest cover, a natural buffer against cyclone surges. Increasing sea levels and accelerated wave action caused subsequent changes in the hydrodynamic regime that led to severe land loss. From 2009 to 2019, the island's area has prominently reduced from 246.76 km^2 to 230.98 km^2; the average decadal percent change in this area for that period accounted for -11.33% [20]. The Bay of Bengal typically experiences 7% of the significant cyclones worldwide, while in the last 120 years, the frequency and intensity of the cyclones have increased between 20% and 26% [21]. Severe cyclones, i.e., Yash (2021), Amphan (2020), Bulbul (2019), and Aila (2009), accompanied by storm surges and flooding, caused large-scale devastation to the coastal regions [22]. The agricultural community depends solely on nature; extreme weather events, cyclone surges, tidal ingression, embankment breaching, saline water intrusion deplete their habitats and livelihoods, forcing them to become environmental refugees [23].

Researchers worldwide are assessing the impacts of climate change, associated vulnerabilities, and risks due to the frequency, magnitude, and tenacity of climate events [24,25]. Globally, research on the human effects associated with climate change and their scale [26] has been used to guide policymaking, with the demarcation of vulnerable areas and the identification of at-risk populations being made according to environmental assessments, along with the introduction of measures aimed at mitigating the impacts of severe climate events [27,28]. Ways to assess risk and vulnerability to the impacts of climate change have been defined by the Inter-Governmental Panel for Climate Change. As of the time of writing, two methodologies have been proposed, one based on the AR4 report and the most recent based on the AR5. As defined by the IPCC (2007), vulnerability is a function of sensitivity, exposure, and adaptive capacity [29]. However, the terminology of risk, as introduced in the fifth Assessment Report (AR5) of the IPCC 2014, defines it as the interplay of exposure, hazard, and vulnerability [2,30,31]. The present work is based on the AR5 methodology, which defines it as the function of exposure, hazard, sensitivity, and adaptive capacity. According to AR5, vulnerability is considered as an internal character and is defined by adaptive capacity and sensitivity. Hence, the effective step of adaptation to the impacts of changing climate is to reduce present exposure and vulnerability. Though the vulnerability indices simplify the intricate aspects of climate change impacts, the merit of this kind of evaluation lies in its instrumentality in the context of policy development and mitigation of climatic risks [32].

Previous studies of Sagar Island have only evaluated climate change vulnerability; hence, there was little effort made to assess risk through the IPCC AR5 framework. The present work is designed to assess how biophysical and socioeconomic variabilities contribute to the risk to the inhabitants of Sagar Island associated with climate change. The principal goal of the present study is to assess socioeconomic and biophysical variability associated with climate change using the vulnerability and risk indices through exploratory factor analysis for the 41 inhabited mouzas of Sagar Island based on the AR5 framework. The primary objectives are: (a) to analyze both the slow and rapid onset climate variabilities and assess related vulnerabilities; (b) to identify the relative contributions of hazard, exposure, sensitivity, and adaptive capacity to the observed risk using statistical analysis; and (c) to enumerate indices to help design strategies for efficient risk management through mapping of hotspot areas. Spatial–temporal changes in weather variables and land-use/land-cover were also analyzed, as these are significant factors in risk management. This context-specific and location-specific geostatistical analysis has been framed on a quantitative scale for policymakers and organizations to determine which areas need specific policy interventions.

2. Study Area and Rationale

Sagar Island is located at the confluence of Ganga at the Bay of Bengal, 100 km south of Kolkata in the western part of Sundarbans. Administratively, Sagar Island (21°36′ N to 21°56′ N; 88°02′ E to 88°11′ E) is a part of Sagar block of south 24 Parganas district (Figure 1). Home to 212,037 people [33], with 41 inhabited mouzas in 9 panchayats, covering an area of 282.11 km^2, this island is composed of the alluvium of the Ganga and Brahmaputra rivers and their tributaries. The low-lying islands of Sundarbans have been considered global climate change hotspots, located as they are in a flood-prone micro-tidal estuary characterized mudflats, creeks, and sandy beaches [34]. The average elevation is 4 m; diurnal tides range between 3.5–5.5 m (Figure 2). Hence, a maximum portion of the island undergoes inundation with saline water periodically from tides and storm surges. In this humid monsoon climate, the average annual temperature and the total precipitation were 27.57 °C and 154.25 mm, respectively, in 2020. Between 1977 and 2017, the island has faced significant changes to its shoreline. Continuous tidal ingression, waves, longshore currents, cyclones, and rising sea levels have been modifying the island's shape. With a 14.22% increase rate, this rural population has an average household size of 4.50, of which 44.46% are below the poverty line. According to the 2011 census, 40.03% of the population are total workers; 43.72% and 24.46% are agricultural laborers and cultivators, respectively [30]. Constant loss of land and expanding salinity reduce opportunities for honey and prawn seed collection. Widespread poverty and lack of development are turning most working populations into daily wage laborers. Agriculture and fishing are the most important economic activities for the local population. The increasing surface temperature of the sea, monsoonal irregularities, and higher sea levels are crucial threats to their livelihoods. Between 1981 and 2020, the frequency and intensity of cyclones striking the Sundarbans increased [35]. During 1891–2016, 232 severe cyclonic storms and 293 cyclonic storms were observed in the Bay of Bengal and surrounding areas [36]. The severe cyclonic storm Aila (27 May 2009) rendered over 5.1 million inhabitants homeless, and thousands of acres of farmland perished from the ingression of saline water. Bulbul (5 November 2019) and Amphan (16 May 2020) caused widespread destruction. This area faced continuous land loss, saline water intrusion, and limited access to resources and livelihoods despite an agricultural economy. Mouzas (administrative units) in the southern part, i.e., Dhablat, Shibpur, Mahismari, Beguakhali, have been heavily eroded. The sediment deficiency, removal of mangroves, sea level increase, unsustainable development, and exploitive mining of clay are the root causes of excessive erosion [37].

Land reclamation started in 1811, long before the accretion–erosion process could reach a stable equilibrium in this newly formed active delta. The construction of the barrage at Farakka, the dying-out of the tributaries, and the subsequent fall in the supply of sediments have changed the accretion rates and altered the vital hydrodynamics of surplus erosion [38]. Higher exposure to climate-related hazards and the over-dependency of inhabitants on the rain-fed agricultural economic system have made the islands, including Sagar, an important example of the climate change-related impacts that are being experienced worldwide [39–44].

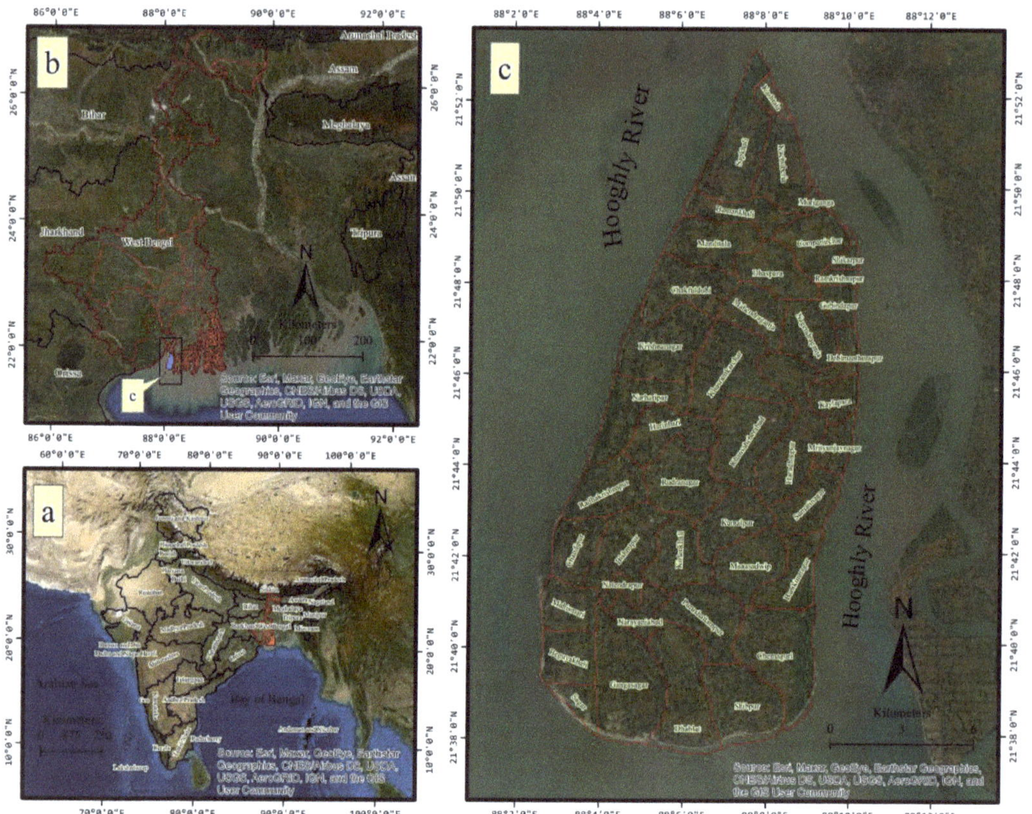

Figure 1. Map showing (**a**) the location of the West Bengal (WB) in relation to India, (**b**) Sagar Island in the southwestern corner of the WB, and (**c**) Sagar Island with mouza boundaries.

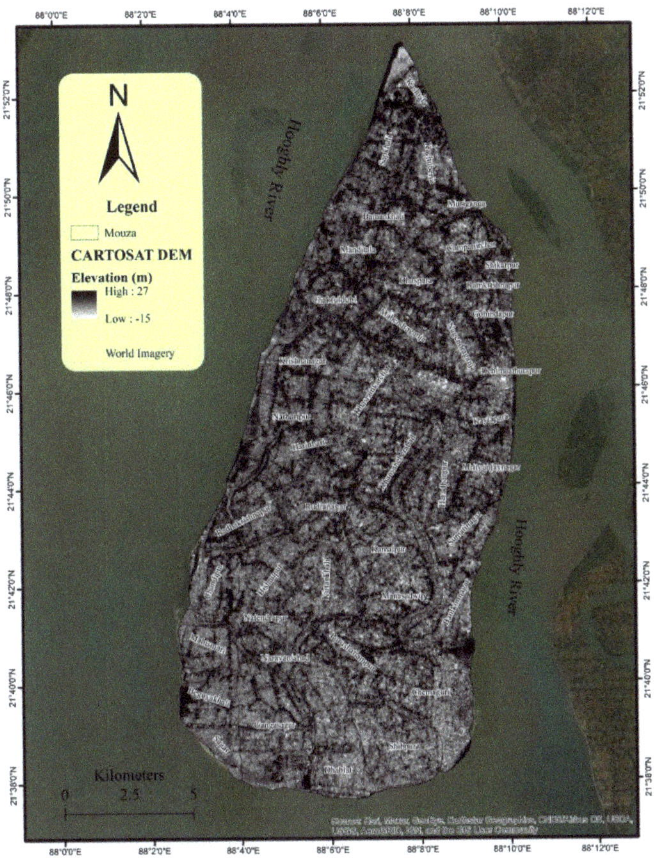

Figure 2. Study area: Sagar Island, administrative setting; Cartosat: 1 DEM; Spatial Resolution: 30 mts; Date stamp: 29 April 2015; Geographic datum: WGS-84.

3. Materials and Methods

3.1. Data

To identify climatic variation and extreme weather situations, the mean monthly temperature and rainfall data for Sagar collected over the past 20 years (2001–2020) along with annual storm data (deep depression, cyclone, and severe cyclones) were obtained from the regional center of IMD. Data for the sea level (1948–2012) of Diamond Harbour (near Sagar) were obtained from the global sea level observing system (GSLO). Data related to the occurrence and intensification of cyclones were also obtained. The Landsat 4-5 TM satellite image (30 m spatial resolution) of MSS for 1990 and OLI for 2020 were acquired from the USGS website.

3.2. Methods

3.2.1. General Framework

Risk is defined as a function of the exposure, hazard, and socioeconomic vulnerability of both resources and communities, according to the IPCC's Fifth Assessment Report (2014) [45]. While the risks are commonly thought of as natural, they can be exacerbated by human-induced variables that speed up or increase the scale of occurrences or processes, or lessen them through interventions and adaptations, such as coastal barriers, embankments, and polders. Overall, in the present work, we first describe how variables characterizing each component of exposure, hazard, and vulnerability are calculated, followed by estimates of the potential risks [45] (Figure 3).

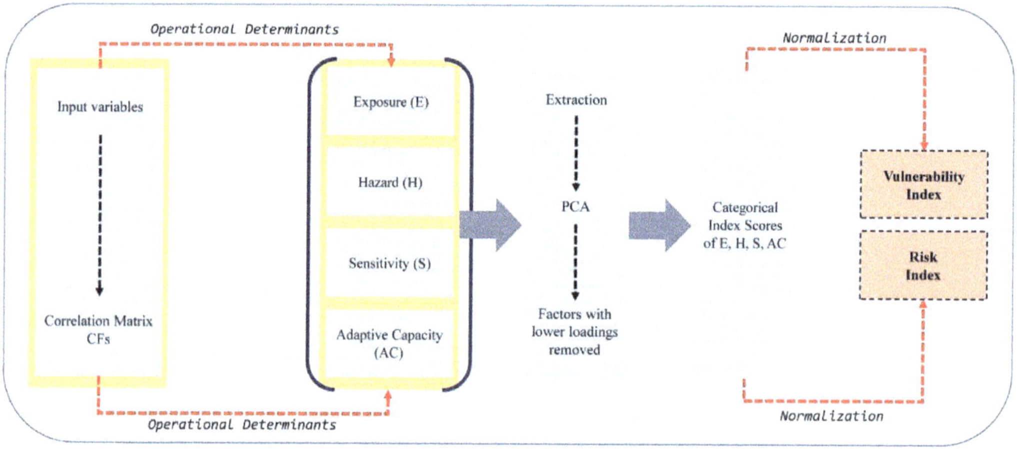

Figure 3. Graphical representation of the overall methodology.

3.2.2. Component Variables

Based on an extensive field survey, literature reviews, official records, and interviews with experts, 28 prominent policy-oriented biophysical and socioeconomic variables of 9 major components of exposure (people, infrastructure, livelihoods), hazard, sensitivity (livelihood activity, demographic profile, and socioeconomic status), and adaptive capacity (human resource, primary facility, infrastructure, and economic security) were selected (Table 1). Climatic variability was measured by the mean monthly temperature and standard precipitation deviation over the last 30 years. Natural hazards were the occurrence of cyclones, floods, and coastal erosion. There were seven components of the internal element identified as adaptive capacity and sensitivity in AR4 and vulnerability and exposure in AR5. In the context of demographic status, socioeconomic, financial security, human resources, and livelihood activity, variables such as household size, number of females and children, disadvantaged individuals, people without land holdings, poverty, literacy, agricultural dependency, marginal worker and non-worker status, work participation, and number of salaried people were computed. Parameters such as access to sanitation and electricity along with the availability of safe drinking water, basic infrastructure, primary control, vital resilience, and adaptive capacity were also assessed among the impacted communities [46,47]. Marginalized rural people are more susceptible to poverty and overdependence on natural resources [48,49]. The present study was carried out on all 41 inhabited mouzas of Sagar Island. All variables were taken at an interval scale, and outliers have been identified through descriptive statistics. To address singularity (perfectly correlated) and multi co-linearity (highly correlated), the list of variables was reduced to 26 by removing redundant variables ($R > 0.8$). The Kaiser–Meyer–Olkin (KMO) test was performed to test sample size suitability and diagnose multi-colinearity. Mann–Kendall parametric rank correlation was computed to detect changes in the time series of seasonal and annual variations of temperature and precipitation. Significant trends were identified by comparing Z values with normal distributions at the selected significance level. Cyclones that affected the island in the last 20 years were plotted in R according to their severity. Land use and land cover (LULC) information was prepared using the above data. For 1990 and 2020, Landsat 4-5 TM (30 m spatial resolution) data were used from the USGS website to compare and identify significant changes under seven classes. A maximum likelihood classifier was chosen for classifying LULC. We obtained an overall accuracy for

1990 and 2020 equal to 80% and 81.45%, respectively. We used ArcGIS 10.2.1 to perform the GIS analysis [50,51].

Table 1. Selected parameters of exposure, hazard, sensitivity and adaptive capacity.

Sl. No.	Components		Variables
1	Exposure		Max. temperature (SD of avg. monthly temperature for the past 30 years)
2			Min. temperature (SD of monthly avg. of past 30 years)
3			Avg. rainfall (SD of monthly avg. of past 30 years
4	Hazard		Flood (percentage of area inundated in past 10 years)
5			Cyclone wind speed (m/s) (interpolated) over past 30 years
6	Sensitivity	Demography	Density of population (no. of persons/sq. km)
7			Avg. size of household
8			Percentage of females relative to total population
9			Percentage of children (0–6 age group) relative to total population
10		Socio-economy	Percentage of SC and ST populations relative to total population
11			Food security (percentage of households with 1 meal/day)
12			Land holding (percentage of households without land)
13			Poverty (percentage of persons under poverty line)
14		Livelihood	Dependency on agriculture (percentage of labourers relative to total population)
15			Percentage of marginal workers
16			Percentage of non-workers
17	Adaptive Capacity	Human resource	Literacy rate
18			Work participation (Percentage workers relative to total population)
19		Economic security	Percentage of salaried persons
20			Home ownership (percentage of households owning a home)
21			Household assets (percentage of households with home assets)
22		Infrastructure	Percentage of pucca houses
23			Road density (km/sq. Km)
24		Basic facilities	Sanitation (percentage of households with sanitation)
25			Electricity (percentage of household with connections)
26			Safe drinking water (percentage of household with access)

3.2.3. Principal Component Analysis (PCA)

Exploratory factor analysis is a multivariate technique widely used in geography and other social research [52]. Principal component analysis (PCA) is the standard statistical data reduction technique for excerpting a smaller and reasoned set of uncorrelated sets (components) amongst many variables. The first set (component) is the most significant variation possible, and each following set accounts for the possible remaining variability. Hence, the variables have been converted into factors, and the coordinate of each variable is computed to ascertain the factor loadings. Factor loadings are values that explicate how strongly the variables are associated with each factor discovered. The sum of the square loadings of each principal component is the component's latent root or Eigenvalue [53].

3.2.4. Calculation of Vulnerability and Risk Indices

Vulnerability and risk indices were calculated at the mouza level using the PCA in the R python Prcomp package. The correlation matrix was used to extract the principal components. To construct the values of contributing factors (CFs) stated in the IPCC framework, coefficients of component scores were multiplied by the ratio of the variance of the corresponding components. The formula used for calculating the contributing factors was Equation (1):

$$CF = \sum \frac{F_i}{TV} \times FS_i \tag{1}$$

where CF is a contributing factor, F_i is the percent variance of component (i), TV is the total variance derived by all the reserved components, FS_i is the coefficient of the component score (i).

It became essential to normalize CFs, as the values can be both positive and negative. The standard normalization equation used is Equation (2):

$$X_{ij} = \frac{(X_i - Min\ X_j)}{(Max\ X_j - Min\ X_j)} \tag{2}$$

where X_{ij} (for mouza) becomes normalized CF (j), X_i is the actual value and Max X_j and Min Xj are the maximum and minimum CF values for complete mouzas, respectively. Then, all normalized CFs were combined into a single composite index ranging from 0 to 1.

From the Fussel and Klein framework [54], exposure (E) and sensitivity (S) are computed together as the potential impact (PI) (3):

$$PI = E \times S \tag{3}$$

A system or community with limited adaptive capacity (AC) becomes more vulnerable according to its sensitivity and exposure to climate change impacts. Hence, vulnerability can be calculated using Equation (4):

$$V = PI(1 - AC) \tag{4}$$

This was also applied in risk calculation, where the function of sensitivity (S), exposure (E), and hazard (H) is PI, as in Equation (5):

$$R = H \times E \times S \times (1 - AC) \tag{5}$$

Vulnerability and risk indexes ranging from 0–1 reflect the current vulnerability and risk quotient of the 41 inhabited mouzas of Sagar Island.

4. Results

4.1. Indicators of Climate Change

The present study performed a trend analysis of temperature and precipitation data over 20 years to identify significant changes in the weather patterns of the study area. The analysis shows a significant rising trend (Z = +2.80 to +2.45) with an average annual rise of (+3.98) for the maximum temperature, while for the minimum temperature a decreasing trend (Z = −0.60 to −0.91) is observed (Table 2 and Figure 4). Furthermore, there was an increase in average annual temperature (26.25 ± 0.57 °C) at the rate of (Z = +2.57) 0.028 °C (Table 3).

Table 2. Mann–Kendall analysis of the temperature of Sagar Island (2001–2020).

Month	T_{max}		T_{min}		T_{mean}	
	Z	Q	Z	Q	Z	Q
Jan	1.58	3.18	−0.60	−0.93	2.24	1.81 *
Feb	1.33	2.09	−0.91	−1.01	0.77	0.58
Mar	1.68	3.08	0.70	0.96	1.54	1.04
Apr	2.73	2.80 *	1.19	1.38	1.23	1.14
May	0.46	0.63	2.00	1.39 *	−0.98	−0.57
June	1.75	1.82	1.61	1.45	0.42	0.15
July	1.16	0.88	2.10	0.94 *	0.77	0.52
August	2.80	2.06 *	1.96	0.91	1.02	0.23
Sep	2.80	3.01 *	1.59	1.04	1.40	0.67
Oct	2.45	3.46 *	1.72	1.73	1.68	1.15
Nov	1.89	1.84	−0.07	−0.06	0.74	0.54
Dec	−0.49	−0.43	0.28	0.41	−0.84	−0.48

* Significant at 95%; Z, Rate; Q, Sen's slope.

Table 3. Intra- and inter-annual trend analysis of the temperature of Sagar Island (2001–2020).

	T_{max}	T_{Min}	T_{Mean}	Summer	Winter
Minimum	27.40	16.80	25.10	-	-
Maximum	33.00	31.90	27.40	-	-
mean	31.84	24.35	26.25	-	-
SD	1.44	0.31	0.57	-	-
CV	4.3%	3.7%	3.8%	-	-
Z	3.98	−2.96	2.57	3.05	0.84
Q	21.66	13.28	3.34	1.138	1.1

SD, Standard deviation; CV, Coefficient of variation; Z, Rate; Q, Sen's slope.

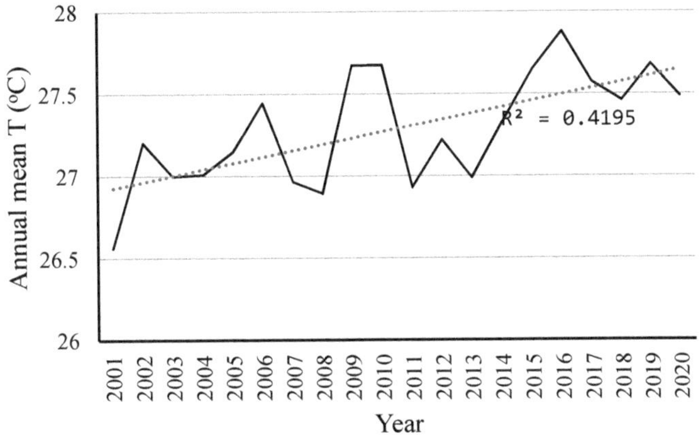

Figure 4. The trend in average annual temperature.

Significant variation in mean monthly rainfall over the past 20 years was observed, with July having the highest (408.86 ± 181.52 mm) and December the lowest (12.01 ± 21.05 mm) monthly rainfall. The analysis reflects a significant trend of increase (5.44 mm/month) in July and of decrease (8.23 mm/month) in June monthly rainfall. The average annual rainfall received is 1797.4 ± 348.9 mm, with 4.3% variation (Table 4). A non-significant trend of decrease in average annual rainfall (Z = −0.07) at a 1.18 mm/year rate is observed (Figure 5). Despite erratic distributions, a significant increasing trend in rainfall (6.19 mm/ year) is evident in monsoon months, while pre-monsoon months show a significant (−1.70 mm/year) decrease in rainfall (Table 5).

Table 4. Mann–Kendall trend analysis of average monthly rainfall for Sagar Island (2001–2019).

Month	Average Monthly Rainfall (mm) (2001–2019)						
	Max	Min	Mean	SD	CV	Z	Q
Jan	87.6	0	12.77	24.37	191%	−0.59	0.00
Feb	195	0	25.49	49.09	193%	1.7	0.52
Mar	127.3	1.6	29.15	33.90	116%	−1.75	−1.00
Apr	133.7	0	43.28	34.19	79%	−0.53	−0.62
May	245.4	29.7	112.94	59.13	52%	−0.42	−1.02
June	533.6	77.5	264.29	133.85	51%	−1.68	−8.23
July	868.9	212.1	408.86	181.52	44%	0.84	5.44
August	905.2	177.3	389.63	186.39	48%	0.63	5.23
Sep	567.6	154.2	303.47	112.01	37%	−0.39	−1.42
Oct	631.3	27.6	173.25	165.81	96%	−0.98	−3.65
Nov	114.8	0	22.32	31.46	141%	0.89	0.26
Dec	75.9	0	12.01	21.05	175%	1.56	0.00

Table 5. Mann–Kendall trend analysis of rainfall distribution for Sagar Island (2001–2019).

Rainfall (mm)	Total8358 (Annual)	Pre-Monsoon	Monsoon	Post-Monsoon
Maximum	2427.10	299.20	2018.70	631.90
Minimum	1289.40	61.60	952.30	81.20
Mean	1797.45	185.36	1366.26	249.70
SD	348.92	61.69	303.82	159.76
CV	19%	33%	22%	64%
Z	−0.07	−0.70	0.35	0.00
Q	−1.18	−1.70	6.19	−0.02

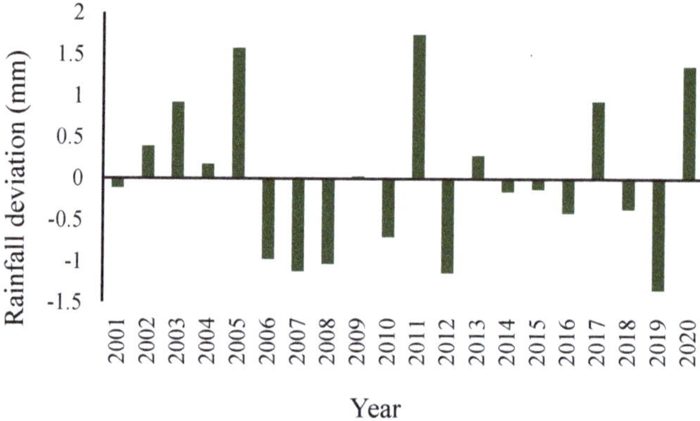

Figure 5. Annual rainfall distribution for Sagar Island.

The occurrence of severe cyclonic storms in the Bay of Bengal often plays an essential role in this low-lying island's climatic vulnerability. A prominent increase in both occurrence and intensity was exhibited in the past ten years. Eight cyclonic storms and seven very severe to highly severe cyclones affected the island from 2010 to 2020 compared with only three severe cyclones from 2000 to 2010 (Figure 6).

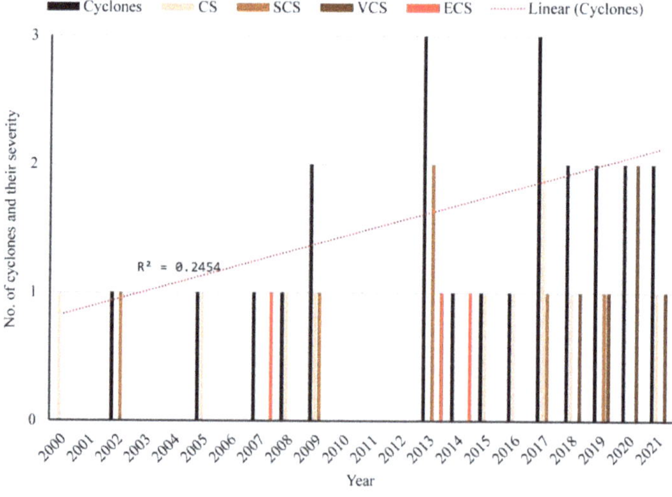

Figure 6. Occurrence and intensity of cyclonic storms. Severity scale: 1, Cyclonic Storm (CS); 2, Severe Cyclonic Storm (SCS); 3, Very Severe Cyclonic Storm (VSCS); 4, Extremely Severe Cyclonic Storm (SSCS).

A crucial outcome of changing climate is sea-level increase which causes severe damage to the coastal ecosystem, infrastructure, and livelihoods [55]. Due to the partial availability of data for Sagar Island, this study considered the sea level data at the nearest station, Diamond Harbor, where a 5.74 mm/year rate of sea level increase has been observed from 1948 to 2015 (Figure 7).

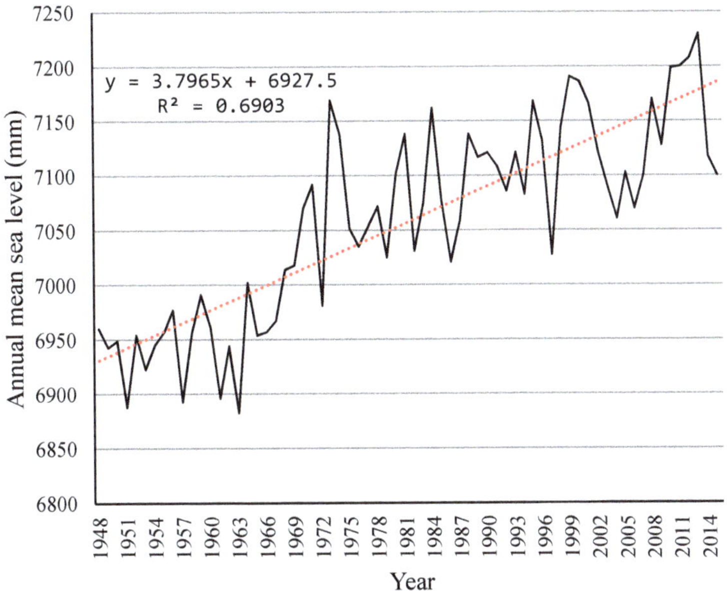

Figure 7. Sea levels at Diamond Harbor Station (near Sagar Island) (1950–2015).

From 1990 to 2020, considerable changes in land use patterns have been associated with coastal erosion. The analysis of the LULC from 1990 and 2020 (Figure 8a,b) shows that significant changes were observed in the water body's sandy areas, which increased by >30% from the year 1990 to 2020. Apart from increases, the built-up area has significantly increased and taken its toll on cultivated land, the area of which shows a decrease of 26.16% (Table 6 and Figure 9).

Table 6. Estimates of the land use and land cover for Sagar Island at two points in time (1990 and 2020).

S No.	Classes	LULC 1990	LULC 2020	Change	% Change
1	Cultivated land	164.36	121.46	−42.90	−26.10
2	Built-up	51.97	84.16	32.19	61.95
3	Mixed open land	0.55	1.04	0.49	88.81
4	Plantation	11.51	17.68	6.17	53.59
5	Sandy areas	0.74	4.55	3.82	518.06
6	Water	4.48	8.87	4.40	98.26
7	Wetland	0.80	1.04	0.23	28.80

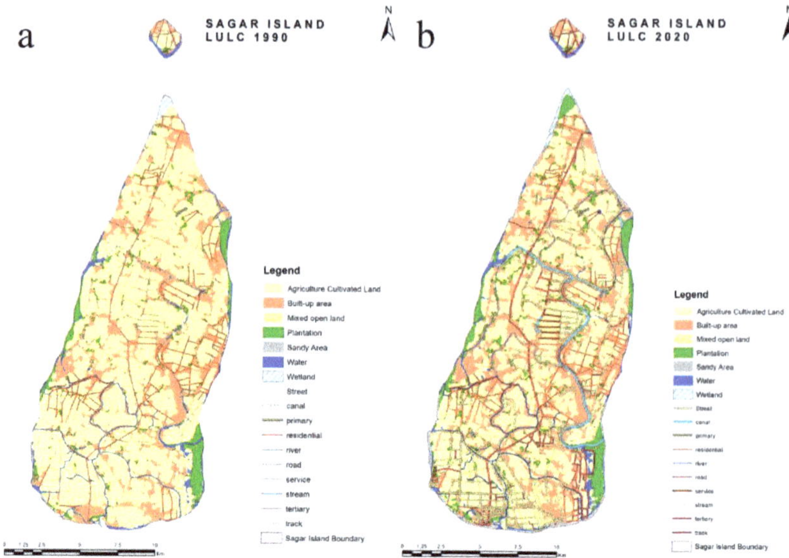

Figure 8. Land use and land cover analysis: (**a**) 1990; (**b**) 2020.

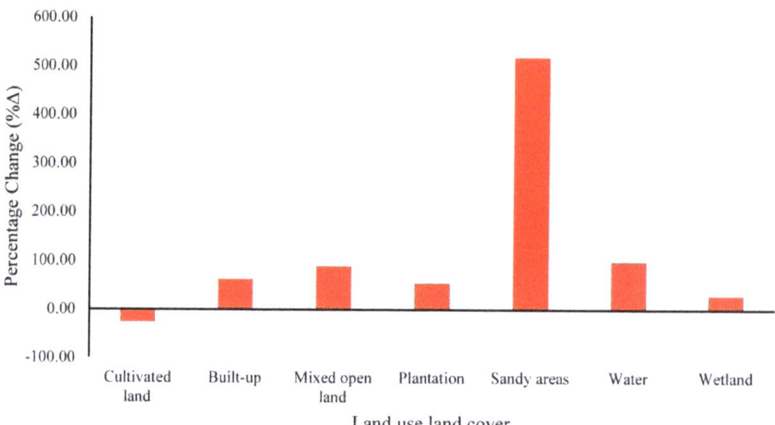

Figure 9. Changes in land use and land cover of Sagar Island (1990 and 2020).

4.2. PCA Results and Construction of Vulnerability and Risk Indices

Spatial assessment of a low-lying deltaic island is crucial as such areas are prone to spatiotemporal variations. The first, second, third, and fourth components accounted for around 53% of the measured variables. For the first component, sensitivity and adaptive capacity variables, i.e., food insecurity, homeownership, and household assets, relatively higher loadings were determined (Table 7).

Two-dimensional plotting (Figure 10) of the first and second components, Dhablat, Shibpur, Beguakhali, and Chemaguri, can be identified as negative outliers due to lower adaptive capacity and higher sensitivity; Rudranagar and Gangasagar are on the positive side.

Table 7. The loading of the principal component analysis (PCA).

Variables	PC 1	PC 2	PC 3	PC 4
Avg. maximum temperature	−0.196	0.019	−0.07	0.377
Avg. minimum temperature	−0.174	0.091	−0.273	0.238
Avg. rainfall	−0.039	−0.063	0.078	0.371
Area under flood	0.288	0.217	−0.095	−0.16
Cyclone wind speed	0.085	0.319	0.18	0.306
Population density	0.046	0.075	0.168	0.173
Avg. household size	0.032	0.235	0.081	−.314
No. of females	−0.107	0.216	0.041	−0.16
No. of children (0–6 years)	−0.083	−0.083	−0.389	0.029
No. of socially backward people	0.248	0.059	−0.099	−0.197
Food insecurity	0.331	0.09	−0.164	−0.06
Without landholding	0.144	0.352	−0.114	0.049
People below poverty line	0.034	−0.234	−0.327	0.012
Agricultural dependency	−0.015	0.046	0.409	−0.132
Marginal worker	0.248	0.207	0.17	0.004
Non-worker	0.24	0.119	−0.109	0.093
Rate of literacy	0.02	0.229	0.319	−0.208
Work participation	−0.256	0.05	0.241	0.166
No. of salaried persons	−0.219	0.306	−0.188	−0.176
No. of homeowners	−0.317	−0.028	0.146	−0.144
Having household assets	−0.353	−0.014	0.059	−0.206
No. of pucca houses	−0.273	0.003	−0.2	−0.318
Road density	−0.081	0.326	−0.118	−0.104
Sanitation	−0.184	0.342	−0.067	−0.077
Electricity	−0.115	0.23	0.039	0.173
Safe drinking water	0.188	0.216	−0.194	0.056
Standard deviation	2.335	2	1.614	1.384
Proportion of variance (Eigenvalues)	0.209	0.154	0.1	0.073
Cumulative proportion	0.209	0.363	0.464	0.537

Statistical test: Kaiser–Meyar–Olkin measures of sampling adequacy ≥ 0.516; determinant of correlation matrix ≥ 0.00001.

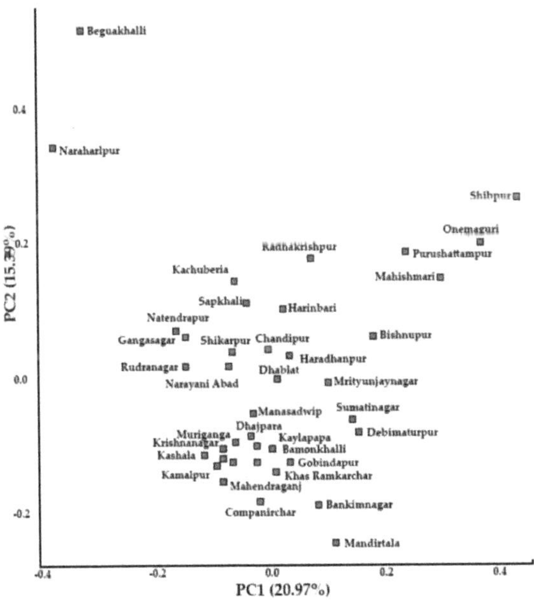

Figure 10. Two-dimensional PCA plot.

The second component explains that the numbers of people without landholdings, road density, and sanitation crucially affect adaptive measures. The third component reflects variables related to sensitivity, i.e., poverty, agricultural dependency, which have a higher significance. The fourth component explains variables of exposure, i.e., Tmax and Tmin, and the variable of the pucca house as a vital parameter. Derived from the outcomes of the PCA, vulnerability and risk values for all the mouzas were calculated (Table 8) and mapped.

Table 8. Specification of severity classes of vulnerability and risk.

Classes	No of Villages (%)		Area (%)		Population in Thousands (%)	
	V	R	V	R	V	R
Very Low (<0.15)	7 (17.07)	9 (21.95)	23.2	25.1	34.81 (16.42)	34.07 (16.07)
Low (>0.15)	9 (21.95)	8 (19.31)	28.6	27.4	38.42 (18.12)	42.84 (20.21)
Moderate (<0.5)	12 (29.26)	11 (26.83)	29.4	28.3	74.81 (32.28)	65.23 (30.82)
High (>0.5)	5 (12.19)	8 (19.51)	10.3	12.9	29.69 (14.0)	40.62 (19.17)
Very High (>0.6)	8 (19.51)	5 (12.19)	8.5	6.3	32.50 (15.33)	23.93 (11.28)

Figure 11 shows mouza hotspots of climate change impacts in terms of vulnerability and risks. This multi-dimensional relative ranking of 41 mouzas indicates that most of the vulnerable communities survive in the marginal areas along the coastline. Shibpur–Dhablat, Beguakhali–Mahismari, and Bankimnagar–Sumatinagar are at high risk, while Kachuberia, Muriganga, Candipur are highly vulnerable but at lower risk of exposure. These location-specific schematic diagrams can effectively target adaptation and mitigation interventions in these geographically homogeneous villages (Figure 12).

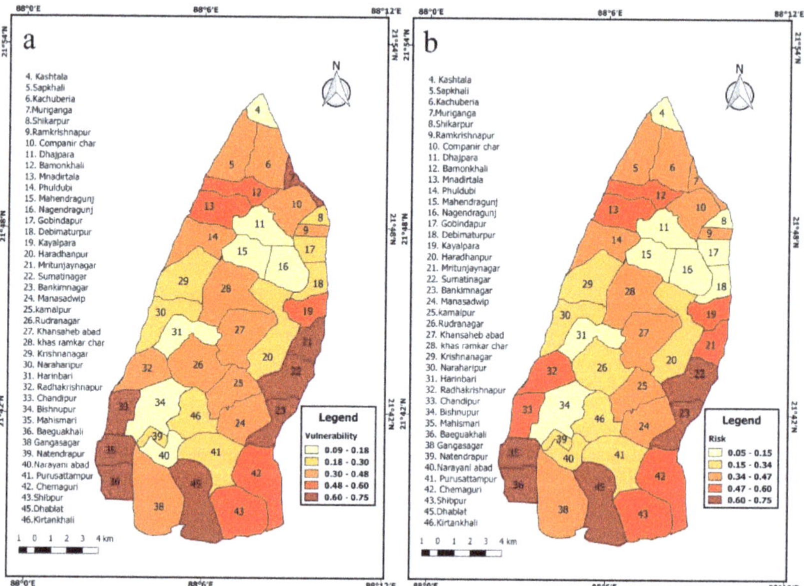

Figure 11. (**a**) Vulnerability and (**b**) risk maps of Sagar Island.

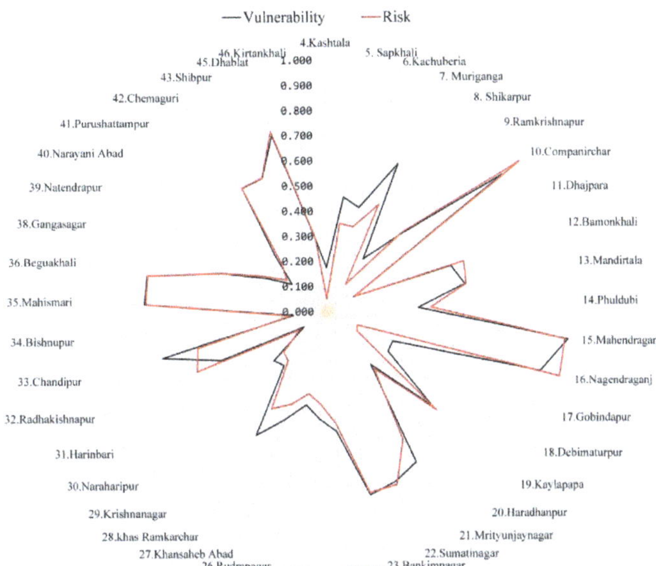

Figure 12. Mouza-level relative ranking of risk and vulnerability for Sagar Island.

5. Discussion

Variations in essential weather and climate parameters have been interconnected with climate change globally [56]. Coastal communities worldwide are threatened by violent cyclones and rising storm surges that cause enormous loss of life and livelihoods [57,58]. Weather parameters play a vital role in the ascription of risk and vulnerability characteristics to any region [59]. Coastal regions are some of the first areas to experience the impacts of a changing climate and are exposed to climate change-related vagaries of nature [60,61]. The climate component of risk analysis in Sagar Island has a uniform influence throughout the region, as spatially explicit climate data information was not used in the study due to the unavailability of such data. Rising storm surges, violent cyclones, and accelerating tidal ingression escalate excessive erosion. The consequences of the overpowering erosion process on human activity and the economy are perfectly portrayed along the coasts of any island system [34,62–65]. The exposure indices of Tmax, Tmin, and precipitation in the risk analysis have also been used by various workers with similar results [66–68]. When examining the vulnerability of the state of Georgia in the United States, Binita et al. (2015) discovered that one of the indications of a changing climate is changes in the intensity and frequency of climate extremes. The study recorded temperature and precipitation anomalies that showed overall trends towards dryness and warming climates correlating well with the recent increase in extreme hydroclimatic events [69]. We also observed a similar correlation with the increase in the intensity and severity of the cyclonic events in the Bay of Bengal. Our study shows a decrease in cultivated land and a consequent increase in the land under built-up areas. This has had tremendous implications for the sensitivity and adaptive capacity indices of the risk analysis. The island has become more vulnerable to agriculture-related climate change impacts that will manifest in the loss of livelihoods related to the island's agricultural sector. A similar finding has also been reported by Kantamaneni et al. (2020). The study suggests that the farmers' economic resources are being harmed by the climate catastrophe, resulting in significant disruptions to social and cultural activities in these coastal communities. The study concluded that climate change calamities, such as floods, cyclones, and strong winds, contribute to higher agricultural vulnerabilities in the investigated areas [70]. Moreover, our analysis of the LULC information from 1990 and 2020 shows that significant changes were observed in the water body's sandy areas, which increased by >30% from 1990 to 2020. This increase is attributed to sea level increases due to climate change.

The vulnerability and risk mapping show a minimal significant difference. However, the overall indices nearly reflect the conceptual linkage and slight advantage of the AR5 framework in identifying specific vulnerable communities through more adequate exposure indices [71,72]. Dhablat, Shibpur, and Beguakhaki are the mouzas facing the maximum potential risk of being adversely affected by climate change. These areas require the immediate intervention of the local government and planning authority. IPCC AR4 did not use the elaborate variables or isolate the variables into the four components of risk: exposure, hazard, sensitivity, and adaptive capacity. The revised AR5 methodology is able to capture this intricate relation [73–75]. The similar exposure, higher hazard, sensitivity, and lower adaptive capacity render the regions inside the island at high risk. The sensitivity and adaptive capacity parameters have been able to be used to locate the at-risk areas within Sagar Island. Overall, the coastal areas are already under serious threat from climate change impacts, and the areas centrally located on the island are also not safe from the consequences [76,77]. Owing to the diverse variables and different methods used, recognizing the components responsible for heightened vulnerability and comparing them with the indices used turns out to be a complex process. Mouzas in high-risk and vulnerability categories, such as Kirtankhali, Dhablat, Shibpur, Chemaguri, and others, require measures to mitigate the effects of catastrophic events. As a result, it is necessary to provide a beneficial infrastructure that can help the inhabitants cope with disasters, e.g., disaster-resistant shelters, early-warning systems, and coastal protection barricades [78,79].

Furthermore, several IT-based inventions have recently been deployed to prevent disasters in many parts of the world, such as WebGIS-based flood simulation scenarios to help cope with an incoming extreme weather phenomenon [80]. Smartphone-based warning applications can be used to alert residents in an area through the supply of real-time information and can help manage rescue operations [81]. Social media and drone-based surveys can be used to locate the people under threat during an extreme event and are thereby proved to be efficient means of saving lives. These strategies need to be made available through government backing to all the risk-prone areas. The Government of India's disaster management policies needs to boost and invest in all such innovative technologies. For efficient mitigation at the local scale, robust policies are required in order to allow access to these technologies to the region's authorities responsible for disaster management. Some of the necessary interventions to assist the mouzas at higher risk and vulnerability are better infrastructural facilities and social benefits, such as health insurance and access to better communication systems [82]. This will ensure that communities live with dignity and a sense of safety that will finally provide them with the required adaptive capacity. For this, administrations have to have a far-sighted approach in devising such policies. This also needs to be collectively handled by various local and international non-governmental organizations involved in adapting and mitigating climate-related hazards.

Programs aimed at improving the socioeconomic health of coastal communities and reducing their overall vulnerability, such as those aimed at providing robust and affordable housing and improving road connectivity, have to be implemented. Modifications to present policies and initiatives to address the particular needs of these coastal areas may further help reduce their risk, as a result of which the communities' adaptive capacities would be enhanced, minimizing their risk. Adaptive capacity is influenced by poverty, housing quality, and education; therefore, measures aiming at overall socioeconomic betterment are needed in vulnerable areas [83]. The growth of coastal economies and livelihood options, such as fishing, mining, tourism, and sea energy, can assist residents in improving their adaptive capacity. This might significantly impact the region's socioeconomic demography while also reducing coastal vulnerability. High-risk mouzas are typically densely populated and have a large built-up area. It is necessary to investigate the possibility of transferring critical companies and economic activities to inland areas with reduced hazards. As risk is the function of four components (exposure, hazard, sensitivity, and adaptive capacity), a multi-dimensional methodology for minimizing disasters in vulnerable wards ought to be a requirement for policymakers [84–86].

6. Conclusions

Coastal regions worldwide are under increasing threat from risk-associated climate change. To assess the spatial dimensions of risk and planning for its aversion, IPCC AR5 constitutes a methodology that takes into account the hazard, exposure, sensitivity, and adaptive capacity of the inhabitants of vulnerable communities. This study was carried out on Sagar Island, West Bengal, India, which is currently trying to cope with multiple challenges from climate hazards, livelihood vulnerability, and underdevelopment. The present study used an index-based approach to assess the island's administrative level (mouza) risk and vulnerability for planning management and mitigation strategies. We observed that hazard parameters, such as cyclonic surges, extensive flooding, embankment breaching, and severe erosion, affected the adaptive capacity of the inhabitants. Furthermore, continuous exploitation of natural resources and unsustainable economic activities increase their sensitivity and risk quotients. The results significantly explained various spatially discrete parameters that determined different degrees of exposure, hazard, sensitivity, and adaptive capacity. Dhablat, Shibpur, and Beguakhaki are the mouzas facing the maximum potential risk of being adversely affected by climate change, and these areas need immediate intervention from the local government and planning authority. We propose providing innovative technologies, better healthcare, and communication, along with robust infrastructural facilities, to the most affected mouzas in order for them to increase their adaptive capacities and ultimately reduce their risk quotients.

Author Contributions: Conceptualization, A.B., G.M. and S.K.S.; methodology, A.B., S.K.S., G.M. and M.F.; software, A.B. and G.M.; validation, A.B., G.M., S.K., S.K.S., P.K. and N.S.; formal analysis, A.B., G.M., S.K.S. and M.F.; investigation, A.B., G.M., S.K., S.K.S. and M.F.; resources, P.K. and N.S.; data curation, A.B.; writing—original draft preparation, A.B., G.M. and S.K.S.; writing—review and editing, G.M., S.K.S., and M.F.; visualization, A.B., G.M. and S.K.S.; supervision, S.K. and S.K.S.; project administration, N.S.; funding acquisition, P.K. All authors have read and agreed to the published version of the manuscript.

Funding: This publication is supported by the Asia Pacific Network for Global Change Research (APN) under Collaborative Regional Research Programme (CRRP) with project reference number CRRP2019-01MY-Kumar.

Institutional Review Board Statement: Not applicable.

Informed Consent Statement: Not applicable.

Data Availability Statement: Data available upon request from the authors.

Acknowledgments: The authors express their gratefulness to the three anonymous reviewers for their valuable comments and suggestions on the earlier version of the manuscript that greatly improved its content and structure. The author G.M. is thankful to Department of Science and Technology, Government of India for providing the Fellowship under Scheme for Young Scientists and Technology (SYST-SEED) [Grant no. SP/YO/2019/1362(G) and (C)].

Conflicts of Interest: The authors declare no conflict of interest.

References

1. Kanga, S.; Meraj, G.; Farooq, M.; Singh, S.K.; Nathawat, M.S. Disasters in the Complex Himalayan Terrains. In *Disaster Management in the Complex Himalayan Terrains*; Springer: Cham, Switzerland, 2022; pp. 3–10.
2. IPCC. *Climate Change 2014: Impacts, Adaptation, and Vulnerability. Part A: Global and Sectoral Aspects. Contribution of Working Group II to the Fifth Assessment Report of the Intergovernmental Panel on Climate Change*; Field, C.B., Barros, V.R., Dokken, D.J., Mach, K.J., Mastrandrea, M.D., Bilir, T.E., Chatterjee, M., Ebi, K.L., Estrada, Y.O., Genova, R.C., et al., Eds.; Cambridge University Press: Cambridge, UK; New York, NY, USA, 2014; 1132p.
3. Ericson, J.P.; Vörösmarty, C.J.; Dingman, S.L.; Ward, L.G.; Meybeck, M. Effective sea-level rise and deltas: Causes of change and human dimension implications. *Glob. Planet Change* **2006**, *50*, 63–82. [CrossRef]

4. Brown, S.; Nicholls, R.J.; Lázár, A.N.; Hornby, D.D.; Hill, C.; Hazra, S.; Addo, K.A.; Haque, A.; Caesar, J.; Tompkins, E.L. What are the implications of sea-level rise for a 1.5, 2 and 3 °C rise in global mean temperatures in the Ganges-Brahmaputra-Meghna and other vulnerable deltas? *Reg. Environ. Chang.* **2018**, *18*, 1829–1842. [CrossRef]
5. Shah, K.U.; Dulal, H.B.; Johnson, C.; Baptiste, A. Understanding livelihood vulnerability to climate change: Applying the livelihood vulnerability index in Trinidad and Tobago. *Geoforum* **2013**, *47*, 125–137. [CrossRef]
6. Dasgupta, P.; Morton, J.F.; Dodman, D.; Karapinar, B.; Meza, F.; Rivera-Ferre, M.G.; ToureSarr, A.; Vincent, K.E. Rural areas. In *Climate Change 2014: Impacts, Adaptation, and Vulnerability. Part A: Global and Sectoral Aspects. Contribution of Working Group II to the Fifth Assessment Report of the Intergovernmental Panel on Climate Change*; Field, C.B., Barros, V.R., Dokken, D.J., Mach, K.J., Mastrandrea, M.D., Bilir, T.E., Chatterjee, M., Ebi, K.L., Estrada, Y.O., Genova, R.C., et al., Eds.; Cambridge University Press: Cambridge, UK; New York, NY, USA, 2014; pp. 613–657.
7. Hahn, M.B.; Riederer, A.M.; Foster, S.O. The livelihood vulnerability index: A pragmatic approach to assessing risks from climate variability and change-a case study in Mozambique. *Glob. Environ. Chang.* **2009**, *19*, 74–88. [CrossRef]
8. Szabo, S.; Brondizio, E.; Renaud, F.G.; Hetrick, S.; Nicholls, R.J.; Matthews, Z.; Tessler, Z.; Tejedor, A.; Sebesvari, Z.; Foufoula-Georgiou, E.; et al. Population dynamics, delta vulnerability and environmental change: Comparison of the Mekong, Ganges–Brahmaputra and Amazon delta regions. *Sustain. Sci.* **2016**, *11*, 539–554. [CrossRef]
9. Mabogunje, A.L. Poverty and environmental degradation: Challenges within the global economy. *Environ. Sci. Policy Sustain. Dev.* **2002**, *44*, 8–19. [CrossRef]
10. Wood, S.A.; Jina, A.S.; Jain, M.; Kristjanson, P.; DeFries, R.S. Smallholder farmer cropping decisions related to climate variability across multiple regions. *Glob. Environ. Chang.* **2014**, *25*, 163–172. [CrossRef]
11. Mortreux, C.; de Campos, R.S.; Adger, W.N.; Ghosh, T.; Das, S.; Adams, H.; Hazra, S. Political economy of planned relocation: A model of action and inaction in government responses. *Glob. Environ. Chang.* **2018**, *50*, 123–132. [CrossRef]
12. Marcinko, C.L.J.; Nicholls, R.J.; Daw, T.M.; Hazra, S.; Hutton, C.W.; Hill, C.T.; Clarke, D.; Harfoot, A.; Basu, O.; Das, I.; et al. The Development of a Framework for the Integrated Assessment of SDG Trade-Offs in the Sundarban Biosphere Reserve. *Water* **2021**, *13*, 528. [CrossRef]
13. Iwasaki, S.; Razafindrabe, B.H.N.; Shaw, R. Fishery livelihoods and adaptation to climate change: A case study of Chilika lagoon, India. *Mitig. Adapt. Strateg. Glob. Chang.* **2009**, *14*, 339–355. [CrossRef]
14. Olsson, L.; Opondo, M.; Tschakert, P.; Agrawal, A.; Eriksen, S.H.; Ma, S.; Perch, L.N.; Zakieldeen, S.A. Livelihoods and poverty. In *Climate Change 2014: Impacts, Adaptation, and Vulnerability. Part A: Global and Sectoral Aspects. Contribution of Working Group II to the Fifth Assessment Report of the Intergovernmental Panel on Climate Change*; Field, C.B., Barros, V.R., Dokken, D.J., Mach, K.J., Mastrandrea, M.D., Bilir, T.E., Chatterjee, M., Ebi, K.L., Estrada, Y.O., Genova, R.C., et al., Eds.; Cambridge University Press: Cambridge, UK; New York, NY, USA, 2014; pp. 793–832.
15. Gopinath, G. Critical coastal issues of Sagar Island, east coast of India. *Environ. Monit. Assess.* **2010**, *160*, 555–561. [CrossRef] [PubMed]
16. Mukherjee, N.; Siddique, G.; Basak, A.; Roy, A.; Mandal, M.H. Climate change and livelihood vulnerability of the local population on Sagar Island, India. *Chin. Geogr. Sci.* **2019**, *29*, 417–436. [CrossRef]
17. Hazra, S.; Mukhopadhyay, A.; Ghosh, A.R.; Mitra, D.; Dadhwal, V.K. (Eds.) *Environment and Earth Observation. Part of the Springer Remote Sensing/Photogrammetry Book Series (SPRINGERREMO)*; Springer: Berlin/Heidelberg, Germany, 2017; pp. 153–172. [CrossRef]
18. Dieng, H.; Cazenave, A.; Meyssignac, B.; Ablain, M. New estimate of the current rate of sea level rise from a sea level budget approach. *Geophys. Res. Lett.* **2017**, *44*, 3744–3751. [CrossRef]
19. WCRP Global Sea Level Budget Group. Global sea level budget 1993-present. *Earth Syst. Sci. Data* **2018**, *10*, 1551–1590. [CrossRef]
20. Bera, A.; Taloor, A.K.; Meraj, G.; Kanga, S.; Singh, S.K.; Đurin, B.; Anand, S. Climate vulnerability and economic determinants: Linkages and risk reduction in Sagar Island, India; A geospatial approach. *Quat. Sci. Adv.* **2021**, *4*, 100038. [CrossRef]
21. Singh, O.P. Long term trends in the frequency of severe cyclone of Bay of Bengal: Observation and simulations. *Mousam* **2007**, *58*, 59–66. [CrossRef]
22. Ghosh, U.; Kjosavik, D.J.; Bose, S. The certainty of uncertainty: Climate change realities of the Indian Sundarbans. In *The Politics of Climate Change and Uncertainty in India*; Routledge: London, UK, 2021; pp. 107–133.
23. IPCC. *Climate Change 2007: Impacts, Adaptation and Vulnerability. Contribution of Working Group II to the Fourth Assessment Report of the Intergovernmental Panel on Climate Change*; Parry, M.L., Canziani, O.F., Palutik, J.P., van der Linden, P.J., Hanson, C.E., Eds.; Cambridge University Press: Cambridge, UK, 2007; 976p.
24. IPCC. *Climate Change 2001: Impacts, Adaptation, and Vulnerability. Contribution of Working Group II to the Third Assessment Report*; Cambridge University Press: Cambridge, UK, 2001.
25. Gbetibouo, G.A.; Ringler, C.; Hassan, R. Vulnerability of the South African farming sector to climate change and variability: An indicator approach. *Nat. Resour. Forum* **2010**, *34*, 175–187. [CrossRef]
26. Eriksen, S.H.; Kelly, P.M. Developing credible vulnerability indicators for climate adaptation policy assessment. *Mitig. Adapt. Strateg. Glob. Chang.* **2007**, *12*, 495–524. [CrossRef]
27. Mondal, A.; Khare, D.; Kundu, S. Spatial and temporal analysis of precipitation and temperature trend of India. *Theor. Appl. Climatol.* **2015**, *122*, 143–158. [CrossRef]

28. Preston, B.L.; Yuen, E.J.; Westaway, R.M. Putting vulnerability to climate change on the map: A review of approaches, benefits, and risks. *Sustain. Sci.* **2011**, *6*, 177–202. [CrossRef]
29. Bera, A.; Singh, S.K. Comparative Assessment of Livelihood Vulnerability of Climate Induced Migrants: A Micro Level Study on Sagar Island, India. *Sustain. Agri Food Environ. Res.* **2021**, *9*, 216–230. [CrossRef]
30. Census of India. Primary Census Abstract. North 24 Parganasand South 24 Parganas: Office of the Registrar General and Commissioner, Government of India. 2011. Available online: https://censusindia.gov.in/pca/cdb_pca_census/houselisting-housing-wb.html (accessed on 25 December 2021).
31. Hajra, R.; Ghosh, A.; Ghosh, T. Comparative Assessment of Morphological and Landuse/Landcover Change Pattern of Sagar, Ghoramara, and Mousani Island of Indian Sundarban Delta Through Remote Sensing. In *Environment and Earth Observation*; Hazra, S., Mukhopadhyay, A., Ghosh, A., Mitra, D., Dadhwal, V., Eds.; Springer: Cham, Switzerland, 2017.
32. Mandal, S.; Choudhury, B.U. Estimation and prediction of maximum daily precipitation at Sagar Island using best fit probability models. *Theor. Appl. Climatol.* **2015**, *121*, 87–97. [CrossRef]
33. Gopinath, G.; Seralathan, P. Rapid erosion of the coast of Sagar Island, West Bengal–India. *Environ. Geol.* **2005**, *48*, 1058–1067. [CrossRef]
34. Kumar, P.; Avtar, R.; Dasgupta, R.; Johnson, B.A.; Mukherjee, A.; Ahsan, M.N.; Nguyen, D.C.H.; Nguyen, H.Q.; Shaw, R.; Mishra, B.K. Socio-hydrology: A key approach for adaptation to water scarcity and achieving human well-being in large riverine islands. *Prog. Disaster Sci.* **2020**, *8*, 100134. [CrossRef]
35. Hazra, S.; Ghosh, T.; Das Gupta, R.; Sen, G. Sea level and associated changes in the Sundarbans. *Sci. Cult.* **2002**, *68*, 309–321.
36. Indian Meteorological Department, 1901–2017. Ministry of Earth Sciences, Government of India. Available online: https://mausam.imd.gov.in/imd_latest/contents/satellite.php (accessed on 25 October 2021).
37. Das, P.; Das, A.; Roy, S. Shrimp fry (meen) farmers of Sundarban Mangrove Forest (India): A tale of ecological damage and economic hardship. *Int. J. Agric. Food Res.* **2016**, *5*, 28–41. [CrossRef]
38. Allen, J.R. A review of the origin and characteristics of recent alluvial sediments. *Sedimentology* **1965**, *5*, 89–191. [CrossRef]
39. Cannon, T.; Twigg, J.; Rowell, J. Social Vulnerability, Sustainable Livelihoods and Disasters. 2003. Available online: https://d1wqtxts1xzle7.cloudfront.net/37233103/CannonTwiggRowellDfidSocialVulnerabilityLivelihoods-with-cover-page-v2.pdf?Expires=1646380822&Signature=gj4QsSOXqSQT4DhYDVooFDkoS7QAJPwyQolXfxGa0PkXlwQj64rk~{}milxSkCZgJGrWJv8ziWgw~{}AWP9lUXDGodmuV0VSuG09BDksmsA1Ra8b9VL1pbUISgqTw03KWIxHhYyDU~{}uRIjm8J4E-eyxrLcj3Fj5cTykrUetcHoaXFF-DtQBot~{}iKC-S~{}l6OpPMKoiPN-UafEOWUYIUqkpJ653iHWdFSlIytbJU-g--tCdCtZLS-iwMBNO2bPb1BBINRgposZuuqRWotOEQPbzyrd0rZ8jy8IUeL1rasTdoa5x83cmwSUtxQXno3YXoe~{}4ArTgODtBm~{}AZKxPxmOs~{}Lczdw__&Key-Pair-Id=APKAJLOHF5GGSLRBV4ZA (accessed on 25 October 2021).
40. Patt, A.G.; Tadross, M.; Nussbaumer, P.; Asante, K.; Metzger, M.; Rafael, J.; Goujon, A.; Brundrit, G. Estimating least-developed countries' vulnerability to climate-related extreme events over the next 50 years. *Proc. Natl. Acad. Sci. USA* **2010**, *107*, 1333–1337. [CrossRef]
41. Bush, M.J. *Climate Change Adaptation in Small Island Developing States*; John Wiley & Sons: Hoboken, NJ, USA, 2018.
42. Hobday, A.J.; Chambers, L.E.; Arnould, J.P. Prioritizing climate change adaptation options for iconic marine species. *Biodivers. Conserv.* **2015**, *24*, 3449–3468. [CrossRef]
43. Hossain, B.; Sohel, M.S.; Ryakitimbo, C.M. Climate change induced extreme flood disaster in Bangladesh: Implications on people's livelihoods in the Char Village and their coping mechanisms. *Prog. Disaster Sci.* **2020**, *6*, 100079. [CrossRef]
44. Singha, P.; Das, P.; Talukdar, S.; Pal, S. Modeling livelihood vulnerability in erosion and flooding induced river island in Ganges riparian corridor, India. *Ecol. Indic.* **2020**, *119*, 106825. [CrossRef]
45. IPCC. Summary for policymakers. In *Climate Change 2014: Impacts, Adaptation, and Vulnerability. Part A: Global and Sectoral Aspects. Contribution of Working Group II to the Fifth Assessment Report of the Intergovernmental Panel on Climate Change*; Field, C.B., Barros, V.R., Dokken, D.J., Mach, K.J., Mastrandrea, M.D., Bilir, T.E., Chatterjee, M., Ebi, K.L., Estrada, Y.O., Genova, R.C., et al., Eds.; Cambridge University Press: Cambridge, UK; New York, NY, USA, 2014.
46. Vincent, K. Creating an index of social vulnerability to climate change for Africa. *Tyndall Cent. Clim. Change Res.* **2004**, *56*, 41.
47. Vijaya, K. *Constructing an Area-Based Socioeconomic Index: A Principal Components Analysis Approach*; Early Child Development Mapping Project: Edmonton, AB, Canada, 2010.
48. Resio, D.T.; Irish, J.L. Tropical cyclone storm surge risk. *Curr. Clim. Change Rep.* **2015**, *1*, 74–84. [CrossRef]
49. Rahman, M.M.; Ghosh, T.; Salehin, M.; Ghosh, A.; Haque, A.; Hossain, M.A.; Das, S.; Hazra, S.; Islam, N.; Sarker, M.H.; et al. Ganges-Brahmaputra-Meghna Delta, Bangladesh and India: A Transnational Mega-Delta. In *Deltas in the Anthropocene*; Nicholls, R., Adger, W., Hutton, C., Hanson, S., Eds.; Palgrave Macmillan: Cham, Switzerland, 2020. [CrossRef]
50. Kanga, S.; Meraj, G.; Das, B.; Farooq, M.; Chaudhuri, S.; Singh, S.K. Modeling the spatial pattern of sediment flow in lower Hugli estuary, West Bengal, India by quantifying suspended sediment concentration (SSC) and depth conditions using geoinformatics. *Appl. Comput. Geosci.* **2020**, *8*, 100043. [CrossRef]
51. Tomar, P.; Singh, S.K.; Kanga, S.; Meraj, G.; Kranjčić, N.; Đurin, B.; Pattanaik, A. GIS-Based Urban Flood Risk Assessment and Management—A Case Study of Delhi National Capital Territory (NCT), India. *Sustainability* **2021**, *13*, 12850. [CrossRef]
52. Mondal, I.; Bandyopadhyay, J.; Dhara, S. Detecting shoreline changing trends using principle component analysis in Sagar Island, West Bengal, India. *Spat. Inf. Res.* **2017**, *25*, 67–73. [CrossRef]

53. Meraj, G.; Romshoo, S.A.; Ayoub, S.; Altaf, S. Geoinformatics based approach for estimating the sediment yield of the mountainous watersheds in Kashmir Himalaya, India. *Geocarto Int.* **2018**, *33*, 1114–1138. [CrossRef]
54. Füssel, H.M.; Klein, R.J. Climate change vulnerability assessments: An evolution of conceptual thinking. *Clim. Chang.* **2006**, *75*, 301–329. [CrossRef]
55. Resurrección, B.P. Persistent women and environment linkages in climate change and sustainable development agendas. *Women's Stud. Int. Forum* **2013**, *40*, 33–43. [CrossRef]
56. Sarkar, S.; Bardhan, S.; Sanyal, P. LULC Change Detection Studies and Its Possible Linkages with IncreasedVisitation in the Sagar Island, W.B. *Int. J. Eng. Trends Technol.* **2021**, *69*, 49–61. [CrossRef]
57. Jayaram, K.S.; Mitra, D.; Mishra, A.K. Coastal geomorphological and land use Land cover study of Sagar Island, Bay of Bengal, India using remotely sensed data. *Int. J. Remote Sens.* **2006**, *27*, 3671–3682. [CrossRef]
58. Saha, C.K. Dynamics of disaster-induced risk in southwestern coastal Bangladesh: An analysis on tropical Cyclone Aila 2009. *Nat. Hazards* **2015**, *75*, 727–754. [CrossRef]
59. Ahmad, N.; Hussain, M.; Riaz, N.; Subhani, F.; Haider, S.; Alamgir, K.S.; Shinwari, F. Flood prediction and disaster risk analysis using GIS based wireless sensor networks, a review. *J. Basic Appl. Sci. Res.* **2013**, *8*, 632–643.
60. Dhiman, R.; VishnuRadhan, R.; Eldho, T.I.; Inamdar, A. Flood risk and adaptation in Indian coastal cities: Recent scenarios. *Appl. Water Sci.* **2019**, *9*, 1–6. [CrossRef]
61. Mukherjee, N.; Siddique, G. Assessment of climatic variability risks with application of livelihood vulnerability indices. *Environ. Dev. Sustain.* **2020**, *22*, 5077–5103. [CrossRef]
62. Barua, P.; Rahman, S.H. Community-based rehabilitation attempt for solution of climate displacement crisis in the coastal area of Bangladesh. *Int. J. Migr. Resid. Mobil.* **2018**, *1*, 358–378. [CrossRef]
63. Swain, D. Tropical Cyclones and Coastal Vulnerability: Assessment and Mitigation. In *Geospatial Technologies for Land and Water Resources Management*; Springer: Cham, Switzerland, 2022; pp. 587–621.
64. von Storch, H.; Fennel, K.; Jensen, J.; Lewis, K.A.; Ratter, B.; Schlurmann, T.; Wahl, T.; Zhang, W. Climate and Coast: Overview and Introduction. *Oxf. Res. Encycl. Clim. Sci.* **2021**.
65. Sharma, D.; Rao, K.; Ramanathan, A.L. A Systematic Review on the Impact of Urbanization and Industrialization on Indian Coastal Mangrove Ecosystem. *Coast. Ecosyst.* **2022**, *38*, 175–199.
66. Sadeqi, A.; Kahya, E. Spatiotemporal analysis of air temperature indices, aridity conditions, and precipitation in Iran. *Theor. Appl. Climatol.* **2021**, *145*, 703–716. [CrossRef]
67. Chen, Y.; Liao, Z.; Shi, Y.; Tian, Y.; Zhai, P. Detectable increases in sequential flood-heatwave events across China during 1961–2018. *Geophys. Res. Lett.* **2021**, *48*, e2021GL092549. [CrossRef]
68. Mie Sein, Z.M.; Ullah, I.; Syed, S.; Zhi, X.; Azam, K.; Rasool, G. Interannual variability of air temperature over Myanmar: The influence of ENSO and IOD. *Climate* **2021**, *9*, 35. [CrossRef]
69. Binita, K.C.; Shepherd, J.M.; Gaither, C.J. Climate change vulnerability assessment in Georgia. *Appl. Geogr.* **2015**, *62*, 62–74.
70. Kantamaneni, K.; Rice, L.; Yenneti, K.; Campos, L.C. Assessing the vulnerability of agriculture systems to climate change in coastal areas: A novel index. *Sustainability* **2020**, *12*, 4771. [CrossRef]
71. Nguyen, K.A.; Liou, Y.A.; Terry, J.P. Vulnerability of Vietnam to typhoons: A spatial assessment based on hazards, exposure and adaptive capacity. *Sci. Total Environ.* **2019**, *682*, 31–46. [CrossRef]
72. Williams, P.A.; Crespo, O.; Abu, M. Assessing vulnerability of horticultural smallholders' to climate variability in Ghana: Applying the livelihood vulnerability approach. *Environ. Dev. Sustain.* **2020**, *22*, 2321–2342. [CrossRef]
73. Viner, D.; Ekstrom, M.; Hulbert, M.; Warner, N.K.; Wreford, A.; Zommers, Z. Understanding the dynamic nature of risk in climate change assessments—A new starting point for discussion. *Atmos. Sci. Lett.* **2020**, *21*, e958. [CrossRef]
74. Akter, M.; Jahan, M.; Kabir, R.; Karim, D.S.; Haque, A.; Rahman, M.; Salehin, M. Risk assessment based on fuzzy synthetic evaluation method. *Sci. Total Environ.* **2019**, *658*, 818–829. [CrossRef] [PubMed]
75. Wang, G.; Liu, Y.; Hu, Z.; Lyu, Y.; Zhang, G.; Liu, J.; Liu, Y.; Gu, Y.; Huang, X.; Zheng, H.; et al. Flood risk assessment based on fuzzy synthetic evaluation method in the Beijing-Tianjin-Hebei metropolitan area, China. *Sustainability* **2020**, *12*, 1451. [CrossRef]
76. Toimil, A.; Losada, I.J.; Nicholls, R.J.; Dalrymple, R.A.; Stive, M.J. Addressing the challenges of climate change risks and adaptation in coastal areas: A review. *Coast. Eng.* **2020**, *156*, 103611. [CrossRef]
77. Abedin, M.; Collins, A.E.; Habiba, U.; Shaw, R. Climate change, water scarcity, and health adaptation in southwestern coastal Bangladesh. *Int. J. Disaster Risk Sci.* **2019**, *10*, 28–42. [CrossRef]
78. Farooq, M.; Singh, S.K.; Kanga, S. Mainstreaming adaptation strategies in relevant flagship schemes to overcome vulnerabilities of climate change to agriculture sector. *Res. J. Agri. Sci. Int. J.* **2021**, *12*, 637–646.
79. Bowen, T.; Del Ninno, C.; Andrews, C.; Coll-Black, S.; Johnson, K.; Kawasoe, Y.; Kryeziu, A.; Maher, B.; Williams, A. *Adaptive Social Protection: Building Resilience to Shocks*; World Bank Publications: Herndon, VA, USA, 2020.
80. Sudipta, C.; Kambekar, A.R.; Arnab, S. Impact of Climate Change on Sea Level Rise along the Coastline of Mumbai City, India. *Int. J. Mar. Environ. Sci.* **2021**, *15*, 164–170.
81. Sun, W.; Bocchini, P.; Davison, B.D. Applications of artificial intelligence for disaster management. *Nat. Hazards* **2020**, *103*, 2631–2689. [CrossRef]
82. Jongman, B. Effective adaptation to rising flood risk. *Nat. Commun.* **2018**, *9*, 1–3. [CrossRef] [PubMed]

3. Devkota, N.; Phuyal, R.K.; Shrestha, D.L. Perception, determinants and barriers for the adoption of climate change adaptation options among Nepalese rice farmers. *Agric. Sci.* **2018**, *9*, 272–298. [CrossRef]
4. Zhang, M.; Liu, Z.; van Dijk, M.P. Measuring urban vulnerability to climate change using an integrated approach, assessing climate risks in Beijing. *PeerJ* **2019**, *7*, e7018. [CrossRef]
5. Mysiak, J.; Torresan, S.; Bosello, F.; Mistry, M.; Amadio, M.; Marzi, S.; Furlan, E.; Sperotto, A. Climate risk index for Italy. Philosophical Transactions of the Royal Society A: Mathematical. *Phys. Eng. Sci.* **2018**, *376*, 20170305.
6. Aznar-Crespo, P.; Aledo, A.; Melgarejo-Moreno, J.; Vallejos-Romero, A. Adapting social impact assessment to flood risk management. *Sustainability* **2021**, *13*, 3410. [CrossRef]

MDPI
St. Alban-Anlage 66
4052 Basel
Switzerland
Tel. +41 61 683 77 34
Fax +41 61 302 89 18
www.mdpi.com

Water Editorial Office
E-mail: water@mdpi.com
www.mdpi.com/journal/water

www.ingramcontent.com/pod-product-compliance
Lightning Source LLC
LaVergne TN
LVHW071017170726
843515LV00006B/1153